Applied Mathematical Sciences
Volume 145

Springer
New York
Berlin
Heidelberg
Barcelona
Hong Kong
London
Milan
Paris
Singapore
Tokyo

Hurricane Fran 1996

Paul K. Newton

The *N*-Vortex Problem
Analytical Techniques

With 79 Figures

 Springer

Paul K. Newton
Department of Aerospace and Mechanical Engineering
  and Department of Mathematics
University of Southern California
Los Angeles, CA 90089-1191
USA
newton@spock.usc.edu

Editors

Jerrold E. Marsden
Control and Dynamical Systems, 107-81
California Institute of Technology
Pasadena, CA 91125
USA

Lawrence Sirovich
Division of Applied Mathematics
Brown University
Providence, RI 02912
USA

Mathematics Subject Classification (2000): 37J99, 37N10, 76B47

Library of Congress Cataloging-in-Publication Data
Newton, Paul K.
    The *N*-Vortex problem : analytical techniques / Paul K. Newton
        p. cm. — (Applied mathematical sciences ; 145)
    Includes bibliographical references and index.
    ISBN 978-1-4419-2916-7
    1. Vortex-motion. I. Title. II. Applied mathematical sciences (Springer-Verlag
New York, Inc.) ; v. 145.
QA1 .A647 no. 145
[QA925]
510 s—dc21
[532′.0595]   00-053198

Printed on acid-free paper.

Production managed by Michael Koy; manufacturing supervised by Jacqui Ashri.
Photocomposed copy prepared from the author's LaTeX2e files.

9 8 7 6 5 4 3 2 1

Springer-Verlag   New York Berlin Heidelberg
A member of BertelsmannSpringer Science+Business Media GmbH

*This book is dedicated
with all my love
to
Lynn and Daniel.*

Preface

This book is an introduction to current research on the N-vortex problem of fluid mechanics in the spirit of several works on N-body problems from celestial mechanics, as for example Pollard (1966), Szebehely (1967), or Meyer and Hall (1992). Despite the fact that the field has progressed rapidly in the last 20 years, no book covers this topic, particularly its more recent developments, in a thorough way. While Saffman's *Vortex Dynamics* (1992) covers the general theory from a classical point of view, and Marchioro and Pulvirenti's *Mathematical Theory of Incompressible Nonviscous Fluids* (1994), Doering and Gibbon's *Applied Analysis of the Navier–Stokes Equations* (1995), and Majda and Bertozzi's *Vorticity and Incompressible Fluid Flow* (2001) cover much of the relevant mathematical background, none of these discusses the more recent literature on integrable and nonintegrable point vortex motion in any depth. Chorin's *Vorticity and Turbulence* (1996) focuses on aspects of vorticity dynamics that are most relevant toward an understanding of turbulence, while Arnold and Khesin's *Topological Methods in Hydrodynamics* (1998) lays the groundwork for a geometrical and topological study of the Euler equations. Ottino's *The Kinematics of Mixing: Stretching, Chaos, and Transport* (1989) is an introductory textbook on the use of dynamical systems techniques in the study of fluid mixing, while Wiggins' *Chaotic Transport in Dynamical Systems* (1992) describes techniques that are of general use, without focusing specifically on vortex motion. All of these books cover aspects of the topics I treat, but none focuses on exactly the same issues.

My goal is to describe the Hamiltonian aspects of vortex dynamics so that graduate students and researchers can use the book as an entry point into the rather large literature on integrable and nonintegrable vortex problems. By describing the N-vortex problem as a branch of dynamical systems theory in the way the N-body problem of celestial mechanics is often presented, I have tried to keep my focus fairly narrow, but deeper than a broader literature survey would be. Despite the fact that problems in celestial mechanics have historically been among the key driving forces behind progress in dynamics, [1] research in this area has fallen out of favor and celestial mechanics is seldom taught in most physics, engineering, and mathematics departments these days. By contrast, the field of vortex dynamics is lively and active, using techniques that have widespread applicability to many general problems in dynamics and modern applied mathematics. These include integrable and nonintegrable Hamiltonian methods, nonlinear stability techniques for fixed and relative equilibria, long-time existence theory, finite-time blow-up (collisions), geometric phase space methods including KAM and singular perturbation theory, geometric phases, transport and mixing, numerical methods and convergence theories relating discrete approximations to their continuum limits, topology and the theory of knots, geometry of curves and surfaces, Lie groups and Lie algebras, statistical mechanics, ergodic theory, and probabilistic methods, to name just a few. While I have not covered all of those topics, my belief is that an extensive study of the N-vortex problem provides an ideal entry into the field of nonlinear dynamics that is physically relevant and mathematically rich.

In the first chapter I present an overview of the two main themes of the book, vorticity dynamics and Hamiltonian systems. The chapter is written more in a review style than are the subsequent ones and, hence, serves the main purpose of introducing many of the background topics necessary for understanding topics covered more thoroughly later in the book. At the end of the chapter, I formulate some of the key questions that will be covered in the chapters that follow. Chapter 2 covers much of what is known concerning the N-vortex problem in the plane with no boundaries. I describe in detail the integrable 3-vortex problem following the classical works of Kirchhoff (1876), Gröbli (1877), Synge (1949), Novikov (1978), and Aref (1979). The 3-vortex problem is fundamental to much of the development of the subject. There are many reasons for this, not the least of which is the fact that the interactions of the general N-vortex problem can

---

[1]The evidence of this can, of course, be found starting with Newton's *Philosophiae Naturalis Principia Mathematica* (1687), Lagrange's *Mécanique Analytique* (1788), Laplace's *Traité de Mécanique Céleste* (1799), Tisserand's *Traité de Mécanique Céleste* (1889), Poincaré's *Les Méthodes Nouvelles de la Mécanique Céleste* (1892), Whittaker's *A Treatise on the Analytical Dynamics of Particles and Rigid Bodies* (1904), Moulton's *An Introduction to Celestial Mechanics* (1902), Birkhoff's *Dynamical Systems* (1927), Wintner's *Analytical Foundations of Celestial Mechanics* (1941), Brouwer and Clemence's *Methods of Celestial Mechanics* (1961), and Moser and Siegel's *Lectures on Celestial Mechanics* (1971). A beautiful, recent book describing part of this history is Barrow-Green (1997).

be written as interacting triads; there are analogues of this decomposition for general nonlinear oscillators and for turbulent wave interactions, where triad dynamics plays a central role. I mention also the fact that the 3-vortex problem is the simplest problem (in terms of smallest N) capable of generating new length and time scales through its dynamics. For the 4-vortex problem I describe the coordinates introduced in Khanin (1982) and Lim (1991a), which conveniently reduce the system from four to two degrees of freedom. I then give a proof of nonintegrability of the restricted 4-vortex problem using Melnikov theory. The original proof is found in Ziglin (1980), followed by Koiller and Carvalho (1985, 1989). However, I follow in more detail the recent proof given in Castilla, Moauro, Negrini, and Oliva (1993).

Chapter 3 contains a discussion of vortex problems in the plane with boundaries. I begin by recalling some classical results from potential theory and the construction of Green's functions for closed domains. Methods for explicitly constructing the relevant Hamiltonian in a closed domain are presented, such as the method of images and conformal mapping techniques. I then explain how integrability is broken in a closed domain without symmetries, following the work of Zannetti and Franchessi (1995). In Chapter 4, I introduce the N-vortex problem on a spherical shell (the one-layer model). The 3-vortex problem is described in detail, following Kidambi and Newton (1998, 1999, 2000a, 2000b). I describe the integrability of this problem, classification of equilibria, as well as the nonequilibrium process of 3-vortex spherical collapse, and contrast the differences between spherical collapse and planar collapse. I discuss what is known about the possible instantaneous streamline topologies that are allowable on the sphere, along with a general classification of topologies for the 3-vortex problem and how this might be relevant in the wider context of atmospheric flows and the "nonlinear decomposition" of weather patterns. I also introduce techniques to treat the case with solid boundaries on a spherical surface, following the recent work of Kidambi and Newton (2000b).

Chapter 5 contains a discussion of geometric phases for vortex problems in the plane, following Newton (1994), and Shashikanth and Newton (1998, 1999). Their role in determining the growth rate of spiral interfaces in vortex-dominated flows is described in the context of several prototypical configurations. In Chapter 6 I present an overview of the statistical mechanics treatment of point vortex motion, which was initiated in a famous paper of Onsager (1949). More recent treatments are described, including those of Pointin and Lundgren (1976), Lundgren and Pointin (1977a,b), Miller and Robert (1991), Miller (1990), Montgomery and Joyce (1974), Robert (1991), Robert and Sommeria (1991), and Lim (1998a, 1999).

Chapters 7 and 8 deal with extensions to the basic theory. In Chapter 7 I relax the assumption that the vorticity is concentrated at singular points and hence describe the dynamics associated with vortex patches in the plane. I explain the work of Kida (1981), in which an exact elliptical patch is derived in the presence of time-independent background strain and vorticity. This work is generalized in the paper of Neu (1984), which is also described. I then present the so-called *moment model* of Melander, Zabusky, and Styczek (1986) in which a self-consistent Hamiltonian system is derived for the interaction of vortex patches under the assumption that the patches are nearly circular and well separated. I also describe a shear layer model introduced and analyzed in Meiburg and Newton (1991) and Newton and Meiburg (1991) based on a viscously decaying, spatially periodic row of vortices. Chapter 8 treats vortex filament dynamics in three dimensions. First, I describe the localized induction equation for the evolution of a thin isolated filament. The DaRios–Betchov equations governing the evolution of the curvature and torsion of the filament are derived, and special solutions are discussed, such as circular rings, helices, torus knots, and solitary wave loops traveling on a filament. Hasimoto's transformation of the localized induction equation to the integrable nonlinear Schrödinger equation is described in some detail, along with discussion of some of the known invariants based on this approach. Higher order theories that include vortex core structure and self-stretching mechanisms are described and a simple model of interacting nearly parallel filaments is presented, following the work of Klein, Majda, and Damodaran (1995). I end with a brief description of the so-called "vorton model," which has been used recently, mostly for numerical calculations. Hence, the book naturally divides into two parts, the first being Chapters 1 through 3, which present much of the basic two-dimensional point vortex theory that is now considered classical. More recent applications and extensions of the basic theory are covered in Chapters 4 through 8, which form the second part of the book.

Despite my best efforts at presenting a comprehensive treatment, there are still many topics I have been forced to ignore, most notably the rather large literature on all numerical aspects associated with vortex dynamics. This includes all the work involving convergence of point vortex decompositions to their continuum Euler limit, which I only briefly summarize. Readers interested in these topics can read the recent book of Cottet and Koumoutsakos (2000). In addition, there are many influential papers on the subject that are primarily computational, which I have not emphasized. In the end I have chosen the topics that I hope will be most useful to the nonlinear dynamicist interested in learning analytical techniques. I have tried to keep the book as self-contained as possible, so that the only training required of the reader is a good background in advanced calculus and ordinary and partial differential equations at the level of a typical undergraduate engineering, physics, or applied mathematics major. An introductory graduate course in fluid mechanics and differential equations would be quite useful, but could be picked up as needed by the diligent reader. Since my experience has been that students come to this subject with many different backgrounds, ultimately, the most important prerequisite is a desire to learn the topics described. At the end of each chapter there

are exercises of varying difficulty which in many cases require the reader
to fill in details of proofs or complete examples whose inclusion would otherwise make the book too long and tedious. Completing these exercises
should help the reader compensate for his or her incomplete background.

By and large, I have avoided extensive discussions of the classical literature of the type found in other books such as Lamb (1932), Batchelor
(1967), or Saffman (1992). References are made to the relevant papers, but
it is assumed that the reader will go to the original sources as needed for
detailed derivations of classical theorems and special solutions. In addition,
I have tried to focus on models that strike a balance between simplicity and
faithfulness to physical reality, describing results that are explicit, sometimes at the expense of generality. My goal is to present tractable models
where one can learn physics along with techniques of nonlinear dynamics
that can be used in other contexts. In all cases I have tried to clearly explain the approximations being used and the subsequent shortcomings of
the model being described.

There are many open problems associated with N-vortex motion; I mention some that are interesting:

- It would be nice to obtain rigorous estimates for the diffusion rate of
  particle motion in flows populated by vortices, particularly in closed
  domains or on compact surfaces, such as spheres. Results of this type
  would be relevant to fluid mixing and transport in turbulent flows.
  It would be particularly interesting to relate the global geometric
  properties of the instantaneous streamline patterns associated with
  the velocity field to the statistical properties of particle transport,
  the key being an analysis of the particle distribution function with
  a velocity field generated by point vortices. Arnold diffusion presum-
  ably also plays a role here, but there are few analytical results in this
  direction.

- Related to the previous problem, rigorous results on the mixing properties of vortex motion in compact regions and on surfaces of general
  curvature would be important; for example, under what conditions
  the motion is ergodic, mixing, Bernouilli ... Techniques borrowed
  from billiard dynamics where similar issues arise might well be useful.

- The general formulation of non-adiabatic geometric phases for vortex
  motion is interesting and more general than in the adiabatic context.
  Results along these lines could well be important for developing a
  control theory based on vortex motion as the importance of geometric
  phases in the control of robotic systems and in other related contexts
  is currently being fleshed out.

- Understanding the dynamics of point vortices on a sphere with additional effects, such as rotation and vertical density stratification, is

crucial for geophysical applications where one is interested in large-scale atmospheric and oceanographic mixing. Vertical density stratification is sometimes treated with N-layer models where vortices interact within each layer, as well as across layers, but not much has been done analytically with these models as compared to what has been done in one layer.

- A general study of the N-vortex collision problem has not yet been carried out. For instance, we do not know if there are any non self-similar collisions of vortices in the plane or on the sphere, and vortex collisions in domains with boundaries are only beginning to be investigated. In general, much more is known about N-body collisions in the context of celestial mechanics, and some of these techniques might prove useful.

- A classification and understanding of all possible instantaneous streamline patterns that are allowable in the plane and on the sphere for an N-vortex problem would be relevant to an understanding of mixing processes. Understanding the dynamical transition from one pattern to another could prove important for these purposes, and further results are needed on the role that these finite-time structures play in the mixing process. A general program for topologically classifying all integrable Hamiltonian systems is at the present time being developed by several groups. The manner in which the N-vortex problem fits within these classifications is currently unclear.

- A general control theory based on knowledge of the vortex motion would be desirable. Controlling the vortex motion by passive or active means near boundaries could have important applications in many technological and environmental flows (for issues like noise supression), and the further development of a Lagrangian theory might prove advantageous over a Hamiltonian one.

- Developing techniques for understanding vortex motion in the far from integrable regime, where the number of vortices is large, but not large enough to warrant the use of statistical mechanics, would be worthwhile. This intermediate regime remains largely unexplored.

- The equations for the interaction of three-dimensional vortex filaments that are nearly parallel have recently been derived as a coupled Schrödinger system (see Chapter 8) and have yet to be thoroughly explored. It would be particularly interesting to analyze the case of 3-vortex filament collapse, using what is known about the collapsing system when the filaments remain perfectly parallel. In addition, using the coupled system as a starting point, one could study the effect of the third dimension on the mixing and transport of particles as they migrate out of plane.

I would like to thank many people who have helped me in learning and presenting the material in this book. Larry Sirovich and Joe Keller were my first teachers of fluid dynamics. They showed me the beauty of the subject and introduced me to some of its challenges. I have learned many things about the Hamiltonian aspects of the field from Jerry Marsden and I thank him for his encouragement and steady advice on all aspects of this project.

I would like to thank my Ph.D. students both at Illinois and USC: Ralph Axel, Banavara Shashikanth, Rangashari Kidambi, and Mohamed Jamaloodeen. I particularly want to thank Rangashari Kidambi and Banavara Shashikanth, whose Ph.D. work forms parts of Chapters 4 and 5, respectively. I have learned much on the subject from my colleagues at USC: Ron Blackwelder, Fred Browand, Andrzej Domaradzki, Tony Maxworthy, Eckart Meiburg, Larry Redekopp, and Geoff Spedding. For reading and commenting on parts of the manuscript special thanks go to Hassan Aref, Bjorn Birnir, Marty Golubitsky, Eugene Gutkin, Darryl Holm, Chjan Lim, Jerry Marsden, Tom Sideris, and Larry Sirovich. Other people who deserve thanks include Pierre-Henri Chavanis, Charlie Doering, David Dritschel, John Gibbon, Paul Goldbart, John Hannay, Jair Koiller, Juan Lasheras, Tony Leonard, Tom Lundgren, Igor Mezić, Mark Nelkin, Rahul Pandit, Sergey Pekarsky, Bhimsen Shivamoggi, Tamas Tél, and George Zaslavsky. Steven Brown deserves my special gratitude for his excellent work with Adobe Illustrator and Adobe Photoshop, rendering many crude figures into polished final products. I particularly want to thank my editor at Springer-Verlag, Achi Dosanjh, for her excellent and professional advice throughout all stages of the project. Finally, I am grateful to my parents for improving both the scientific aspects of the manuscript, as well as the writing style.

The following figures have been reprinted from their original journal articles (as cited in the text) with special permission from the publishers: Figure 1.5 (Institute of Physics); Figures 2.4, 2.5, 4.35 through 4.39, 7.1, 7.3 (American Institute of Physics); Figure 2.6 (Canadian Mathematical Society); Figures 2.10, 7.4 through 7.6 (Cambridge University Press); Figure 7.2 (SIAM Publishing); Figures 4.1 through 4.34 and Figure 8.2 (Elsevier Press); Figures 5.3 through 5.8 (Springer-Verlag).

The book manuscript was completed while I was a long-term visitor at the Institute for Theoretical Physics at UC Santa Barbara, participating in the workshop on "The Physics of Hydrodynamic Turbulence".[2] I thank the organizers of this workshop, K. Sreenivasan, P. Constantin, I. Procaccia, and B. Shraiman for inviting me to participate, and the directors of the ITP, David Gross and Daniel Hone for making the ITP such a stimulating place to spend a sabbatical. The project was supported by the National Science Foundation through grant NSF-ITP-00-74 to the ITP, as well as

---

[2]Downloadable lectures are available at http://online.itp.ucsb.edu/online/hydrot00/

continued support to the author through the Division of Mathematical Sciences (Applied Mathematics).

Santa Barbara, California, USA   PAUL K. NEWTON
May 2001

Contents

1
Introduction

In this chapter the two main themes of this book are introduced:

1. Vorticity dynamics for the Euler equations and the point vortex decomposition.

2. Hamiltonian methods.

The discussion is largely synthetic and serves the important purpose of setting the tone for what follows. Instead of providing proofs for much of this well-known material, I refer the reader to appropriate references. After reviewing the basics of vorticity dynamics for incompressible flows in the first section, and Hamiltonian theory in the second section, the chapter ends with a summary of the main questions of interest.

1.1 Vorticity Dynamics

1.1.1 Basics

Given a flowfield with velocity distribution $\mathbf{u} = (u, v, w) \in \mathbb{R}^3$, the associated vorticity field is

$$\boldsymbol{\omega} = \nabla \times \mathbf{u}. \tag{1.1.1}$$

In this book we deal exclusively with inviscid, incompressible flows with constant (normalized) density $\rho = 1$. The incompressibility condition is given by

$$\nabla \cdot \mathbf{u} = 0. \tag{1.1.2}$$

Taking the divergence and curl of (1.1.1) gives the relations

$$\nabla \cdot \boldsymbol{\omega} = 0,$$
$$\nabla^2 \mathbf{u} = -\nabla \times \boldsymbol{\omega}.$$

The first equation shows that the components of vorticity are related to each other in the same way as the components of velocity, while the second shows that each component of the velocity satisfies a Poisson equation. If the divergence of the vorticity is integrated over a finite volume $V \in \mathbb{R}^3$, the divergence theorem yields

$$\int_V \nabla \cdot \boldsymbol{\omega} dV = \int_S \boldsymbol{\omega} \cdot \mathbf{n} dS = 0, \tag{1.1.3}$$

where S is the surface bounding the volume V with outward unit normal $\mathbf{n}$. Thus, the vorticity flux through a closed surface is zero. The same statement holds for the flux of velocity.

The flow is said to be irrotational if $\nabla \times \mathbf{u} = 0$, i.e., if the vorticity is zero. In this case there exists a scalar function ϕ, called the **velocity potential**, such that $\mathbf{u} = \nabla \phi$ as long as the flow is in all of $\mathbb{R}^3$ or a simply connected region. It follows from (1.1.2) that ϕ satisfies Laplace's equation $\nabla^2 \phi = 0$ and we write $\mathbf{u} = \mathbf{u}_\phi$ for potential flow. In general, the velocity field can be decomposed in terms of this velocity potential plus a solenoidal vector potential (sometimes referred to as a **Helmholtz**, or **Hodge decomposition** — see, for example, Lamb (1932) or Oates (1983)) $\boldsymbol{\psi}$, where $\nabla \cdot \boldsymbol{\psi} = 0$ and

$$\mathbf{u} \equiv \mathbf{u}_\phi + \mathbf{u}_\omega = \nabla \phi + \nabla \times \boldsymbol{\psi}. \tag{1.1.4}$$

The second term is generated from the vorticity distribution.

It follows from (1.1.1) that

$$\nabla^2 \boldsymbol{\psi} = -\boldsymbol{\omega}. \tag{1.1.5}$$

Hence, the vector potential satisfies a Poisson equation with vorticity on the right hand side. By standard techniques (see, for example, Courant and Hilbert, Volume II (1962)), one can write the solution of this Poisson equation in terms of the Green's function for the Laplacian (in free space)

$$\boldsymbol{\psi}(\mathbf{x}) = \int G(\mathbf{x} - \mathbf{z}) \boldsymbol{\omega}(\mathbf{z}) d\mathbf{z}, \tag{1.1.6}$$

where

$$G(\mathbf{x}) = \begin{cases} -\frac{1}{2\pi} \log \|\mathbf{x}\| & (\text{in } \mathbb{R}^2) \\ \frac{1}{4\pi} \frac{1}{\|\mathbf{x}\|} & (\text{in } \mathbb{R}^3) \end{cases} \tag{1.1.7}$$

with

$$\nabla^2 G(\mathbf{x}) + \delta(\mathbf{x}) = 0. \tag{1.1.8}$$

Then, since $\mathbf{u}_\omega(\mathbf{x}) = \nabla \times \psi$, we have

$$\mathbf{u}_\omega(\mathbf{x}) = \nabla \times \int G(\mathbf{x} - \mathbf{z})\omega(\mathbf{z})d\mathbf{z}$$

$$= \int K(\mathbf{x} - \mathbf{z})\omega(\mathbf{z})d\mathbf{z}$$

where K is the singular Biot–Savart kernel defined as

$$K(\mathbf{x}) = \begin{cases} \frac{1}{2\pi}\frac{1}{\|\mathbf{x}\|^2}(-y, x) \quad (\text{in } \mathbb{R}^2) \\[2mm] \frac{1}{4\pi}\frac{1}{\|\mathbf{x}\|^3}\begin{pmatrix} 0 & z & -y \\ -z & 0 & x \\ y & -x & 0 \end{pmatrix} \quad (\text{in } \mathbb{R}^3) \end{cases} \tag{1.1.9}$$

Often, the Biot–Savart formula in $\mathbb{R}^3$ is written:

$$\mathbf{u}_\omega(\mathbf{x}) = -\frac{1}{4\pi}\int \frac{(\mathbf{x} - \mathbf{z}) \times \omega(\mathbf{z})d\mathbf{z}}{\|\mathbf{x} - \mathbf{z}\|^3}. \tag{1.1.10}$$

Remark.   On the sphere, ∇^2 in (1.1.5) is the Laplace–Beltrami operator whose Green's function will be described in Chapter 4.   ◇

A fundamental scalar quantity associated with the vorticity is the circulation $\Gamma = \oint_C \mathbf{u} \cdot d\mathbf{s}$ of the fluid around a simple closed curve. Stokes's theorem gives

$$\Gamma = \oint_C \mathbf{u} \cdot d\mathbf{s} = \int_A \omega \cdot \mathbf{n}dS, \tag{1.1.11}$$

where A is any open surface bounded by the closed curve C. This formula shows that the circulation is the flux of vorticity through an open surface A bounded by the curve. The principal result concerning the circulation is **Kelvin's circulation theorem**, which says that in an ideal, incompressible fluid, acted on by conservative forces, the circulation $\Gamma(t)$ around a closed material curve $C(t)$ moving with the fluid is constant, i.e., $d\Gamma/dt = 0$ (see Chorin and Marsden (1979) or Saffman (1992) for proofs and discussion). Vorticity cannot be created unless one of the conditions of the theorem is violated by, for example, the introduction of viscous boundary layers or the presence of stratification, both of which are important physical effects discussed extensively in the literature. If the circulation is zero for all closed curves, then the flow is irrotational. The converse of this is true only if the fluid is contained in a simply connected region.

Curves in the fluid drawn parallel to the local vorticity vector at each point are called **vortex lines**, while the collection of vortex lines passing through the points of a closed curve in the flow is called a **vortex tube** . By construction, a vortex tube has the property that vorticity is everywhere parallel to its surface, i.e., $\boldsymbol{\omega} \cdot \mathbf{n} = 0$ on the surface of a vortex tube, hence the flux of vorticity through any cross section of the tube is a constant. The flux of vorticity along the tube is equal to the circulation around any closed curve on the tube wall circling the tube once, which is called the tube strength. A vortex tube that is surrounded by irrotational fluid is called a **vortex filament**. Typically, a vortex filament is taken to have a small cross section compared to its curvature. The fact that the circulation does not change along a tube means that vortex tubes are either closed, go off to infinity, or terminate at solid boundaries or fluid-fluid interfaces. (See Truesdell (1954) and Moffatt (1969, 1990) for more discussions on the geometry of vortex tubes.)

For ideal incompressible flows in three dimensions acted on by conservative body forces, the basic equations of motion, based on Newton's law $\mathbf{F} = m\mathbf{a}$, are given by the incompressible Euler equations

$$\frac{D\mathbf{u}}{Dt} \equiv \mathbf{u}_t + \mathbf{u} \cdot \nabla \mathbf{u} = -\nabla p + \mathbf{f}. \tag{1.1.12}$$

Here, $D/Dt \equiv (\partial/\partial t + \mathbf{u} \cdot \nabla)$ is the usual material derivative operator, p is the pressure field, and $\mathbf{f}$ represents the external body forces per unit mass. For inviscid flows the boundary conditions are given by $\mathbf{u} \cdot \mathbf{n} = 0$ on the boundary. Upon taking the curl of this equation (see Saffman (1992) or Green (1995) for details and the appropriate vector identities), one arrives at the most important dynamical equation governing the vorticity evolution,

$$\frac{D\boldsymbol{\omega}}{Dt} = \boldsymbol{\omega} \cdot \nabla \mathbf{u}. \tag{1.1.13}$$

The advantage of this formulation is the absence of the pressure term in the equations — the evolution of vorticity depends only on the local values of velocity and vorticity of the fluid. The term on the right is commonly referred to as the **vortex-stretching** term, which is crucial to many aspects of three-dimensional turbulent flows (see Majda (1986, 1988, 1991) for interesting discussions of these and other related points). Note also that if $\boldsymbol{\omega}(\mathbf{x}, 0) = 0$, then $\boldsymbol{\omega}(\mathbf{x}, t) = 0$ for $t > 0$, which means that if the vorticity is zero initially, it stays zero for all time. If a passive particle is placed in the flow field at position $\mathbf{x}(t)$, it is advected by the velocity field according to the equation

$$\dot{\mathbf{x}} = \mathbf{u}(\mathbf{x}, t).$$

By differentiating this formula (see Batchelor (1967)), one can conclude that an infinitesimal line element δl satisfies the evolution equation

$$\frac{D\delta l}{Dt} = \delta l \cdot \nabla \mathbf{u}.$$

Since this equation is identical to the vorticity evolution equation (1.1.13) with δl and $\boldsymbol{\omega}$ interchanged, we can conclude that vortex lines move precisely as material lines do. In particular, the right hand side of this equation shows nicely how, in three dimensions, material lines are stretched and compressed by the flow. For viscous flows there is an additional term on the right hand side of (1.1.13) causing diffusion of vorticity given by $R^{-1}\nabla^2\boldsymbol{\omega}$, where R is the Reynolds number. In this case, Kelvin's circulation theorem becomes

$$\frac{d\Gamma}{dt} = -R^{-1} \oint (\nabla \times \boldsymbol{\omega}) \cdot dS,$$

showing that the circulation is no longer constant. For flowfields where the velocity is confined to a two dimensional surface, the vorticity vector has only one component in the direction perpendicular to the surface, hence the vortex-stretching term is absent. Primarily for this reason, two dimensional flows are quite different from three dimensional flows. In fact, the basic long time existence theory for the Euler equations in $\mathbb{R}^3$ has not yet been settled (see Ebin and Marsden (1970), Kato (1972), Temam (1975, 1976), Beale, Kato, Majda (1984), Ponce (1985)). It is believed by some that the vortex-stretching term produces a finite-time singularity (see, for example, Grauer and Sideris (1991, 1995), Kerr (1993), and Pumir and Siggia (1992a,b), Pelz (1997)) and we know that the key to this is the quantity $\int_0^T \|\boldsymbol{\omega}\|_{L^\infty} dt$, where $\|\cdot\|_{L^\infty}$ is the L^∞ norm, which must blow up if there is a finite-time singularity, as shown by Beale, Kato, and Majda (1984) and discussed more recently in the books of Doering and Gibbon (1995), P.L. Lions (1996), and Majda and Bertozzi (2001). In $\mathbb{R}^2$ when the vortex-stretching term is absent, long time existence has been established, see Wolibner (1933), Yudovich (1963), Kato (1967), and Bardos (1972).

1.1.2 Invariants of the Euler Equations

The inviscid equations have several invariants which will play an important role in future discussions. Assuming the velocity field decays sufficiently rapidly at infinity and that there are no external body forces, the total kinetic energy E in the flow and the total vorticity W_1 are conserved

$$E(t) = \frac{1}{2} \int_{\mathbb{R}^3} \|\mathbf{u}\|^2 d\mathbf{x} = E(0)$$

$$W_1(t) = \int_{\mathbb{R}^3} \boldsymbol{\omega}(\mathbf{x}) d\mathbf{x} = W_1(0).$$

For viscous flows, kinetic energy is no longer conserved since viscosity causes it to be transformed into internal energy. In general, one has

$$\frac{dE}{dt} = -R^{-1} \int_{\mathbb{R}^3} \|\boldsymbol{\omega}\|^2 d\mathbf{x} \leq 0.$$

However, since the macroscopic equations do not include microscopic effects, there is no accounting for increase in the internal energy to balance the decrease in kinetic energy.

The Euler equations are invariant with respect to spatial translations and rotations, so the "linear momentum" L, and "angular momentum" I, of the fluid are conserved

$$L(t) = \frac{1}{2} \int_{\mathbb{R}^3} (\mathbf{x} \times \boldsymbol{\omega}) d\mathbf{x} \;\; = \;\; L(0) \tag{1.1.14}$$

$$I(t) = \frac{1}{3} \int_{\mathbb{R}^3} \mathbf{x} \times (\mathbf{x} \times \boldsymbol{\omega}) d\mathbf{x}$$

$$= -\frac{1}{2} \int_{\mathbb{R}^3} \|\mathbf{x}\|^2 \boldsymbol{\omega} d\mathbf{x} \;\; = \;\; I(0). \tag{1.1.15}$$

These quantities are also conserved for viscous flows. An interesting topological invariant for inviscid flows, as shown by Moffatt (1969), is the so-called **helicity** J where

$$J = \int_{\mathbb{R}^3} \mathbf{u} \cdot \boldsymbol{\omega} d\mathbf{x} = J(0).$$

Its conservation follows essentially from Kelvin's circulation theorem. The helicity measures the degree of linkage of groups of closed vortex filaments. For example, if we have a collection of n linked vortex filaments C_k, where $k = 1, \ldots, n$, each of strength Γ_k then

$$J = \sum_{i,j} \alpha_{ij} \Gamma_i \Gamma_j,$$

where α_{ij} is the winding number of C_i and C_j (i.e., the integral number of times the filaments wind around each other). This is seen by assuming the vorticity distribution $\boldsymbol{\omega}$ is everywhere zero except on the two (or more) closed filaments C_1, C_2 (see Chapter 8). Then the volume integral defining J degenerates to

$$J = \Gamma_1 \oint_{C_1} \mathbf{u} \cdot d\mathbf{x} + \Gamma_2 \oint_{C_2} \mathbf{u} \cdot d\mathbf{x}.$$

By Stokes's theorem, the first integral is equal to the vorticity flux across the surface spanning C_1, which is given by $\pm n\Gamma_2$, with n being the linking number (number of times C_2 pierces the spanning surface of C_1), and +

(−) is chosen if the linkage is right (left) handed. One argues similarly for the second integral, obtaining

$$J = \pm n\Gamma_1\Gamma_2 \pm n\Gamma_2\Gamma_1 = \pm m\Gamma_1\Gamma_2.$$

This result can be generalized in the obvious way for arbitrarily many filaments. For example, as shown below in Figure 1.1, two unlinked filaments have $\alpha_{12} = 0$, two singly linked filaments have $\alpha_{12} = \pm 1$, etc. A knotted vortex filament can be decomposed into unknotted but linked filaments by inserting pairs of equal and opposite vortex segments. Therefore, the helicity can also be used to classify the degree of knottedness of a single knotted filament, as described in Moffatt (1969, 1985, 1990). The helicity alone, however, is not sufficient to provide an unambiguous classification for knotted and linked vortex filaments, as seen below for the configuration (c) which has zero helicity, yet is topologically distinct from the case of unlinked filaments. Hence, a more complete classification scheme is desirable, such as, for example, the concept of the "energy spectrum" of a knot. For more on these developments, we refer the reader to Moffatt (1990), Freedman and He (1991a,b), or Arnold and Khesin (1992, 1998). For a nice description of the effect of viscosity on the helicity invariant, see the article by Aref and Zawadzki (1991). More generally, however, the vorticity field of a given turbulent or complex flow cannot be explicitly written as a collection of linked vortex tubes. However, one can always decompose $\boldsymbol{\omega}$ into the sum of three fields,

$$\boldsymbol{\omega} \equiv \boldsymbol{\omega}_1 + \boldsymbol{\omega}_2 + \boldsymbol{\omega}_3,$$

where

$$\boldsymbol{\omega}_1 = \left(0, \frac{\partial u_1}{\partial z}, -\frac{\partial u_1}{\partial y}\right)$$

$$\boldsymbol{\omega}_2 = \left(-\frac{\partial u_2}{\partial z}, 0, \frac{\partial u_2}{\partial x}\right)$$

$$\boldsymbol{\omega}_3 = \left(\frac{\partial u_3}{\partial y}, -\frac{\partial u_3}{\partial x}, 0\right).$$

Since the vortex lines for the field $\boldsymbol{\omega}_1$ are the closed curves given by $u_1 = const$, $x = const$, the self-helicity is zero. This is true also for $\boldsymbol{\omega}_2$ and $\boldsymbol{\omega}_3$, and the helicity invariant can be written

$$J = \sum_{n \neq m} J_{nm}, \qquad J_{nm} = \int \mathbf{u}_n \cdot \boldsymbol{\omega}_m dV,$$

where J_{nm} is the degree of linkage of the vorticity fields $\boldsymbol{\omega}_n$, $\boldsymbol{\omega}_m$. See the review article of Moffatt and Tsinobar (1992) for more thorough discussions of the role of the helicity invariant in general flows.

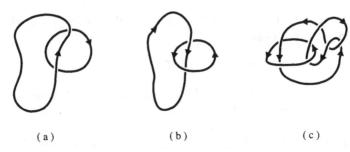

(a)   (b)   (c)

FIGURE 1.1. Geometry of vortex tubes : (a) $\alpha_{12} = 0$, (b) $\alpha_{12} = -1$, (c) $\alpha_{12} = 0$.

Although not conserved in three dimensions, an important integral quantity is the **enstrophy** W_2,

$$W_2 = \frac{1}{2} \int_{\mathbb{R}^3} \|\boldsymbol{\omega}\|^2 d\mathbf{x}.$$

As we have already seen, the energy and enstrophy are related by

$$\frac{dE}{dt} = -2R^{-1}W_2.$$

Another common integral quantity is the **palinstrophy** P,

$$P = \frac{1}{2} \int_{\mathbb{R}^3} (\nabla \times \boldsymbol{\omega})^2 d\mathbf{x},$$

which can be related to the rate of change of enstrophy via

$$\frac{dW_2}{dt} = -2R^{-1}P + \int_{\mathbb{R}^3} \boldsymbol{\omega} \cdot (\boldsymbol{\omega} \cdot \nabla)\mathbf{u} d\mathbf{x}.$$

The helicity, energy, and enstrophy of the flow are related via the Schwarz inequality to give a lower bound on the enstrophy

$$W_2 \geq \frac{J^2}{E}.$$

See Moffatt (1969), Arnold (1974), and Moffatt and Tsinobar (1992) for more discussions of this, and Chapter 3 of Saffman (1994) for complete derivations of the invariants. For more general discussions of the Euler invariants see also Khesin and Chekanov (1989), as well as the paper of Howard (1957).

1.1.3   Two Dimensions

In two dimensions to which we now specialize, since the right hand side of (1.1.13) is zero, the vorticity evolution equation becomes a conservation law for the *scalar* vorticity

$$\frac{D\omega}{Dt} \equiv 0, \tag{1.1.16}$$

where $\omega = \left(\dfrac{\partial v}{\partial x} - \dfrac{\partial u}{\partial y}\right)$. For two-dimensional flows, there is an infinite hierarchy of invariant quantities W_k ($k = 1, 2, 3 \ldots$) called the **vorticity moments**, where

$$W_k = \frac{1}{k} \int_{\mathbb{R}^2} \omega^k(\mathbf{x}) d\mathbf{x}.$$

Thus, in two dimensions both the energy and enstrophy are conserved (a fact that is usually said to prohibit a direct cascade of energy to small scales), while in three dimensions the enstrophy can increase due to vortex tube stretching, as discussed, for example, in Kraichnan and Montgomery (1980). In fact, it is easily verified that any smooth function of the vorticity $f(\omega)$ is conserved under the flow. To see this, consider a closed path C enclosing area A that moves with the flowfield. Then

$$\frac{d}{dt} \int_C f(\omega(\mathbf{x})) d\mathbf{x} = \int_C f'(\omega(\mathbf{x})) \left[\frac{\partial \omega}{\partial t} + \mathbf{u} \cdot \nabla \omega\right] d\mathbf{x} = 0.$$

An additional feature of two-dimensional flows not available for 3-D flows (unless they possess certain symmetries — see the papers of Mezić and Wiggins (1994), and Haller and Mezić (1998) for more on this) is that the vector potential ψ becomes the scalar streamfunction

$$\mathbf{u} = \nabla \times \psi = \nabla^\perp \psi \equiv \begin{pmatrix} 0 & 1 \\ -1 & 0 \end{pmatrix} \begin{pmatrix} \partial_x \\ \partial_y \end{pmatrix} \psi = (\psi_y, -\psi_x).$$

The equation for a fluid particle then becomes

$$\dot{\mathbf{x}} = \mathbf{u} = (\psi_y, -\psi_x)$$

which takes the form of *Hamilton's* canonical equations

$$\dot{x} = \frac{\partial \mathcal{H}}{\partial y}(x, y; t)$$

$$\dot{y} = -\frac{\partial \mathcal{H}}{\partial x}(x, y; t).$$

The time-dependent Hamiltonian plays the role of the streamfunction $\mathcal{H} \equiv \psi$, with the position coordinates (x, y) acting as the canonically conjugate variables. When the flow is time-independent, the level curves of the Hamiltonian are the streamlines that the fluid particles (so called "passive tracers") follow, their velocity being tangent to the streamline. This can be understood by noting that the normal vector $\mathbf{n}$ to a level curve $\psi = const$ is given by

$$\mathbf{n} = \nabla \psi = \nabla \mathcal{H}. \tag{1.1.17}$$

Since $\mathbf{u} \cdot \mathbf{n} = \mathbf{u} \cdot \nabla \psi = 0$, the local fluid velocity is tangent to each level curve. Also, since $\mathbf{u} \cdot \mathbf{n} = 0$, each streamline can be viewed as a solid boundary, which no fluid can penetrate. For time-dependent flows the streamline pattern evolves dynamically, and it is of interest to understand the possible instantaneous topological patterns achievable by a given flow. Therefore, in two dimensions the "vorticity- streamfunction" or "vorticity-Hamiltonian" form of the Euler equations becomes

$$\omega_t + \mathcal{J}(\omega, \mathcal{H}) = 0$$
$$\nabla^2 \mathcal{H} = -\omega,$$

where $\mathcal{J}$ is the Jacobian

$$\mathcal{J}(\omega, \mathcal{H}) \equiv \left(\frac{\partial \omega}{\partial x} \frac{\partial \mathcal{H}}{\partial y} - \frac{\partial \omega}{\partial y} \frac{\partial \mathcal{H}}{\partial x} \right).$$

Hence, in general, one must solve a Poisson equation for the Hamiltonian, coupled nonlinearly to the evolution equation for the vorticity profile. An alternative and sometimes useful formulation arises by uncoupling these equations and writing a single evolution equation for the Hamiltonian as

$$\Delta \mathcal{H}_t + (\nabla^\perp \mathcal{H}) \cdot \nabla(\Delta \mathcal{H}) = 0,$$

where

$$\omega = \nabla \times (\nabla^\perp \mathcal{H}).$$

Since we are interested in solutions to this equation such that $\mathcal{H} = const$, we refer to this equation as the **level set** equation for vortex motion. In particular, level sets of this equation for a given vorticity distribution will give rise to streamline patterns generated by the flowfield and will be discussed further in Chapter 4 in the context of spherical atmospheric weather patterns. It is worth mentioning here that several naturally occuring flows, such as Hele–Shaw flows, flows on soap films, and many large scale atmospheric and oceanographic flows, can be considered two-dimensional. See Kraichnan (1975) and Kraichnan and Montgomery (1980) for an overview of 2-D turbulence theory.

The power of the vorticity formulation is now clear — vorticity is a conserved quantity that acts as a fluid marker. In principle, one can experimentally track a patch of vorticity just as one would track a particle introduced in the fluid. Two dimensional vorticity is simply transported by the velocity field, while in 3-D vortex tubes can stretch or shrink. This "Lagrangian" concept is quite appealing and gives a complementary approach to the "Eulerian" idea of measuring the velocity field $\mathbf{u}(\mathbf{x}, t)$ at a fixed station.

Finally, we mention that in two dimensions, one can define a complex quantity called the complex potential.

Definition 1.1.1. *The complex analytic function*

$$w(z) \equiv \phi + i\psi$$

is called the **complex potential** *for the flow, where* $z = x + iy$. *The real part,* ϕ, *is the* **velocity potential**, *while the imaginary part,* ψ, *is the* **streamfunction**. *Since the function* $w(z)$ *is analytic,* ϕ *and* ψ *are related to each other via the Cauchy–Riemann equations*

$$\frac{\partial \phi}{\partial x} = \frac{\partial \psi}{\partial y}$$

$$\frac{\partial \phi}{\partial y} = -\frac{\partial \psi}{\partial x}.$$

The **complex velocity** *is defined by*

$$\frac{dw}{dz} = u - iv.$$

Occasionally, it is useful to write the velocity and vorticity in terms of polar coordinates. The radial and azimuthal velocity components, (u_r, u_θ) are given by

$$u_r = \frac{1}{r}\frac{\partial \psi}{\partial \theta}, \qquad u_\theta = -\frac{\partial \psi}{\partial r},$$

while the vorticity is given by the formula

$$\omega = -\frac{1}{r}\frac{\partial}{\partial r}\left(r\frac{\partial \psi}{\partial r}\right) - \frac{1}{r^2}\frac{\partial^2 \psi}{\partial \theta^2}.$$

We list here several examples of complex potentials:

Point source flow

$$w(z) = \frac{m}{2\pi}\log(z - z_0)$$

$$\phi(z) = \frac{m}{2\pi}\log|z - z_0|$$

$$\psi(z) = \frac{m}{2\pi}arg(z - z_0)$$

$$\frac{dw}{dz} = \frac{m}{2\pi}\cdot\frac{(z - z_0)^*}{|z - z_0|^2}$$

Point vortex flow

$$w(z) = \frac{\Gamma}{2\pi i}\log(z - z_0)$$

$$\phi(z) = \frac{\Gamma}{2\pi}arg(z - z_0)$$

$$\psi(z) = -\frac{\Gamma}{2\pi}\cdot\log|z - z_0|$$

$$\frac{dw}{dz} = \frac{\Gamma}{2\pi i}\cdot\frac{(z - z_0)^*}{|z - z_0|^2}$$

Corner flow at origin, with angle π/m

$$w(z) = Az^m$$
$$\phi = Ar^m \cos m\theta$$
$$\psi = Ar^m \sin m\theta$$
$$\frac{dw}{dz} = mAz^{m-1}$$

1.1.4 Point Vortex Decomposition

We next turn our attention to the limiting case of thin parallel vortex filaments and formulate the equations for point vortex dynamics on two-dimensional surfaces. When the vorticity is concentrated along a perfectly straight vortex filament, a slice of this filament by a two-dimensional surface gives us a vorticity distribution in the plane that is highly localized around the centerline of the filament, i.e., the vorticity function has small support. In general, for a discrete vortex representation in two dimensions, one assumes a vorticity distribution in the form

$$\omega(\mathbf{x}) = \sum_{i=1}^{N} \frac{\Gamma_i}{2\pi} \phi_\epsilon(\mathbf{x} - \mathbf{x}_i)$$
$$\phi_\epsilon(\mathbf{x}) = \frac{1}{\epsilon^2} \phi\left(\frac{\mathbf{x}}{\epsilon}\right)$$

with $\epsilon << 1$ and ϕ any suitably normalized radially symmetric function so that $\int \phi d\mathbf{x} = 1$. A limiting ($\epsilon \to 0$) case of such a representation, called a **point vortex**, is given by the choice

$$\phi_\epsilon(\mathbf{x}) = \delta(\mathbf{x}),$$

where δ is the usual Dirac delta function. The velocity field and vorticity associated with an isolated point vortex is given by

$$u_\theta(r, \theta, z) = \frac{\Gamma}{2\pi r}$$
$$u_r = u_z = 0$$
$$\omega_z = \delta(r)$$

This flow is sometimes referred to as an **irrotational vortex**, since $\omega = 0$ except at the location of the point vortex, which has zero measure. Physically, this discrete vortex approximation is analogous to the "point-mass" approximation used in gravitational N-body problems where the internal structure and deformation of the local vorticity is ignored and one is interested only in tracking the vorticity maximum or center point of the vorticity region. There are, in fact, rigorous results of Marchioro and Pulverinti (1983), which show that point vortices faithfully track the centers of vorticity of smoothed out vorticity distributions (such as vortex patches) for sufficiently short times. The paper of Turkington (1987) treats the case of a single vortex in a bounded domain and obtains the result that as the initial vorticity distribution tends to a Dirac measure, the corresponding solution to the Euler equation converges weakly to the solution of the corresponding point vortex problem, a result that is valid over arbitrarily long time intervals. A global result of this type was obtained by Marchioro (1988) for like-signed vortices.

Detailed questions involving the internal dynamics of the local vorticity distribution, while clearly important particularly for understanding the generation of small-scale structures, are not included in the model unless one superposes a cluster of point vortices and uses this to represent a distributed vorticity region. The usefulness of the technique ultimately rests on the physical fact that many fluid flows of interest have compact vorticity distributions, making it appealing to directly model the vorticity as opposed to the velocity. There are, however, some shortcomings of the point vortex approximation to real-life flows, and we refer the reader to Aref (1983, 1985) and Saffman and Baker (1979) for some discussion on this topic.

To derive the equations for the dynamics of a collection of N vortices located at $\mathbf{x}_\alpha = (x_\alpha(t), y_\alpha(t))$ where $\alpha = 1, \ldots, N$, we start with the velocity equations and note that the velocity induced by an isolated point vortex is

$$\dot{\mathbf{x}} = \mathbf{u}(\mathbf{x}_\alpha, t)$$
$$= \nabla^\perp \psi_\alpha(\mathbf{x}_\alpha, t).$$

From (1.1.6)

$$\psi_\alpha(\mathbf{x}, t) = -\frac{1}{2\pi} \int \Gamma_\alpha \log(\mathbf{x} - \mathbf{z}) \delta(\mathbf{x}_\alpha - \mathbf{z}) dz$$
$$= -\frac{\Gamma_\alpha}{2\pi} \log(\mathbf{x} - \mathbf{x}_\alpha).$$

The velocity field due to a collection of N such vortices is obtained by linear superposition (since the velocity-vorticity relation is a linear one)

$$\dot{\mathbf{x}} = \sum_{\alpha=1}^{N} \nabla^\perp \psi_\alpha(\mathbf{x}, t).$$

Then, since each point vortex moves with the local velocity of the fluid and the vorticity is conserved, we have the equations for a collection of N-vortices

$$\dot{\mathbf{x}}_\beta = \sum_{\alpha \neq \beta}^{N} \nabla^\perp \psi_\alpha(\mathbf{x}_\beta, t), \tag{1.1.18}$$

where $\beta = 1, \dots, N$.

There are many alternative ways of writing the equations of motion for N-vortices, most of which will be discussed in later chapters. We mention here two of the principal systems to be studied.

1. Vortex dynamics in the plane:

$$\dot{x}_\alpha = -\frac{1}{2\pi} \sum_{\beta \neq \alpha}^N \frac{\Gamma_\beta (y_\alpha - y_\beta)}{l_{\beta\alpha}^2} \qquad (1.1.19)$$

$$\dot{y}_\alpha = \frac{1}{2\pi} \sum_{\beta \neq \alpha}^N \frac{\Gamma_\beta (x_\alpha - x_\beta)}{l_{\beta\alpha}^2} \qquad (1.1.20)$$

where $l_{\beta\alpha} = \|\mathbf{x}_\beta - \mathbf{x}_\alpha\|$ are the intervortical distances. The system is more compactly expressed in the complex notation $z_\alpha(t) = x_\alpha(t) + iy_\alpha(t)$,

$$\dot{z}_\alpha = \frac{i}{2\pi} \sum_{\beta \neq \alpha}^N \Gamma_\beta \frac{z_\alpha - z_\beta}{|z_\alpha - z_\beta|^2}. \qquad (1.1.21)$$

Equivalently,

$$\dot{z}_\alpha^* = \frac{1}{2\pi i} \sum_{\beta \neq \alpha}^N \frac{\Gamma_\beta}{z_\alpha - z_\beta}, \qquad (1.1.22)$$

assuming all vortex strengths Γ_β are real, and $\alpha = 1, \dots, N$. One can find a nice review of many aspects of this planar system in Aref (1982) and it is treated in detail in Chapter 2.

2. Vortex dynamics on spheres:

$$\dot{\theta}_\alpha = -\frac{1}{2\pi} \sum_{\beta \neq \alpha}^N \frac{\Gamma_\beta \sin\theta_\beta \sin(\phi_\alpha - \phi_\beta)}{l_{\beta\alpha}^2} \qquad (1.1.23)$$

$$\sin\theta_\alpha \dot{\phi}_\alpha = \frac{1}{2\pi} \sum_{\beta \neq \alpha}^N \frac{\Gamma_\beta \gamma_{\alpha\beta}}{l_{\beta\alpha}^2} \qquad (1.1.24)$$

using standard spherical coordinates, and where

$$\gamma_{\alpha\beta} = \sin\theta_\alpha \cos\theta_\beta - \cos\theta_\alpha \sin\theta_\beta \cos(\phi_\alpha - \phi_\beta).$$

The details can be found in Chapter 4.

3. Vortex dynamics on general curved surfaces:

$$\dot{z}_\alpha^* = h^{-2}(z_\alpha, z_\alpha^*) \left(\frac{1}{2\pi i} \sum_{\beta \neq \alpha}^N \frac{\Gamma_\beta}{z_\alpha - z_\beta} \right.$$
$$\left. - \frac{\Gamma_\alpha}{2\pi i} \frac{\partial \log[h(z_\alpha, z_\alpha^*)]}{\partial z_\alpha} \right) \qquad (1.1.25)$$

where

$$g_{ij}(x,y) \equiv \delta_{ij} h^2(x,y)$$

is the Riemannian metric characterizing the surface, and

$$ds^2 = h^2(x,y)(dx^2 + dy^2)$$

is the line element associated with the surface. For example, surfaces of revolution are characterized by $h(x,y) = h(r)$ where $r = (x^2 + y^2)^{1/2}$. A special case is a sphere, where

$$h(x,y) = h(r) = 2R(1+r^2)^{-1},$$

and $r = \tan(\theta/2)$. The line element is

$$ds^2 = R^2(d\theta^2 + \sin^2\theta d\phi^2)$$
$$= 4R^2(1+r^2)^{-2}(dr^2 + r^2 d\phi^2).$$

Derivations of these formulas as well as discussions of conservation laws and related issues can be found in Hally (1980). In a recent paper of Kimura (1999) the equations of motion for point vortices on general surfaces of *constant* curvature are derived in a unified way that incorporates the system on a sphere (positive constant curvature), plane (zero curvature), and the hyperbolic plane (negative constant curvature).

Remark. Mathematically, one could take the vortex strengths Γ_α to be pure imaginary, in which case the system corresponds to a collection of interacting sources and sinks. In general, when $\Gamma_\alpha \in \mathbb{C}$, the system combines rotation due to vorticity along with source-sink behavior (called "spiral vortices" in Kochin, Kibel, and Rose (1965)).   ◊

We can unify the planar and spherical formulas by writing the velocity field at an arbitrary point $\mathbf{x}$ due to a point vortex of strength Γ_α located at position $\mathbf{x}_\alpha$ as

$$\dot{\mathbf{x}} = \frac{\Gamma_\alpha}{2\pi} \cdot \frac{\mathbf{n}_\alpha \times (\mathbf{x} - \mathbf{x}_\alpha)}{\|\mathbf{x} - \mathbf{x}_\alpha\|^2},$$

where $\mathbf{n}_\alpha$ is the unit normal vector to the surface at the vortex location $\mathbf{x}_\alpha$. The velocity due to a collection of N such vortices is then

$$\dot{\mathbf{x}} = \sum_{\alpha=1}^{N} \frac{\Gamma_\alpha}{2\pi} \cdot \frac{\mathbf{n}_\alpha \times (\mathbf{x} - \mathbf{x}_\alpha)}{\|\mathbf{x} - \mathbf{x}_\alpha\|^2},$$

which gives the equations for a collection of N-vortices

$$\dot{\mathbf{x}}_\beta = \sum_{\alpha \neq \beta}^N \frac{\Gamma_\alpha}{2\pi} \cdot \frac{\mathbf{n}_\alpha \times (\mathbf{x}_\beta - \mathbf{x}_\alpha)}{\|\mathbf{x}_\beta - \mathbf{x}_\alpha\|^2}, \tag{1.1.26}$$

where $\beta = 1, \dots, N$. The characteristic feature of the surface enters only through specifying the normal vector at each vortex location.

Example 1.1 For the planar problem, $\mathbf{n}_\alpha = \mathbf{e}_z$, and the equations become

$$\dot{\mathbf{x}}_\beta = \sum_{\alpha \neq \beta}^N \frac{\Gamma_\alpha}{2\pi} \cdot \frac{\mathbf{e}_z \times (\mathbf{x}_\beta - \mathbf{x}_\alpha)}{l_{\beta\alpha}^2}. \tag{1.1.27}$$

On the sphere with radius R, $\mathbf{n}_\alpha = \mathbf{x}_\alpha / R$ and the equations are

$$\dot{\mathbf{x}}_\beta = \sum_{\alpha \neq \beta}^N \frac{\Gamma_\alpha}{2\pi} \cdot \frac{(\mathbf{x}_\alpha / R) \times (\mathbf{x}_\beta - \mathbf{x}_\alpha)}{l_{\beta\alpha}^2}. \tag{1.1.28}$$

One, of course, has the additional constraint on the sphere that $\|\mathbf{x}_\alpha\| = R$. $\Diamond$

Remark. For numerical purposes, the vorticity distribution is often desingularized into "vortex blobs," whose centers evolve according to the equations

$$\dot{\mathbf{x}}_\beta = \sum_{\alpha \neq \beta}^N \frac{\Gamma_\alpha}{2\pi} \cdot \frac{\mathbf{n}_\alpha \times (\mathbf{x}_\beta - \mathbf{x}_\alpha)}{l_{\beta\alpha}^2 + \delta_c^2},$$

where δ_c^2 is a parameter measuring the core size. $\delta_c^2 = 0$ corresponds to a point vortex system , while $\delta_c^2 > 0$ gives vortex blobs with vorticity profiles

$$\omega(\mathbf{x}) = \frac{\Gamma_\alpha}{\pi} \cdot \frac{\delta_c^2}{\left(\delta_c^2 + (\mathbf{x} - \mathbf{x}_\alpha)^2\right)^2}.$$

Discussions of this system in the plane can be found in Leonard (1985), Winckelmans and Leonard (1993), and Cottet and Koumoutsakos (2000), but no work has been done on this system for the spherical case. $\Diamond$

We finish this section by listing several other well-studied and useful vorticity profiles that the reader should be familiar with, as points of comparison to the point vortex profiles emphasized in this book.

1. Solid body rotation

$$u_\theta(r, \theta, z) = \Omega r$$
$$u_r = u_z = 0$$
$$\omega_z = \frac{1}{r}\left(\frac{\partial(r u_\theta)}{\partial r} - \frac{\partial u_r}{\partial \theta}\right) = 2\Omega$$
$$\Gamma = 2\pi r^2 \Omega \to \infty \quad \text{as} \quad r \to \infty$$

2. Rankine vortex

$$u_\theta(r, \theta, z) = \begin{cases} \left(\dfrac{\Gamma}{2\pi R^2}\right) r, & r \le R \\ \dfrac{\Gamma}{2\pi r}, & r > R \end{cases}$$

$$\omega_z = \begin{cases} \dfrac{\Gamma}{\pi R^2}, & r \le R \\ \delta(r), & r > R \end{cases}$$

$$u_r = u_z = 0$$

The flow corresponds to solid body rotation inside a circle of radius R and point vortex flow outside the circle. The Rankine vortex is a special degenerate member of a family of elliptical vortex patches known as Kirchhoff elliptical patches. See Saffman (1992) for further discussion.

3. Lamb–Oseen vortex (diffusive)

$$u_\theta(r, \theta, z) = \frac{\Gamma}{2\pi r}\left[1 - \exp\left(-\frac{r^2}{4\nu t}\right)\right]$$
$$u_r = u_z = 0$$
$$\omega_z = \frac{\Gamma}{4\pi \nu t}\exp\left(-\frac{r^2}{4\nu t}\right)$$

The flow corresponds to the evolution of an initial point vortex in the presence of diffusion ν. The so-called vortex core, in this case, increases like $\sqrt{\nu t}$.

4. Burgers vortex (3D stretching)

This is a three-dimensional steady flow in which diffusion of vorticity counterbalances convection and stretching in the z direction.

$$u_\theta(r, \theta, z) = \frac{\Gamma}{2\pi r} \left[1 - \exp\left(-\frac{r^2}{2\delta^2} \right) \right]$$
$$u_r = -\sigma r$$
$$u_z = 2\sigma z$$
$$\omega_z = \frac{\Gamma}{\pi \delta^2} \exp\left(-\frac{r^2}{2\delta^2} \right)$$

The azimuthal component of the flow corresponds to the Lamb–Oseen vortex at fixed time, but includes the effects of stretching in the z direction, as well as an inward radial component. See the original articles of Rott (1958) and Burgers (1948). The core size is $\delta = \sqrt{\nu/\sigma}$.

Most of these examples can be found in the standard texts on fluid mechanics, such as Lamb (1932), Landau and Lifschitz (1959), or Batchelor (1967). The book of Saffman (1992) also contains many examples as do the recent books of Leal (1992) and Pozrikidis (1997).

1.1.5 Bibliographic Notes

There are, of course, many texts where one can get a general introduction to classical incompressible flows (see Lamb (1932), Batchelor (1967), Serrin (1959), Sommerfeld (1964), Landau and Lifschitz (1959), or Kochin, Kibel and Rose (1965), for example). A nice overview of the general mathematical theory can be found in the classic book of Ladyzhenskaya (1969), the article by Caflisch (1988), and the books of Marchioro and Pulvirenti (1994), Doering and Gibbon (1995), Lions (1996), and Majda and Bertozzi (2001). In particular, Chapter 4 of Lions (1996) contains a nice account of the current status of the mathematical theory. A concise summary of the two-dimensional theory is given in Sulem and Sulem (1983) and a general description of the mathematical status of the Navier–Stokes equations (authored by C. Fefferman) can be downloaded from http://www.claymath.org/.

One can also find a large literature on the mathematical aspects of the discrete vortex approximation on which numerical methods are based. Because of the singular nature of the Biot–Savart kernel (1.1.9), as two vortices approach each other, their velocity becomes unbounded. Since this behavior is unphysical and causes problems with numerical implementation, a wide range of techniques for smoothing the singularity in the Biot–Savart formula have been proposed — a good entry into this discussion is the review article by Hald (1991). The choice of a smoothing function ϕ_ϵ to represent vorticity, as well as the method to smooth the singularity in the

Biot–Savart kernel, affect the accuracy of the method and dictate its convergence properties. In general, for a given choice of smoothing function ϕ_ϵ, one would like to prove that as $\epsilon \to 0$ (along with other discretizing parameters such as mesh size and timestep) the solutions of the discretized problem converge to solutions of the Euler equations. The literature on this topic is large and begins with the papers by Hald and del Prete (1978), Hald (1979), followed by Beale and Majda (1982), which were the first to show that vortex methods could be designed in both two and three dimensions to rigorously approximate smooth inviscid flows. The field has developed rapidly since then, with a large number of papers significantly refining the method — we refer the readers to Sethian (1991), Hald (1991), as well as the review by Hou (1991) and the book of Cottet and Koumoutsakos (2000). For some recent convergence results in both two and three dimensions, see Goodman, Hou, and Lowengrub (1990), and Hou and Lowengrub (1990). For periodic domains and other recent developments, see Schochet (1995, 1996). A survey of several recent topics related to vortex methods can be found in Puckett (1993) and a good mathematical introduction can be found in the book of Marchioro and Pulvirenti (1994).

Related at least in spirit, not in detail, to these ideas is a growing body of work on the "averaged-Euler," or "Euler- α " equations which are written

$$(1 - \alpha^2 \Delta)\partial_t \mathbf{u} + \mathbf{u} \cdot \nabla(1 - \alpha^2 \Delta)\mathbf{u} - \alpha^2(\nabla \mathbf{u})^t \cdot \Delta \mathbf{u} = -\nabla p,$$

along with the usual incompressibility condition. The parameter $\alpha > 0$ can be thought of as a "cut-off" length scale below which spatial information is averaged (see Shkoller (2000) for discussions). One can take the curl of this equation in the usual way to see the vorticity evolution equation, which in two dimensions is

$$q_t + \mathbf{u} \cdot \nabla q = 0$$
$$\mathbf{u} = K^\alpha * q,$$

where

$$q = (1 - \alpha^2 \Delta)\nabla \times \mathbf{u}.$$

One can see from this that q is a potential vorticity term, and K^α is the integral kernel associated with the inverse of $(1 - \alpha^2 \Delta)\nabla \times$, hence generalizes the kernel (1.1.9) and reduces to it when $\alpha = 0$. Interesting discussions of this equation and its relation to the vortex blob method (using Bessel function smoothing) and the equations for non-Newtonian fluids can be found in Oliver and Shkoller (1999), where it is shown that the vortex blob method generates unique global weak solutions to the averaged Euler equations. Holm, Marsden, and Ratiu (1998a) first wrote the equation as an approximation to the Euler equations and the mathematical structure, including variational forms, are developed in Holm, Marsden, and Ratiu (1998b). A nice review of the Euler-α equation can be found in Marsden and Shkoller (2000).

1.2 Hamiltonian Dynamics

1.2.1 Canonical Formulation

The starting point for our discussion are the canonical Hamiltonian equations in the form $\mathcal{H}(\mathbf{q}, \mathbf{p}) \equiv \mathcal{H}(\mathbf{z})$, where $\mathbf{z} \equiv (q_1, \ldots, q_n; p_1, \ldots, p_n) \in \mathbb{R}^{2n}$, and

$$\dot{\mathbf{z}} = \mathbb{J} \cdot \nabla \mathcal{H}(\mathbf{z}). \qquad (1.2.1)$$

Here, $\nabla \equiv (\partial_{z_1}, \ldots, \partial_{z_{2n}})$ is the gradient operator on $\mathbb{R}^2$, and $\mathbb{J}$ is the $2n \times 2n$ symplectic matrix

$$\mathbb{J} = \begin{pmatrix} 0 & \mathbf{1} \\ -\mathbf{1} & 0 \end{pmatrix}$$

where $\mathbf{1}$ is a $n \times n$ unit matrix.

Definition 1.2.1. *A matrix* $\mathbb{T}$ *is* **symplectic** *if* $\mathbb{T}^T \mathbb{J} \mathbb{T} = \mathbb{J}$, *and (1.2.1) is said to be in* **symplectic form.**

Notice that $\mathbb{J}$ is symplectic since $\mathbb{J}^{-1} = \mathbb{J}^T$. The right hand side of (1.2.1) is called the **symplectic gradient** of $\mathcal{H}$.

Example 1.2 For the N-vortex problem in the plane we can introduce the new variables

$$q_i = \sqrt{|\Gamma_i|} sgn(\Gamma_i) x_i$$
$$p_i = \sqrt{|\Gamma_i|} sgn(\Gamma_i) y_i$$
$$\mathbf{z} = (q_1, \ldots, q_n; p_1, \ldots, p_n) \in \mathbb{R}^{2n},$$

$\mathbb{J}$ is the $2n \times 2n$ symplectic matrix and the Hamiltonian is

$$\mathcal{H}(\mathbf{q}, \mathbf{p}) = -\frac{1}{4\pi} \sum_{i \neq j} \Gamma_i \Gamma_j \log(q_i - p_j). \qquad \Diamond$$

We summarize here the main facts from classical mechanics that take us from physical principles to these equations. (See, for example Landau and Lifshitz (1959), Goldstein (1980), or Chetaev (1989).)

1. One starts by writing down a suitable Lagrangian $L(\mathbf{q}, \dot{\mathbf{q}}; t) = T - V$ in terms of the kinetic energy T and the potential energy V. From this we can define the action integral

$$W = \int_{t_1}^{t_2} L(\mathbf{q}, \dot{\mathbf{q}}; t),$$

the vanishing of whose first variation with fixed endpoints gives the
Lagrange equations of motion

$$\frac{d}{dt}\left(\frac{\partial L}{\partial \dot{q}_i}\right) - \frac{\partial L}{\partial q_i} = 0.$$

It is straightforward to show that if L does not explicitly depend on
time, the energy of the system

$$E = T + V = \sum_i \dot{q}_i \frac{\partial L}{\partial \dot{q}_i} - L$$

is conserved.

2. One then makes a change of variables (called a Legendre transformation) by introducing the generalized momenta $p_i = \partial L/\partial \dot{q}_i$ from which the energy of the system can be written $E = \sum_{i=1}^{n} p_i \dot{q}_i - L$. To explicitly write the $\dot{q}_i$ in terms of the p_i requires that $|\frac{\partial^2 L}{\partial \dot{q}_i \partial \dot{q}_j}| \neq 0$. The space defined by the variables $(\mathbf{q}, \mathbf{p})$ is called the $2n$-dimensional phase space of the system. The Hamiltonian is defined as the energy,

$$\mathcal{H}(\mathbf{q}, \mathbf{p}, t) = \sum_{i=1}^{n} p_i \dot{q}_i - L(\mathbf{q}, \dot{\mathbf{q}}, t)$$

and typically is written as the sum of kinetic and potential energies
$\mathcal{H} = T + V$.

3. From the definition of $\mathcal{H}$, one gets

$$d\mathcal{H} = \sum (\dot{q}_i dp_i - \dot{p}_i dq_i) - \frac{\partial L}{\partial t}$$

to obtain Hamilton's equations in canonical form,

$$\dot{q}_i = \frac{\partial \mathcal{H}}{\partial p_i}, \qquad \dot{p}_i = -\frac{\partial \mathcal{H}}{\partial q_i}. \tag{1.2.2}$$

4. Liouville's theorem states that volumes in phase space are preserved

$$\nabla \cdot (\dot{\mathbf{z}}) = \nabla \cdot (\mathbb{J} \cdot \nabla \mathcal{H})$$

$$= \sum_{i=1}^{n} \frac{\partial}{\partial q_i}\left(\frac{\partial \mathcal{H}}{\partial p_i}\right) + \frac{\partial}{\partial p_i}\left(-\frac{\partial \mathcal{H}}{\partial q_i}\right) = 0$$

Remarks.

1. For fluid flows Liouville's theorem is a simple consequence of incompressibility.

2. The phase space velocity $\dot{z}$ is perpendicular to the gradient of the Hamiltonian, i.e., $(\mathbb{J} \cdot \nabla \mathcal{H}) \cdot (\nabla \mathcal{H}) = 0$. Therefore, contours of a time-independent Hamiltonian correspond to particle paths. Instantaneous contours of a time-dependent Hamiltonian correspond to instantaneous particle paths.

3. Since the Hamiltonian for the point vortex model **is not** the sum of potential and kinetic energies, it cannot be obtained from a Lagrangian in the standard way. It is given by the Green's function of ∇^2 (e.g., (1.1.7)) and has a singularity at each vortex site .

4. The phase space for the point vortex model is made up entirely of position variables $(\mathbf{x}, \mathbf{y})$, which form a coordinate chart on the actual 2D physical surface. $\Diamond$

A fundamental result useful for classifying as well as actually "solving" a given system is **Noether's Theorem**, which states that corresponding to every continuous transformation group that leaves the Lagrangian invariant, there exists a conserved quantity. (See, for example Marsden and Ratiu (1999) for discussions of this along with the closely related momentum map idea. Bluman and Kumei (1989) and Olver (1993) also have nice discussions of the relation between symmetries and conservation laws.) From this one can draw the following conclusions:

- Invariance under time translations $\Leftrightarrow$ conservation of energy.

- Invariance under translations in space $\Leftrightarrow$ conservation of linear momentum.

- Invariance under rotations $\Leftrightarrow$ conservation of angular momentum.

By direct inspection one can write a Lagrangian,

$$\mathcal{L}(z_\alpha, z_\alpha^*; \dot{z}_\alpha, \dot{z}_\alpha^*) = \frac{1}{2i} \sum_{\alpha=1}^{N} \Gamma_\alpha (z_\alpha^* \dot{z}_\alpha - z_\alpha \dot{z}_\alpha^*) - \frac{1}{2} \sum_{\beta \neq \alpha}^{N} \Gamma_\alpha \Gamma_\beta \log |z_\beta - z_\alpha|^2,$$

whose Euler–Lagrange equations

$$\frac{d}{dt} \left(\frac{\partial \mathcal{L}}{\partial \dot{z}_\alpha} \right) - \frac{\partial \mathcal{L}}{\partial z_\alpha} = 0$$

give the equations of motion (1.1.22). The second term in the Lagrangian is, of course, the Hamiltonian, while the first is the invariant

$$
\begin{aligned}
I_0 &\equiv \frac{1}{2i} \sum_{\alpha=1}^{N} \Gamma_\alpha (z_\alpha^* \dot{z}_\alpha - z_\alpha \dot{z}_\alpha^*) \\
&= \frac{1}{2} \sum \Gamma_\alpha \Gamma_\beta \\
&= \Im(\sum p_i \dot{q}_i),
\end{aligned}
$$

where the variables $q_\alpha = z_\alpha$, $p_\alpha = \Gamma_\alpha z_\alpha^*$ are treated as canonical. As shown in Chapman (1978), I_0 is generated from the scale invariance $t \to \eta^2 t$, $z \to \eta z$ in the system (1.1.22).

The number of functionally independent conserved quantities in a given system determines what fraction of the phase space a particular trajectory explores over long times and whether or not the system is completely integrable. Consider the time derivative of a C^1 function $f(\mathbf{q}, \mathbf{p}; t)$ defined on the $2n$ dimensional phase space. By the chain rule

$$
\begin{aligned}
\frac{df}{dt} &= \sum_{i=1}^{n} \left(\frac{\partial f}{\partial q_i} \frac{\partial q_i}{\partial t} + \frac{\partial f}{\partial p_i} \frac{\partial p_i}{\partial t} \right) + \frac{\partial f}{\partial t} \\
&= \sum_{i=1}^{n} \left(\frac{\partial f}{\partial q_i} \frac{\partial \mathcal{H}}{\partial p_i} - \frac{\partial f}{\partial p_i} \frac{\partial \mathcal{H}}{\partial q_i} \right) + \frac{\partial f}{\partial t} \\
&\equiv \{f, \mathcal{H}\} + \frac{\partial f}{\partial t}.
\end{aligned} \tag{1.2.3}
$$

The last step defines the *canonical* Poisson bracket [1] of two functions

$$
\{f, g\} \equiv \sum_{i=1}^{n} \left(\frac{\partial f}{\partial q_i} \frac{\partial g}{\partial p_i} - \frac{\partial f}{\partial p_i} \frac{\partial g}{\partial q_i} \right). \tag{1.2.4}
$$

The formula (1.2.3) shows that if the function f is time-independent and if its Poisson bracket with the Hamiltonian vanishes, then it is a conserved quantity or integral of the motion. As we will see, how one defines the Poisson bracket is fundamental to understanding the underlying structure of the Hamiltonian system and in generalizing it to incorporate a much wider class of equations.

Using this definition, the following Poisson bracket identities hold:

1. $\{f, g\} = -\{g, f\}$

2. $\{f + g, h\} = \{f, h\} + \{g, h\}$

---

[1] Every student and practitioner of mechanics should read the wonderful article of Jost (1964) on this subject.

   3. $\{f, \{g, h\}\} + \{g, \{h, f\}\} + \{h, \{f, g\}\} = 0$ (Jacobi's identity)

   4. $\{fg, h\} = f\{g, h\} + g\{f, h\}$ (Leibniz identity)

Identities (1),(2), and (3) define a **Lie algebra**.

Example 1.3 $\{\mathcal{H}, \mathcal{H}\} = 0$; hence, the Hamiltonian is a conserved quantity $\mathcal{H} = E$ for time-independent systems.   ◊

From Jacobi's identity, it is straightforward to prove the following:

Theorem 1.2.1. *If two functions f and g are conserved quantities, then $\{f, g\}$ is also a conserved quantity.*

Proof. Use Jacobi's identity with f, g, and $\mathcal{H}$ to show $\{\mathcal{H}, \{f, g\}\} = 0$.   □

Remark. Sometimes this computation can be used to generate new conserved quantities, although there is no guarantee that $\{f, g\}$ will be functionally independent from f and g.   ◊

Definition 1.2.2. *Two functions whose Poisson bracket vanishes are said to be **involutive** or **in involution**. Equivalently, they are said to **Poisson commute**.*

See Ratiu (1980) for a general discussion of involution theorems, as well as Abraham, Marsden, and Ratiu (1988).

    Hamilton's equations can now be compactly rewritten and interpreted by making the following observations. Given any C^∞ function F defined on the phase space $(\mathbf{q}, \mathbf{p})$ of the Hamiltonian system, the directional derivative of F along the flow of Hamilton's equations is

$$\dot{F} = \sum_{i=1}^{n} \left(\frac{\partial F}{\partial q_i} \dot{q}_i + \frac{\partial F}{\partial p_i} \dot{p}_i \right)$$

$$= \sum_{i=1}^{n} \left(\frac{\partial F}{\partial q_i} \frac{\partial \mathcal{H}}{\partial p_i} - \frac{\partial F}{\partial p_i} \frac{\partial \mathcal{H}}{\partial q_i} \right)$$

$$= \{F, \mathcal{H}\}.$$

If we let F be the canonical variables q_i, p_i in turn, we get Hamilton's canonical equations

$$\dot{q}_i = \{q_i, \mathcal{H}\} \quad \dot{p}_i = \{p_i, \mathcal{H}\} \tag{1.2.5}$$

for the evolution of the phase space variables along the Hamiltonian flow.

    In order to generalize the discussion so that, as a first step, the canonical machinery can incorporate systems that are spatially extended, hence governed by partial differential equations, we must introduce the notion of a variational derivative of a functional. We start with a simple example to illustrate the main concept.

Example 1.4   Consider the kinetic energy functional $T(u(x))$ associated with a one-dimensional incompressible fluid in the domain $x \in \mathbb{R}^1$, defined by

$$T(u(x)) \equiv \frac{1}{2} \int_a^b u^2 dx,$$

where $a \leq x \leq b$. How does the the value of $T(u(x))$ vary if we perturb $u(x)$ so that

$$u(x) \to u(x) + \epsilon \delta u(x), \qquad 0 < \epsilon << 1$$

and $\delta u(a) = \delta u(b) = 0$? The first variation of $T(u(x))$, denoted $\delta T(u; \delta u)$, is computed

$$\begin{aligned}
\delta T(u; \delta u) &= \lim_{\epsilon \to 0} \frac{1}{\epsilon} [T(u + \epsilon \delta u; \delta u) - T(u)] \\
&= \frac{d}{d\epsilon} T(u + \epsilon \delta u) \mid_{\epsilon=0} \\
&= \int_a^b \delta u \frac{\delta T}{\delta u(x)} dx \\
&= \left\langle \frac{\delta T}{\delta u}, \delta u \right\rangle,
\end{aligned}$$

where $\langle \cdot, \cdot \rangle$ defines an inner product over all admissable variations δu (see Gelfand and Fomin (1963)). In this case we have

$$\delta T(u; \delta u) = \int_a^b u \cdot \delta u dx,$$

hence, the functional derivative $\delta T/\delta u$ is

$$\frac{\delta T}{\delta u} = u,$$

where the inner product is defined by

$$\langle v(x), w(x) \rangle \equiv \int_a^b v(x) \cdot w(x) dx. \qquad \Diamond$$

The idea can be generalized further by considering a more general functional

$$F(u) \equiv \int_a^b \mathcal{F}(x, u, u_x, u_{xx}, \ldots) dx$$

where we assume that $\mathcal{F}$ is differentiable in all of its variables. The first variation of F is easily computed as

$$\delta F(u; \delta u) \equiv \int_a^b \left(\frac{\partial \mathcal{F}}{\partial u} \delta u + \frac{\partial \mathcal{F}}{\partial u_x} \delta u_x + \dots \right) dx.$$

The terms on the right can be integrated by parts

$$\delta F(u; \delta u) = \int_a^b \delta u \left(\frac{\partial \mathcal{F}}{\partial u} - \frac{d}{dx} \frac{\partial \mathcal{F}}{\partial u_x} + \frac{d^2}{dx^2} \frac{\partial \mathcal{F}}{\partial u_{xx}} - \dots \right) dx$$
$$+ \left(\frac{\partial \mathcal{F}}{\partial u_x} \delta u + \dots \right) \Big|_a^b$$

where the last expression is the boundary contribution. If we choose $\delta u(a) = \delta u(b) = 0; \delta u_x(a) = \delta u_x(b) = 0$, etc., as is typically the case, then the boundary terms vanish and we have

$$\delta F(u, \delta u) = \left\langle \frac{\delta F}{\delta u}, \delta u \right\rangle,$$

with

$$\frac{\delta F}{\delta u} \equiv \frac{\partial \mathcal{F}}{\partial u} - \frac{d}{dx} \frac{\partial \mathcal{F}}{\partial u_x} + \frac{d^2}{dx^2} \frac{\partial \mathcal{F}}{\partial u_{xx}} - \dots$$

We are now in a position to generalize Hamilton's canonical equations (1.2.2) to the case of fields. Adopting the notation of Marsden and Ratiu (1999), we consider Hamiltonians of the form

$$\mathcal{H}(\phi^1, \dots, \phi^m; \pi_1, \dots, \pi_m)$$

where $(\phi^1, \dots, \phi^m)$ are functions of both space and time and $(\pi_1, \dots, \pi_m)$ are the associated conjugate momentum variables. The analogue of the canonical system (1.2.2) is written

$$\frac{\partial \phi^\alpha}{\partial t} = \frac{\delta \mathcal{H}}{\delta \pi_\alpha} \tag{1.2.6}$$

$$\frac{\partial \pi_\alpha}{\partial t} = -\frac{\delta \mathcal{H}}{\delta \phi^\alpha}, \tag{1.2.7}$$

for $\alpha = 1, 2, 3, \dots, m$. We can then generalize our previous definition of the canonical Poisson bracket (1.2.4) to

$$\{F, G\} \equiv \sum_{\alpha=1}^m \int_{\mathbb{R}^n} \left(\frac{\delta F}{\delta \phi^\alpha} \frac{\delta G}{\delta \pi_\alpha} - \frac{\delta F}{\delta \pi_\alpha} \frac{\delta G}{\delta \phi^\alpha} \right) dx$$

for any two functionals $F(\phi^1, \dots, \phi^m; \pi_1, \dots, \pi_m)$, $G(\phi^1, \dots, \phi^m; \pi_1, \dots, \pi_m)$. As before, the Hamiltonian system of partial differential equations can then be conveniently expressed in Poisson bracket form

$$\dot{F} = \{F, \mathcal{H}\}.$$

In particular, if we let F be the canonical functions ϕ^α, π_α, in turn, we obtain the canonical equations

$$\dot{\phi}^\alpha = \{\phi^\alpha, \mathcal{H}\}$$
$$\dot{\pi}_\alpha = \{\pi_\alpha, \mathcal{H}\}.$$

1.2.2 Noncanonical Formulation

While the previous canonical formalism goes a long way toward the development of a general theory for many classical mechanical systems, it does not by any means tell the whole story. In particular, the main system we will be concerned with, the Euler equations for incompressible fluids, does not fit comfortably within this framework, nor do the general equations for geophysical fluid mechanics, or, in fact, the simpler system of rigid body equations. An important generalization to these ideas is the formulation of *noncanonical mechanics*, which we briefly describe next.

The main idea behind the development of noncanonical mechanics starts with generalizing the definition of the Poisson bracket. Consider the example of rigid body dynamics:

Example 1.5   The equations of motion for rigid body dynamics are

$$I_1\dot{\Omega}_1 = (I_2 - I_3)\Omega_2\Omega_3$$
$$I_2\dot{\Omega}_2 = (I_3 - I_1)\Omega_3\Omega_1$$
$$I_3\dot{\Omega}_3 = (I_1 - I_2)\Omega_1\Omega_2,$$

where $\boldsymbol{\Omega} \equiv (\Omega_1, \Omega_2, \Omega_3)$ is the angular velocity of the rigid body from a body-fixed frame of reference, and $\mathbf{I} \equiv (I_1, I_2, I_3)$ are the principal moments of inertia of the body. Since the system is made up of three equations, it is clearly not immediately in canonical form. However, if we allow for generalizations of the Poisson bracket definition, we can put them in noncanonical form as follows. Define the angular momentum vector

$$\boldsymbol{\Pi} \equiv (\Pi_1, \Pi_2, \Pi_3) = (I_1\Omega_1, I_2\Omega_2, I_3\Omega_3).$$

Then the rigid body equations can be written compactly as

$$\dot{\boldsymbol{\Pi}} = \boldsymbol{\Pi} \times \boldsymbol{\Omega}. \tag{1.2.8}$$

We now define the rigid body Poisson bracket, which acts on functions of the angular momentum vector as

$$\{F, G\} = -\boldsymbol{\Pi} \cdot (\nabla F \times \nabla G).$$

Then it can be verified that the system (1.2.8) is equivalent to

$$\dot{F} = \{F, \mathcal{H}\}$$

with Hamiltonian $\mathcal{H}(\mathbf{\Pi})$ given by

$$\mathcal{H} = \frac{1}{2} \left(\frac{\Pi_1^2}{I_1} + \frac{\Pi_2^2}{I_2} + \frac{\Pi_3^2}{I_3} \right),$$

taking F to be the variables Π_1, Π_2, and Π_3, in turn. Notice also that if we take the total angular momentum

$$\mathcal{C}(\mathbf{\Pi}) \equiv \frac{1}{2} \left(\Pi_1^2 + \Pi_2^2 + \Pi_3^2 \right)$$

and use it in the above, we obtain

$$\begin{aligned}
\dot{\mathcal{C}} &= \{\mathcal{C}, \mathcal{H}\} \\
&= -\mathbf{\Pi} \cdot (\nabla \mathcal{C} \times \nabla \mathcal{H}) \\
&= -\mathbf{\Pi} \cdot (\mathbf{\Pi} \times \nabla \mathcal{H}) \\
&= 0,
\end{aligned}$$

a statement of conservation of angular momentum. In fact, for any function G one can prove that

$$\{\mathcal{C}, G\} = 0,$$

i.e., $\mathcal{C}$ Poisson commutes with all functions.   ◊

This defines what are called *Casimir* functions:

Definition 1.2.3. *A **Casimir function** $\mathcal{C}$ is a differentiable function that Poisson commutes with any function G, i.e.,*

$$\{\mathcal{C}, G\} = 0.$$

Remarks.

1. If we take the canonical Poisson bracket (1.2.4), then the Casimirs are simply constants.

2. Typically, the Casimir invariants are not related to explicit symmetries of the system the way the invariants generated by Noether's theorem are.   ◊

These ideas need to be generalized further in order to incorportate the Euler equations (1.1.12). To do this we introduce the so-called *ideal fluid bracket*

$$\{F, G\}(\mathbf{u}) = \int_\Omega \mathbf{u} \cdot \left[\frac{\delta F}{\delta \mathbf{u}}, \frac{\delta G}{\delta \mathbf{u}} \right]_L d\mathbf{x}, \tag{1.2.9}$$

where the variational derivative $\delta F/\delta \mathbf{u}$ is defined as

$$\frac{\delta F}{\delta \mathbf{u}} \equiv \lim_{\epsilon \to 0} \frac{1}{\epsilon} [F(\mathbf{u} + \epsilon \delta \mathbf{u}) - F(\mathbf{u})] \equiv \int_{\Omega} \left(\delta \mathbf{u} \cdot \frac{\delta F}{\delta \mathbf{u}} \right) d\mathbf{x}$$

and $[\cdot, \cdot]_L$ is the Jacobi–Lie bracket defined for vectors $\mathbf{v}$, $\mathbf{w}$ as

$$[\mathbf{v}, \mathbf{w}]_L^i \equiv \sum_{j=1}^{n} \left(w^j \frac{\partial v^i}{\partial x^j} - v^j \frac{\partial w^i}{\partial x^j} \right).$$

Based on these broadened definitions, the Euler equations (1.1.12) are expressible in the familiar form

$$\dot{F} = \{F, \mathcal{H}\},$$

with the kinetic energy Hamiltonian

$$\mathcal{H}(\mathbf{u}) = \frac{1}{2} \int_{\Omega} \|\mathbf{u}\|^2 d\mathbf{x}.$$

Note that in checking this, the constraint of divergence-free $\mathbf{u}$ and $\delta F/\delta \mathbf{u}$ must be enforced. The pressure term, which seems missing from this discussion, arises when projecting the equations onto the divergence-free part, using the decomposition (1.1.4). See Marsden and Ratiu (1999) for more details on this calculation.

For two-dimensional flows the Hamiltonian structure for the incompressible Euler equations are elegantly formulated in terms of the vorticity, whose basic equation we know to be

$$\dot{\omega} = [\omega, \psi],$$

where

$$[f, g] \equiv \left(\frac{\partial f}{\partial y} \frac{\partial g}{\partial x} - \frac{\partial f}{\partial x} \frac{\partial g}{\partial y} \right).$$

The Hamiltonian energy is written

$$\mathcal{H}(\omega) = \frac{1}{2} \int_{\mathcal{D}} \|\mathbf{u}\|^2 d\mathbf{x}$$

$$= \frac{1}{2} \int_{\mathcal{D}} |\nabla \psi|^2 \, d\mathbf{x}$$

$$= -\frac{1}{2} \int_{\mathcal{D}} \omega \psi d\mathbf{x}.$$

This last expression is obtained after integration by parts, taking $\mathcal{D}$ to be a finite box with periodic boundary conditions to avoid complications with boundary terms. If we now define the *noncanonical Lie–Poisson bracket* as

$$\{F, G\} = \int_{\mathcal{D}} \omega \left[\frac{\delta F}{\delta \omega}, \frac{\delta G}{\delta \omega} \right] d\mathbf{x}, \qquad (1.2.10)$$

and use the fact that

$$\int_{\mathcal{D}} f [g, h] \, d\mathbf{x} = - \int_{\mathcal{D}} g [f, h] \, d\mathbf{x}$$

to write

$$\{F, G\} = - \int_{\mathcal{D}} \frac{\delta F}{\delta \omega} \left[\omega, \frac{\delta G}{\delta \omega} \right] d\mathbf{x},$$

then the equations of motion in vorticity-streamfunction form are

$$\dot{\omega} = \{\omega, \mathcal{H}\},$$

where

$$\frac{\delta \mathcal{H}}{\delta \omega} \equiv -\psi.$$

The brackets defined in (1.2.9) and (1.2.10) are actually related through the isomorphism between the space of divergence-free vector fields $\mathbf{u}$ and the space of vorticity fields ω. The brackets above are special cases of general Lie–Poisson brackets defined on the dual of any Lie algebra. Lie–Poisson systems are known to have invariant subsystems called co-adjoint orbits (see the rigid body angular momentum sphere of Example 1.5) and on these orbits the Hamiltonian system (at least locally) becomes canonical. In fluid systems examples of these coadjoint orbits are point vortices , vortex patches, and vortex filaments. To a large extent, this general procedure and the examples are detailed in Marsden and Weinstein (1983). It can be shown that any smooth function of the vorticity $\Phi(\omega(\mathbf{x}))$ is a Casimir, i.e.,

$$\mathcal{C}(\omega) = \int \Phi(\omega(\mathbf{x})) d\mathbf{x}.$$

This is a simple manifestation of Kelvin's circulation theorem in two dimensions.

One could alternatively base the starting point for discussions of the noncanonical generalizations on the symplectic form (1.2.1). To generalize this expression one allows for any skew-symmetric tensor J_{ij} such that the so-called Jacobi condition is satisfied —

$$\epsilon_{ijk} J_{im} \frac{\partial J_{jk}}{\partial u_m} = 0, \qquad (1.2.11)$$

where standard tensor notation and summation conventions are being used, i.e., ϵ_{ijk} is the alternating tensor. The conservation of $\mathcal{H}$ follows as before from the fact that

$$\frac{d\mathcal{H}}{dt} = \frac{\partial\mathcal{H}}{\partial u_i}\frac{du_i}{dt} = \frac{\partial\mathcal{H}}{\partial u_i}J_{ij}\frac{\partial\mathcal{H}}{\partial u_j} = 0.$$

Note that in this formulation $\mathbb{J}$ can be a singular matrix, which significantly generalizes the types of systems one can treat. Typically, singular $\mathbb{J}$'s appear after a transformation from canonical coordinates to a reduced set of coordinates by exploiting symmetries, a process generally known as **reduction** (see Marsden (1992)). When $\mathbb{J}$ is nonsingular, it can be shown that locally one can make a change of coordinates to bring the system into canonical form (Darboux's theorem — see Salmon (1988), Marsden (1992), Olver (1993), or Marsden and Ratiu (1999)). In this way the equations for rigid body dynamics can be written in the form (1.2.1), with

$$\mathbf{u} = (\Pi_1, \Pi_2, \Pi_3)^T$$

$$\mathcal{H}(\mathbf{u}) = \frac{1}{2}\left(\frac{\Pi_1^2}{I_1} + \frac{\Pi_2^2}{I_2} + \frac{\Pi_3^2}{I_3}\right)$$

$$\mathbb{J} = \begin{pmatrix} 0 & \Pi_3 & -\Pi_2 \\ -\Pi_3 & 0 & \Pi_1 \\ \Pi_2 & -\Pi_1 & 0 \end{pmatrix}.$$

Clearly this formulation is noncanonical since $\mathbb{J}$ is singular and there is an odd number of independent variables. Also, it is easy to show that $\mathbb{J}$ satisfies the Jacobi condition.

Based on the fact that $\mathbb{J}$ is singular, one can then interpret the Casimirs as differentiable functions $\mathcal{C}(\mathbf{u})$ that satisfy $\mathbb{J}\cdot\nabla\mathcal{C} = 0$ for any skew-symmetric tensor $\mathbb{J}$ satisfying the Jacobi condition (1.2.11). It is straightforward to verify that $d\mathcal{C}/dt = 0$ using the properties of $\mathbb{J}$. If the system is in canonical form, hence, $\mathbb{J}$ is invertible, the Casimirs are constants. However, if $\mathbb{J}$ is singular, there exist k independent nontrivial Casimirs, where k is the corank of $\mathbb{J}$ (see Salmon (1988), Littlejohn (1982)). For the rigid body equations, as shown before, the angular momentum $\frac{1}{2}(\Pi_1^2 + \Pi_2^2 + \Pi_3^2)$ is a Casimir.

Noncanonical partial differential equations in Hamiltonian form can be written

$$u_t = J\frac{\delta\mathcal{H}}{\delta u}$$

for a given Hamiltonian functional $\mathcal{H}$. The skew-adjoint differential operator, J satisfies the Jacobi condition along with the inner product condition

$$\langle u, Jv\rangle = -\langle Ju, v\rangle.$$

By analogy with the definition earlier, a Casimir invariant $\mathcal{C}$ satisfies

$$J\frac{\delta\mathcal{C}}{\delta u} = 0.$$

Using this structure, the Euler equations (1.1.13) can be put in Hamiltonian form by letting

$$\mathbf{u} = \omega$$
$$\mathcal{H} = \frac{1}{2} \int \|\mathbf{u}\|^2 d\mathbf{x}.$$

In two dimensions,

$$J = \left(\frac{\partial \omega}{\partial y} \frac{\partial}{\partial x} - \frac{\partial \omega}{\partial x} \frac{\partial}{\partial y} \right).$$

For the three-dimensional case J becomes the operator

$$J = (\omega \cdot \nabla - (\nabla\omega)\nabla\times). \tag{1.2.12}$$

Remarks.

1. The previous discussion only scratches the surface of a much more general and encompassing theory, which is described most recently in Marsden and Ratiu (1999). For our purposes, we describe the theory only for appropriate background. One can read more specialized articles such as that of Morrison (1998) or Shepherd (1990) for more specific treatments of ideal fluids and geophysical flows.

2. The original observation that the Euler equations for incompressible flow could be put in Lagrangian and Hamiltonian form is due to Arnold (1966a, 1969). It was subsequently generalized to a more abstract setting in Ebin and Marsden (1970). For other discussions of the Hamiltonian structure for the three-dimensional Euler equations, see Olver (1993) (Example 7.10 and Exercise 7.5), Bretherton (1970), Roberts (1972), Olver (1982), Benjamin and Olver (1982), Marsden and Weinstein (1983), Benjamin (1984), Serre (1984), Salmon (1988), and Arnold and Khesin (1998).

3. One can read more about the noncanonical approach in Greene (1982), Shepherd (1990), Holm (1997), Marsden (1992), Lewis, Marsden, Montgomery, and Ratiu (1986), or Marsden and Ratiu (1999). A tutorial style introduction can be found in Morrison (1998).   ◇

1.2.3   Geometric Fluid Mechanics

There is a geometric formulation of ideal fluid mechanics originating in the works of Arnold (1966) and Ebin and Marsden (1970), a formulation that allows one to view Lagrangian trajectories as geodesics, and, as an added benefit, work on general Riemannian manifolds. We summarize this theory here and point to the recent books of Marsden and Ratiu (1999) and Arnold and Khesin (1998) for more in-depth discussions.

Written first in Eulerian form, the velocity field $\mathbf{u}$ of an ideal, incompressible, homogeneous fluid moving on a C^∞ compact, oriented n-dimensional Riemannian manifold $\mathcal{M}$ (with C^∞ boundary $\partial\mathcal{M}$), endowed with metric g and Riemannian volume form μ, is written in the usual way:

$$\mathbf{u}_t + \mathbf{u} \cdot \nabla\mathbf{u} = -\nabla p$$
$$\nabla \cdot \mathbf{u} = 0,$$

with auxiliary boundary condition that $\mathbf{u}$ is tangent to the boundary $\partial\mathcal{M}$. On a general manifold $\mathcal{M}$ the nonlinear term $\mathbf{u} \cdot \nabla\mathbf{u}$ is interpreted as the covariant derivative of $\mathbf{u}$ along $\mathbf{u}$, which we denote $\nabla_u\mathbf{u}$, and one considers vector fields $\mathbf{u}$ in the Sobolev space H^s for $s > (n/2) + 1$.

In Lagrangian form, the flow of the vector field $\mathbf{u}(t,x)$ is denoted by $\eta(t,x)$, i.e.,

$$\frac{\partial}{\partial t}\eta(t,x) = \mathbf{u}(t,\eta(t,x)),$$

where $\eta(0,x) \equiv x$ for all $x \in \mathcal{M}$. Thus, passively advected particles and point vortices evolve under the flow map $x \mapsto \eta_t(x)$, where η_t denotes the map $\eta(t,\cdot)$ for each time t. Incompressibility then manifests itself in the statement that η_t is a volume-preserving diffeomorphism, and η_t is in the group of H^s volume-preserving diffeomorphisms $D^s_\mu(\mathcal{M})$, i.e., $\eta_t \in D^s_\mu(\mathcal{M})$. The result of Arnold (1966) and Ebin and Marsden (1970) can be stated as

Theorem 1.2.2. *The time-dependent velocity field* $\mathbf{u}$ *satisfies the Euler equations if and only if the Lagrangian trajectory* η_t *is a geodesic of the right invariant* L^2*-metric on* $D^s_\mu(\mathcal{M})$.

One can find further detail in the book of Marsden and Ratiu (1999).

In terms of Lie derivatives, the Euler equations can be written as

$$\frac{\partial u^b}{\partial t} + \pounds_u u^b = d(\frac{1}{2}|u|^2 + p) = dp', \tag{1.2.13}$$

where u^b is the 1-form associated with the vector field $\mathbf{u}$ via the metric g, and $\pounds_u u^b$ denotes the Lie derivative of this 1-form along u. One then obtains the vorticity formulation by taking the exterior derivative of (1.2.13)

$$\frac{\partial \omega}{\partial t} + \pounds_u \omega = 0, \tag{1.2.14}$$

where $\omega \equiv du^b$ is now a 2-form whose solution is written

$$\omega_t = (\eta_t) * \omega_0.$$

In this way, both the two-dimensional and the three-dimensional equations are formally written in the same way. In two dimensions, however, ω is a scalar, while in three dimensions, using the volume form μ, ω is a vector field, which in both cases is the curl of **u**.

This geometric point of view has several concrete benefits that have been exploited over the past 20 years. Local existence and uniqueness theorems for the Euler equations are reduced essentially to Picard iteration arguments, convergence proofs of vortex methods and of Navier–Stokes to Euler limits as $\nu \to 0$ (with no boundaries) are simplified, to name only a few. With regard to modeling, the geometric point of view has also been instrumental in the development of the Euler-α equations (also known as the Lagrange Averaged Euler (LAE) and Lagrange Averaged Navier–Stokes (LANS) equations) discussed earlier. Fluid stability methods such as the energy Casimir or Arnold method, which generalize the classical linear Rayleigh criteria to nonlinear settings, were also developed with this framework. The reader can learn this point of view from the recent books of Marsden and Ratiu (1999) and Arnold and Khesin (1998).

1.2.4   Integrable Systems and Action-Angle Coordinates

We can now define the notion of complete integrability for a Hamiltonian system based on the number of conserved quantities, or symmetry groups of the system:

Definition 1.2.4. *In a Hamiltonian system with n degrees of freedom (2n-dimensional phase space) and k functionally independent involutive conserved quantities, one can reduce the dimension of the phase space to $2(n - k)$. If $k = n$, the system is said to be* **completely integrable** *and in principle can be solved by quadrature.*

See Whittaker (1937), Arnold (1978), or Abraham, Marsden, and Ratiu (1988) for further discussion.

Explicitly carrying out the integration procedure for a given system, even if one knows all the conserved quantities, can, of course, be complicated and generally requires clever use of transformation and reduction theory. These classical topics can be found in several texts, for example, Lichtenberg and Lieberman (1983), Percival and Richards (1982), Goldstein (1980), or Whittaker (1937) are good starting points.

For completely integrable systems one can find special "action-angle" coordinates $(\mathbf{I}, \boldsymbol{\theta})$, usually defined in subregions of the phase space, and a new Hamiltonian $\hat{\mathcal{H}}(\mathbf{I})$ independent of the angle variables so that the equations of motion become

$$\dot{I}_k = -\frac{\partial \hat{\mathcal{H}}}{\partial \theta_k} = 0$$

$$\dot{\theta}_k = \frac{\partial \hat{\mathcal{H}}}{\partial I_k} = \omega_k(\mathbf{I}).$$

In this case, the n actions $\mathbf{I}$ are conserved and the angles $\theta \in [0, 2\pi]$ increase linearly in time:

$$I_k = const \tag{1.2.15}$$

$$\theta_k = \omega_k(\mathbf{I}) \cdot t + \theta_k(0). \tag{1.2.16}$$

The motion is said to take place on an n-dimensional torus, which globally foliates the phase space. An n-torus has the property that one can define n topologically independent closed loops $\mathcal{C}_k$ on its surface.

One way to understand the meaning of action-angle variables is to think of them as nonlinear polar coordinates, i.e., a separable coordinate system for closed regions, which need not be circular. Consider the Hamiltonian $\mathcal{H}(q, p) = p^2/2 + V(q) = E$, with equations of motion given by

$$\dot{q} = \frac{\partial \mathcal{H}}{\partial p} = p$$

$$\dot{p} = -\frac{\partial \mathcal{H}}{\partial q} = -\frac{dV}{dq}.$$

The potential well and phase curves in the (q, p)-plane are shown in Figure 1.2 for a potential with one local minimum. For each fixed energy value E, a solution of the system corresponds to a trajectory on a fixed closed phase curve in the (q, p)-plane. We now seek a new set of coordinates $(q, p) \mapsto (\theta, I)$ with the following properties:

1. The transformation from (q, p) to (θ, I) is **canonical**[2], i.e., the Jacobian determinant has value one:

$$J(q(\theta, I), p(\theta, I)) \equiv \frac{\partial q}{\partial \theta}\frac{\partial p}{\partial I} - \frac{\partial q}{\partial I}\frac{\partial p}{\partial \theta} = 1. \tag{1.2.17}$$

This means that closed regions in the (q, p)-plane are transformed to closed regions in the (θ, I)-plane with the same area A, since by the change of variable formula,

$$A = \int\int dq dp = \int\int J(q, p) d\theta dI = \int\int d\theta dI.$$

---

[2] A general necessary and sufficient condition for a transformation to be canonical is that its Jacobian be symplectic. Greenwood (1997) has a nice discussion of this.

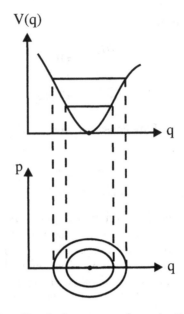

FIGURE 1.2. Potential well and phase curves for potential with one local minimum.

2. In the new coordinates, the action I is constant along each phase curve, so that it can be used as a label to specify each curve.

3. On a phase curve of constant action the angle variable $\theta \in [0, 2\pi]$ can be used to parametrize the position.

Example 1.6 Consider the quadratic Hamiltonian $\mathcal{H}(q, p) = p^2/2 + \omega^2 q^2/2$ corresponding to a harmonic oscillator with equations of motion

$$\dot{q} = \frac{\partial \mathcal{H}}{\partial p} \quad = p$$

$$\dot{p} = -\frac{\partial \mathcal{H}}{\partial q} \quad = -\omega^2 q.$$

Make the canonical transformation $(q, p) \mapsto (\theta, r)$

$$q = (\frac{2r}{\omega})^{1/2} \sin \theta$$

$$p = (2r\omega)^{1/2} \cos \theta.$$

The equations in the new variables are obtained using the chain rule. One gets

$$\mathcal{M} \begin{bmatrix} \dot{r} \\ \dot{\theta} - \omega \end{bmatrix} = 0,$$

where

$$\mathcal{M} = \begin{pmatrix} (2\omega r)^{-1/2} \sin\theta & (\frac{2r}{\omega})^{1/2} \cos\theta \\ (\frac{\omega}{2r})^{1/2} \cos\theta & -(2\omega r)^{1/2} \sin\theta \end{pmatrix}.$$

Since $\det\mathcal{M} \neq 0$, the new equations decouple,

$$\dot{r} = 0$$
$$\dot{\theta} = \omega.$$

This gives the solution by inspection

$$r = r(0)$$
$$\theta = \omega t + \theta(0).$$

The Hamiltonian transforms to $\mathcal{H}(q(r,\theta), p(r,\theta)) = r\omega \equiv \hat{\mathcal{H}}(r)$. Hence, the coordinates (r, θ) (where $r = E/\omega$) are the action-angle variables for this system. $\diamond$

This can be seen more geometrically: We specialize the discussion, for the moment, to a two-dimensional phase space. For a fixed energy value E the area enclosed by a closed trajectory (i.e., level curve of the Hamiltonian) in the (q, p)-plane is given by

$$A(E) = \oint p\,dq.$$

Since the action-angle coordinates are canonical, the corresponding area in the (I, θ)-plane is identical and is given by

$$A(E) = \int_0^{2\pi} I\,d\theta = 2\pi I.$$

This simple formula offers a nice geometrical interpretation of the action variable I (see Figure 1.3),

$$I = \frac{A(E)}{2\pi}.$$

The action I is the (normalized) area enclosed by a fixed energy contour in the phase space.

Example 1.7 Consider the motion of a particle positioned at (x, y) in the flow field of an isolated point vortex of strength Γ located at the origin. The Hamiltonian is given by $\mathcal{H} = \frac{\Gamma}{2\pi} \log(\sqrt{x^2 + y^2})$ with equations of motion

$$\dot{x} = \frac{\Gamma}{2\pi} \cdot \frac{y}{\sqrt{x^2 + y^2}}$$

$$\dot{y} = -\frac{\Gamma}{2\pi} \cdot \frac{x}{\sqrt{x^2 + y^2}}.$$

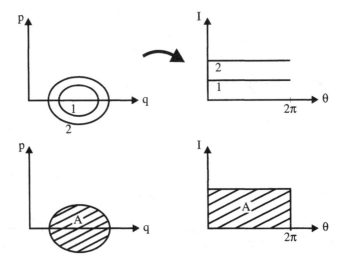

FIGURE 1.3. Closed phase curves in (q,p) and (I,θ)-coordinates.

The transformation $(x, y) \mapsto (\sqrt{2r}\sin\theta, \sqrt{2r}\cos\theta)$ is canonical and leads to the action-angle decomposition

$$\dot{r} = 0$$
$$\dot{\theta} = \omega(r),$$

where $\omega(r) = (\Gamma/2\pi)\left(1/\sqrt{2r}\right)$. ◊

Example 1.8 Consider the motion of two planar point vortices with strength (Γ_1, Γ_2), where $\Gamma_1 + \Gamma_2 \neq 0$. Their governing equations are written in complex form $z_1(t) = (x_1(t), y_1(t))$, $z_2(t) = (x_2(t), y_2(t))$,

$$\dot{z}_1(t) = i\frac{\Gamma_2}{2\pi} \cdot \frac{z_1 - z_2}{|z_1 - z_2|^2}$$
$$\dot{z}_2(t) = i\frac{\Gamma_1}{2\pi} \cdot \frac{z_2 - z_1}{|z_1 - z_2|^2}.$$

From these equations, it is straightforward to derive the two conserved quantities (D, C), where

$$D^2 = |z_1(t) - z_2(t)|^2 = |z_1(0) - z_2(0)|^2$$
$$C = \frac{\Gamma_1 z_1(t) + \Gamma_2 z_2(t)}{\Gamma_1 + \Gamma_2} = \frac{\Gamma_1 z_1(0) + \Gamma_2 z_2(0)}{\Gamma_1 + \Gamma_2}.$$

C is called the **center of vorticity** (or barycenter) and will play an important role in the following chapters. After some manipulations, the equations can be decoupled to give

$$\begin{pmatrix} z_1 \\ z_2 \end{pmatrix}' = i\frac{(\Gamma_1 + \Gamma_2)}{2\pi D^2}\begin{pmatrix} 1 & 0 \\ 0 & 1 \end{pmatrix}\begin{pmatrix} z_1 - C \\ z_2 - C \end{pmatrix},$$

then put in action-angle form by the canonical transformation

$$z_1 - C = \sqrt{2R_1(t)} \exp[i\theta_1(t)]$$
$$z_2 - C = \sqrt{2R_2(t)} \exp[i\theta_2(t)],$$

giving

$$\dot{R}_1 = 0 \quad \dot{\theta}_1 = \omega$$
$$\dot{R}_2 = 0 \quad \dot{\theta}_2 = \omega,$$

where $\omega = (\Gamma_1 + \Gamma_2)/2\pi D^2$. The two vortices rotate on concentric circles about their center of vorticity. When $\Gamma_1 = \Gamma_2$ the vortices move on the same circle. When $\Gamma_1 + \Gamma_2 = 0$ the center of vorticity is at ∞ and the vortices translate in parallel with velocity $[\frac{1}{2}(\Gamma_1^2 + \Gamma_2^2)]^{1/2}/2\pi D$. We show in Figure 1.4 the two distinct streamline topologies that can occur for the case $\Gamma_1 + \Gamma_2 \neq 0$. Note that in the plane the two figures are homotopically distinct, i.e., one cannot be continuously deformed into the other. The case where $\Gamma_1 + \Gamma_2 = 0$, called a "vortex couple," is treated in more detail in Chapter 2. $\Diamond$

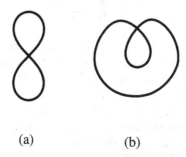

(a)   (b)

FIGURE 1.4. Streamline topologies for the 2-vortex problem: (a) Figure eight. (b) Limacon.

Definition 1.2.5. *A **vortex configuration** refers to the geometrical object formed by the collection of vortices. If the configuration neither translates nor rotates, it is called a **fixed equilibrium**. Otherwise, it is a **relative equilibrium**.*

Remarks.

1. For the planar 2-vortex problem there are no fixed equilibria ($\omega \neq 0$) and no finite-time collapse ($D \neq 0$). All solutions form relative equilibria.

2. It turns out that the action variable is remarkably stable under perturbations — this is described in the next subsection.

3. In the examples, where the closed trajectories were circular, the transformation $(x, y) \mapsto (\sqrt{2I}\sin\theta, \sqrt{2I}\cos\theta)$ was canonical and led to action-angle variables (I, θ). In standard polar coordinates the radial variable $r = \sqrt{2I}$, or $I = r^2/2 = \pi r^2/2\pi =$ area of circle $/2\pi$. $\Diamond$

Typically, one performs the transformation from (q, p) coordinates to action-angle coordinates (I, θ) by introducing a generating function. See Goldstein (1980), Tabor (1989), or Lichtenberg and Lieberman (1982) for more on this approach.

For systems with n degrees of freedom these ideas can be generalized to form the basis of a large literature on integrable systems. In general, the action variables are written as contour integrals

$$I_k = \frac{1}{2\pi} \oint_{C_k} \sum_{i=1}^{n} p_i dq_i,$$

where the closed loops are the n independent loops on the surface of an n-torus. The above equation says that the sum of the areas given by the projections of the closed curve C_k onto the n planes spanned by $(q_i, p_i), i = 1, \ldots, n$ is conserved. Extensions of the action-angle idea to infinite-dimensional integrable systems have been and are currently being developed. See, for example, the large literature on soliton systems (Newell (1985), Faddeev and Takhtajan (1987)) and more recently, the contribution of McKean and Vaninsky (1997).

A particular n-torus is picked out in the $2n$-dimensional phase space by specifying an initial condition for the flow, $(\mathbf{q}(0), \mathbf{p}(0))$. Since energy is conserved, each trajectory lies on an energy surface of dimension $2n - 1$. Hence, for systems with only one degree of freedom the torus and the energy surface are the same. For systems with two degrees of freedom, the two-dimensional tori lie on a three-dimensional energy surface which therefore is divided into two distinct regions by each torus. For $n = 3$ the three-dimensional tori lie on a five-dimensional energy surface, which topologically is no longer separated into two distinct regions — the motion on the energy surface can be nontrivial. This is particularly true when the equations are subjected to small perturbations, as will be described in the next section.

For motion lying on an invariant n-torus in the $2n$-dimensional phase space, one typically expresses the multiperiodic position coordinates in terms of Fourier series in the angle variables

$$q_i(t) = \sum_{k_1} \cdots \sum_{k_n} a^{(i)}_{k_1, k_2, \ldots, k_n} \exp[i(\mathbf{k} \cdot \boldsymbol{\theta})]$$
$$= \sum \cdots \sum a^{(i)}_{k_1, k_2, \ldots, k_n} \exp\{i[(\mathbf{k} \cdot \boldsymbol{\omega})t + \mathbf{k} \cdot \boldsymbol{\theta}(0)]\}.$$

The Fourier coefficients $a^{(i)}_k$ are functions of the action $\mathbf{I}$

$$a^{(i)}_k(\mathbf{I}) = \int_0^{2\pi} d\theta_1 \cdots \int_0^{2\pi} d\theta_n q_i(\mathbf{I}, \boldsymbol{\theta}) \exp[i(\mathbf{k} \cdot \boldsymbol{\theta})].$$

It is important to distinguish between two types of motions based on properties of the frequency vector. $\boldsymbol{\omega}$ is called **nonresonant** if $\mathbf{k} \cdot \boldsymbol{\omega} = 0$ (where the k_i are integers) implies all $k_i = 0$. In this case, the frequencies are incommensurate, or not rationally related, and the quasiperiodic orbit on the n-torus will densely cover the surface. Additional requirements are necessary for the flow to be **ergodic** on it (for more details, see Chapter 6), i.e., for time averages to equal space averages (see Arnold and Avez (1989), Halmos (1958), Petersen (1982), or most recently Katok and Hasselblatt (1996) for a general discussion in the context of differentiable dynamics). The proof of these statements can be found in Arnold (1978). If, however, the frequencies are rationally related, i.e., if $\mathbf{k} \cdot \boldsymbol{\omega} = 0$ does not imply $k_i = 0$, then each orbit will be closed and the motion periodic.

Example 1.9 For a 2-torus if $n\omega_1 + m\omega_2 = 0$ does not imply $m = n = 0$, then we can express the frequency ratio as

$$\frac{\omega_1}{\omega_2} = -\frac{m}{n}.$$

In this case the orbit will close after n cycles of θ_1 and m cycles of θ_2.   ◊

Every $2n$-dimensional phase space corresponding to an integrable system is foliated by n-dimensional invariant tori.

Definition 1.2.6. *The system is called **nondegenerate** if*

$$\det\left(\frac{\partial \omega_i(\mathbf{I})}{\partial I_j}\right) \equiv \det\left(\frac{\partial^2 \mathcal{H}(\mathbf{I})}{\partial I_i \partial I_j}\right) \neq 0. \tag{1.2.18}$$

Some of the tori will have closed periodic orbits (the tori with rationally related frequencies), while others will have quasiperiodic orbits (the tori with irrationally related frequencies). Since the rationals form a zero-measure set in the space of real numbers and the irrationals form a set of full measure, the quasiperiodic orbits far outnumber the closed periodic orbits, even though the latter form a dense set.

If the integrable system is perturbed, depending on the form of the perturbation, either the n conserved quantities will persist or $m \leq n$ will be broken. If they persist, the perturbed system is still completely integrable. We next describe what happens if some or all of the conserved quantities are lost under perturbation.

1.2.5 Near-Integrable Systems: KAM Theory and the Poincaré–Melnikov Method

We now suppose that our integrable system is subjected to Hamiltonian perturbations, so that the perturbed system has a governing Hamiltonian given by

$$\mathcal{H}(\mathbf{I}, \boldsymbol{\theta}; \epsilon) = \mathcal{H}_0(\mathbf{I}) + \epsilon \mathcal{H}_1(\mathbf{I}, \boldsymbol{\theta}; \epsilon).$$

In what follows we will assume that $\mathcal{H}$ is analytic in all of its arguments, although this is not strictly necessary for the theory. The perturbation $\mathcal{H}_1$ only needs to be C^r where $r \geq 2n + 2$ for analytic $\mathcal{H}_0$.

The perturbed equations of motion are

$$\dot{I}_k = -\epsilon \frac{\partial \mathcal{H}_1}{\partial \theta_k} \qquad \dot{\theta}_k = \omega_k(\mathbf{I}) + \epsilon \frac{\partial \mathcal{H}_1}{\partial I_k}.$$

It is natural to ask if any of the unperturbed ($\epsilon = 0$) invariant tori survive under perturbation. This question was answered in the KAM theorem (Kolmogorov–Arnold–Moser), about which much has been written since the original papers by Kolmogorov (1954), Arnold (1963), Moser (1962). One can read the various proofs in many places and we refer the reader to Lichtenberg and Lieberman (1982) and Arnold (1978) for a start. Our statement of the theorem follows the one given in Arnold (1978). For this we first need to define the notion of an isoenergetically nondegenerate system.

Definition 1.2.7. *The unperturbed Hamiltonian $\mathcal{H}_0(\mathbf{I})$ is called **isoenergetically nondegenerate** if*

$$\det \begin{pmatrix} \dfrac{\partial^2 \mathcal{H}_0}{\partial I_i \partial I_j} & \dfrac{\partial \mathcal{H}_0}{\partial I_i} \\ \dfrac{\partial \mathcal{H}_0}{\partial I_j} & 0 \end{pmatrix} \neq 0.$$

We can now state the KAM theorem as follows:

Theorem 1.2.3 (KAM Theorem). *If the unperturbed Hamiltonian $\mathcal{H}_0$ is nondegenerate, then for a sufficiently small ϵ, most nonresonant invariant tori persist. The phase space of the perturbed system has invariant tori densely filled with quasiperiodic trajectories, in which the number of independent frequencies are equal to the number of degrees of freedom of the*

*system. The invariant tori form a majority in the sense that the measure
of the complement of their union is small when $\epsilon \ll 1$. If $\mathcal{H}_0$ is isoenergetically nondegenerate, the invariant tori form a majority on each fixed
energy surface.*

Remarks.

1. The basic idea of the proof of the existence of perturbed tori has two
   main ingredients:

   - Linearization around an approximate solution, which introduces
     the so-called problem of small divisors, causing divergences to
     appear in the infinite series representation of the solution.

   - Inductive improvement of the approximation using a "supercon-
     vergent" Newton iteration method to overcome the small divisor
     problem.

   The recent review article by Wayne (1996) describes the proof in more
   detail.

2. The nondegeneracy condition ensures that the frequencies of the unperturbed system are functionally independent. The isoenergetic nondegeneracy condition ensures that they are functionally independent
   on each fixed energy surface. These two conditions are independent
   of each other (Arnold (1978)).

3. For systems with two degrees of freedom the isoenergetic nondegeneracy condition guarantees that the action variables remain close to
   their initial values for all time, provided $\epsilon \ll 1$. This is because
   the nested two-dimensional invariant tori partition each 3-D energy
   surface into regions in which the nonintegrable trajectories are confined. For $n \geq 3$ this is no longer guaranteed since the n-dimensional
   KAM tori no longer partition the $2n - 1$-dimensional energy manifolds. Trajectories not lying exactly on the invariant tori may "leak
   out," causing the actions to deviate significantly over long times, a
   phenomenon known as "Arnold diffusion" (Arnold (1964)). See Lichtenberg and Lieberman (1982), Arnold (1964), Nekhoroshev (1971),
   Chirikov (1978), Holmes and Marsden (1982), Dumas (1993) for discussions of this phenomenon. An upper bound of this diffusion rate
   (due to Nekhoroshev (1971)) is $O(\exp(-1/\epsilon^d))$, where $0 < d < 1$.
   There are ongoing attempts to get sharper estimates on this phase
   space diffusion rate (see, for example, Dumas (1993), Lochak and
   Neishtadt (1992), and Lochak (1990) for more discussion of this literature).

4. The term "most" means a Cantor set of positive measure, hence given a KAM torus for $\epsilon = 0$, there exists another one (for $0 < \epsilon \ll 1$) arbitrarily close to it. For example, with $n = 2$ the preserved tori satisfy the irrationality condition

$$\left| \frac{\omega_1}{\omega_2} - \frac{r}{s} \right| > \frac{K(\epsilon)}{s^{5/2}}$$

for all integer pairs (r, s), where $K(\epsilon) \to 0$ as $\epsilon \to 0$. To see that this preserves a set of positive measure as $\epsilon \to 0$, consider the interval $[0, 1]$. Delete from this interval zones of width $K(\epsilon) \cdot s^{-\mu}$ for each integer s. The total deleted length is then

$$D \equiv \sum_{s=1}^{\infty} \left(\frac{K(\epsilon)}{s^{\mu}} \right) \cdot s$$

$$= K(\epsilon) \sum_{s=1}^{\infty} s^{1-\mu}.$$

As long as $\mu > 2$, the sum converges and $D \sim K(\epsilon) \to 0$. For more on this, as well as discussions of KAM theory for finitely differentiable Hamiltonians, see Pöschel (1982). KAM provides the existence result ensuring that when one uses perturbation theory to approximate a perturbed orbit, one is approximating something that actually exists. See Treshchev (1991) for an interesting discussion of the fate of the resonant tori.

5. KAM ideas have recently been extended to infinite dimensions. We refer the interested reader to Wayne (1990), Craig (1996), and Kuksin (1993).   ◊

Coincident with the regions of periodic and quasiperiodic orbits, in typical perturbed integrable systems, are regions where the dynamics are chaotic. Our goal here is to describe what has become the standard analytical tool for checking when a given system exhibits chaotic behavior. Called the **Poincaré–Melnikov** method, or more simply the Melnikov method, it has been applied in a wide variety of contexts. Since our goal is to use it as a tool only in the Hamiltonian context, we will limit our description somewhat, and then comment at the end on where one can find much more general treatments.

Start by considering the planar perturbed Hamiltonian system

$$\mathcal{H}(\mathbf{z}) = \mathcal{H}_0(\mathbf{z}) + \epsilon \mathcal{H}_1(\mathbf{z}; t) \quad (0 < \epsilon \ll 1)$$

with corresponding equations

$$\dot{\mathbf{z}} = \mathbf{f}_0(\mathbf{z}) + \epsilon \mathbf{f}_1(\mathbf{z}; t),$$

where $\mathbf{z} \in \mathbb{R}^2$ and $\mathcal{H}_1$ is T-periodic in t, i.e.,

$$\mathcal{H}_1(\mathbf{z}; t + T) = \mathcal{H}_1(\mathbf{z}; t).$$

We do not necessarily assume that the system is in action-angle form; however the unperturbed system must have a special structure, which we now describe.

1. We assume that $\mathbf{z} = \mathbf{z}_0$ is a *hyperbolic fixed point* of the unperturbed system

$$\mathbf{J} \cdot \nabla \mathcal{H}_0(\mathbf{z}_0) = 0.$$

This implies that the eigenvalues $\lambda^{\pm}$ of the linearized system around $\mathbf{z}_0$ have the property $Re(\lambda^+) > 0$, $Re(\lambda^-) < 0$. The corresponding eigenvectors are denoted $\mathbf{z}^{\pm}$.

2. The phase plane associated with the unperturbed problem has either a *homoclinic orbit* $\overline{\mathbf{z}}_{ho}(t)$ or a *heteroclinic orbit* $\overline{\mathbf{z}}_{he}(t)$, where we define these orbits as follows.

Definition 1.2.8. *A **homoclinic orbit** $\overline{\mathbf{z}}_{ho}(t)$ has the property that*

$$\overline{\mathbf{z}}_{ho}(t) \to \mathbf{z}_0 \text{ as } t \to \pm\infty,$$

*i.e., the orbit globally connects the hyperbolic fixed point back to itself doubly asymptotically. A **heteroclinic orbit** $\overline{\mathbf{z}}_{he}(t)$ globally connects two different hyperbolic fixed points $\mathbf{z}_0^{(1)}, \mathbf{z}_0^{(2)}$, i.e., $\overline{\mathbf{z}}_{he}(t) \to \mathbf{z}_0^{(1)}$ as $t \to \infty$, and $\overline{\mathbf{z}}_{he}(t) \to \mathbf{z}_0^{(2)}$ as $t \to -\infty$. Locally, the eigenvectors $\mathbf{z}^{\pm}$ associated with the fixed point $\mathbf{z}_0$ are tangent to the homoclinic orbit at $\mathbf{z}_0$. The global manifold tangent to $\mathbf{z}^+$ at $\mathbf{z}_0$ is called the **unstable manifold**, denoted W_0^u, while the one tangent to $\mathbf{z}^-$ at $\mathbf{z}_0$ is called the **stable manifold**, denoted W_0^s.*

For the unperturbed problem, the stable and unstable manifolds coincide (i.e., the "distance" between them is zero). See, for example, Chapter 3 of Wiggins (1988) for further discussion.

Under generic perturbations, the homoclinic (or heteroclinic) connection splits transversely, a sufficient condition for what is now commonly referred to as **homoclinic chaos** (Wiggins (1988)). For problems of this type it is useful to introduce the **Poincaré map** $\mathcal{P}_\epsilon : \Sigma \mapsto \Sigma$, associated with the augmented time-periodic system

$$\dot{\mathbf{z}} = \mathbf{f}_0(\mathbf{z}) + \epsilon \mathbf{f}_1(\mathbf{z}; t) \qquad (1.2.19)$$
$$\dot{t} = 1,$$

where the phase space is extended to $(\mathbf{z}, t)$ and Σ denotes the cross section $\Sigma = \{(\mathbf{z}, t)|t = t_0\}$ (called the Poincaré section). Using the fact that $\mathbf{f}_1(\mathbf{z}; t)$ is T-periodic, we know that for all future times $t = t_0 \pm nT$, the phase trajectory again intersects Σ. Hence, if at time $t = t_0$ the point on Σ is denoted $\mathbf{z}^{(0)}$, then we can denote the point of intersection at time $t = t_0 + nT$ by $\mathbf{z}^{(n)}$. The Poincaré map is then defined as follows.

Definition 1.2.9. *The **Poincaré map** $\mathcal{P}_\epsilon(\mathbf{z}^{(n)}) : \Sigma \mapsto \Sigma$ associated with (1.2.19) is given by*

$$\mathcal{P}_\epsilon(\mathbf{z}^{(n)}) = \mathbf{z}^{(n+1)}.$$

The Nth iterate of the map, denoted $\mathcal{P}_\epsilon^N(\mathbf{z}^{(0)})$, is defined by

$$\mathcal{P}_\epsilon^N(\mathbf{z}^{(0)}) = \mathcal{P}_\epsilon \circ \mathcal{P}_\epsilon \circ \cdots \circ \mathcal{P}_\epsilon(\mathbf{z}^{(0)})$$

where one composes $\mathcal{P}_\epsilon$ with itself N times.

Thus, to each initial condition $\mathbf{z}^{(0)}$ there is a collection of points on the cross section Σ that are all future iterates of that point under the map. See Wiggins (1990) for further discussion of Poincaré maps and their uses. We show in Figure 1.5 (reprinted with permission from Neufeld and Tél (1997)) the contrasting integrable and nonintegrable cases for a 3-vortex problem and particle. This configuration was used by Ziglin (1980) in his proof of nonintegrability. Figure 1.5(a) shows the level curves for the 3-equal-vortex problem placed on an equilateral triangle in rigid rotation, corotating at the system frequency. The level curves confirm the integrable structure of the configuration. Notice also the hyperbolic fixed points and heteroclinic orbits. Contrast this with Figure 1.5(b) in which the three vortices no longer lie exactly on an equilateral triangle. The figure is a Poincaré section showing iterates of a passive particle under the flow generated by the system of three vortices in a perturbed equilateral configuration. The chaotic region is seen surrounding the heteroclinic orbits of the unperturbed problem. The general nonequilateral triangle motion can be decomposed into a superposition of relative vortex motion along with global rotation of the whole system around the center of vorticity. As in Neufeld and Tél (1997), call the period of the relative motion T, with frequency $\omega_{rel} = 2\pi/T$. Then measure the angular displacement $\Delta\phi_0$ about the center of vorticity between two configurations separated by time T. The frequency of the global rotation can be defined as

$$\omega_{glob} = \frac{\Delta\phi_0}{T}.$$

Depending on whether the ratio $\omega_{rel}/\omega_{glob}$ is rational or irrational, the overall vortex motion will be periodic or quasiperiodic.

The Poincaré–Melnikov method amounts to checking under what conditions one has a transverse intersection of the stable and unstable manifolds

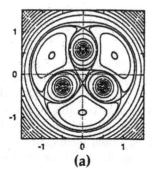
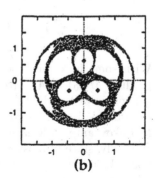

(a)   **(b)**

FIGURE 1.5. 3-vortex configuration in corotating frame. (a) Equilateral triangle configuration. (b) Poincaré map for the perturbed equilateral triangle.

associated with a particular fixed point $\mathbf{z_0}$. The distance $\mathbf{d}(t)$ between the two orbits, under perturbation is generically nonzero. However, if there are transverse intersections, there will be isolated points $t = t_0$, at which the distance function has a zero, namely, $\mathbf{d}(t_0) = 0$. To leading order in ϵ, this distance function is given by the Poincaré–Melnikov formula

$$M(t_0) = \int_{-\infty}^{\infty} \{\mathcal{H}_0, \mathcal{H}_1\}(\bar{\mathbf{z}}(t - t_0), t)dt. \qquad (1.2.20)$$

Based on this, we have:

Theorem 1.2.4 (Poincaré–Melnikov Theorem). *If $M(t_0)$ has a simple zero, then for sufficiently small ϵ, the system (1.2.19) has homoclinic chaos in the sense of transversal intersecting separatrices.*

Remarks.

1. Particularly complete discussions of the proof of the theorem can be found in Wiggins (1988, 1990, 1993). The method is also described in Guckenheimer and Holmes (1983) and Marsden and Ratiu (1999). The foundations of the method are based on work of Poincaré (1890), who was the first to realize that the transverse splitting of separatrices precludes the existence of analytic integrals of motion (see also Cherry (1924)). The paper by Melnikov (1963) was influential in that it helped turn the idea into a concrete technique for checking integrability. Other important papers include Holmes (1980), Holmes and Marsden (1982a,b), Chow, Hale, and Mallet-Paret (1980), and Robinson (1988).

2. It is usually more convenient to make the change of variable $t \mapsto t+t_0$ in (1.2.20), putting it in the equivalent form

$$M(t_0) = \int_{-\infty}^{\infty} \{\mathcal{H}_0, \mathcal{H}_1\}(\overline{\mathbf{z}}(t), t + t_0)dt.$$

With some luck, the integral can sometimes be separated into the product $M(t_0) = \mathcal{A} \cdot \mathcal{T}(t_0)$, where the Melnikov amplitude $\mathcal{A}$ is independent of t_0, while the function $\mathcal{T}(t_0)$ is oscillatory, having easily computable isolated zeros. It then simply remains to prove that the Melnikov amplitude is nonzero, which is typically carried out numerically, since in most applications, closed form expressions for the homoclinic or heteroclinic orbits are unobtainable or too complicated for practical use.

3. There is much more to say about the method and the type of chaos associated with it. For our purposes, suffice it to say that a simple zero of the Poincaré–Melnikov function precludes the existence of enough analytic integrals or conserved quantities for the system to be completely integrable. Thus, proving that the function $M(t_0)$ has a simple zero means, one can end the search for n integrals in involution. The question of exactly how many conserved quantities do survive the perturbation is not addressed by the method and is a more involved problem for high dimensional systems. The perturbed problem has infinitely many periodic solutions of arbitrarily high period, while the Poincaré section possesses so-called "Smale horseshoes." Good places to read more about these ideas include Guckenheimer and Holmes (1983), Wiggins (1988, 1990), and many related ideas in the classic book of Moser (1973). We will develop the Melnikov method further in Chapters 2 and 3 in the context of specific examples. Strictly speaking, the case of a heteroclinic orbit is more complicated than that of a homoclinic orbit for the following reason. Although the Melnikov calculation is identical, one needs a recurrence mechanism to conclude that horseshoes are present in the map, and hence that the system is nonintegrable. In the homoclinic case this is automatically satisfied, while in the heteroclinic case one must deal with this separately.

4. The method is by no means limited to planar Hamiltonian systems. It can be used for dissipative perturbations, as well as higher dimensional systems. The literature discussing applications of the method to physical problems is large and growing. See, for example, Beige, Leonard and Wiggins (1994) and Ottino (1989) for applications to fluid mixing and transport problems, certainly one of the richest application areas to date. Generalizations to infinite-dimensional systems include the work of Holmes and Marsden (1981), as well as the more recent work of McLaughlin and Overman (1995), Li, McLaugh-

lin, Shatah, and Wiggins (1996), and McLaughlin, and Shatah (1996).
◊

Example 1.10   (Marsden and Ratiu (1999)) Consider the forced pendulum system

$$\ddot{\theta} + \sin\theta = \epsilon\cos\omega t,$$

written in first order form as

$$\frac{d}{dt}\begin{pmatrix} \theta \\ \dot{\theta} \end{pmatrix} = \begin{pmatrix} \dot{\theta} \\ -\sin\theta \end{pmatrix} + \epsilon\begin{pmatrix} 0 \\ \cos\omega t \end{pmatrix}.$$

In our previous notation, $\mathbf{z} \equiv (\theta, \dot{\theta})$, and

$$\mathcal{H}_0(\theta, \dot{\theta}) = \frac{1}{2}(\dot{\theta})^2 - \cos\theta$$

$$\mathcal{H}_1(\theta, \dot{\theta}; t) = \theta\cos\omega t.$$

The system has hyperbolic fixed points at $\theta_0 = \pm(n+\frac{1}{2})\pi$. The homoclinic orbits can be written conveniently as

$$\theta(t) = \pm 2\tan^{-1}(\sinh t).$$

This gives

$$\bar{\mathbf{z}}(t) = \begin{pmatrix} \theta(t) \\ \dot{\theta}(t) \end{pmatrix} = \begin{pmatrix} \pm 2\tan^{-1}(\sinh t) \\ \pm\operatorname{sech} t \end{pmatrix}.$$

The Poincaré–Melnikov function $M(t_0)$ is given by

$$\begin{aligned}
M(t_0) &= \int_{-\infty}^{\infty}\left(\frac{\partial\mathcal{H}_0}{\partial\theta}\frac{\partial\mathcal{H}_1}{\partial\dot{\theta}} - \frac{\partial\mathcal{H}_0}{\partial\dot{\theta}}\frac{\partial\mathcal{H}_1}{\partial\theta}\right)(\bar{\mathbf{z}}(t-t_0), t)dt \\
&= -\int_{-\infty}^{\infty}\dot{\theta}(t-t_0)\cos\omega t\, dt \\
&= \pm\int_{-\infty}^{\infty}[\operatorname{sech}(t-t_0)\cos\omega t]\, dt \\
&= \mathcal{A}\cdot\cos\omega t_0,
\end{aligned}$$

where

$$\mathcal{A} = \pm 2\int_{-\infty}^{\infty}\operatorname{sech} t\cos\omega t\, dt.$$

The last step is obtained via the change of variable $t \mapsto t+t_0$. The function $\mathcal{A}$ is the Melnikov amplitude, and one must prove that $\mathcal{A} \neq 0$, since $\cos(\omega t_0)$

clearly has isolated zeros at $t_0 = (n+\frac{1}{2})\pi/\omega$. The evaluation of this integral is a straightforward residue calculation, and one can show there is a nonzero contribution from the simple pole $z = i\pi/2$. In many applications the final step of evaluating the Melnikov amplitude (or Melnikov integral) must be carried out numerically.   ◇

1.2.6   Adiabatic Invariants and Geometric Phases

For perturbed systems, it is important to ask what are the precise rates of change of the action and angle variables over long times? We first consider the action variable **I** for "slowly varying" Hamiltonians $\mathcal{H}(\mathbf{q}, \mathbf{p}; \tau)$, where $\tau = \epsilon t$, with $0 < \epsilon << 1$. It is well-known that, for ϵ sufficiently small, the action variable is *approximately* conserved. We will describe this result in the context of an example, which will lead to the formal definition of *adiabatic invariance.*

Example 1.11   Consider the harmonic oscillator with slowly varying frequency (see Littlewood (1963),Wasow (1973))

$$\mathcal{H}_\epsilon(q, p; \tau) = \frac{p^2}{2} + \frac{1}{2}\omega(\tau)^2 q^2$$
$$\tau = \epsilon t \quad 0 < \epsilon << 1.$$

Define the "frozen" system as that corresponding to the $\epsilon = 0$ limit

$$\mathcal{H}_0(q, p; 0) \equiv \frac{p^2}{2} + \frac{1}{2}\omega(0)^2 q^2 \equiv E,$$

where $\omega(0) = const.$ We know that the frozen system conserves energy, i.e., $E = const.$ For this limit, one has the action-angle variables $(E/\omega, \theta)$ defined by

$$\frac{E}{\omega} = \frac{1}{2}(\frac{p^2}{\omega} + \omega q^2),$$
$$\theta = \tan^{-1}(\frac{\omega q}{p}).$$

Consider the transformation $(q, p) \mapsto (E, \theta)$ for the slowly varying problem $\epsilon \neq 0$. Using the change of variables

$$q = \frac{(2E)^{1/2}}{\omega} \sin\theta,$$
$$p = (2E)^{1/2} \cos\theta,$$

and taking into account the time dependence of energy, gives the new system

$$\dot{E} = \epsilon \frac{(d\omega/d\tau)}{\omega} E(1 - \cos 2\theta),$$
$$\dot{\theta} = \omega(\tau) + \frac{\epsilon}{2} \frac{(d\omega/d\tau)}{\omega} \sin 2\theta,$$

which shows that the energy is no longer conserved. In fact, because of the first term on the right hand side of the energy equation, even if we average the angle variable θ over a 2π interval, there is still an $O(\epsilon)$ drift term (we assume ω is bounded away from zero). To eliminate this term we make one more transformation to the action coordinate $I = E/\omega$. Then

$$\frac{dE}{dt} = \dot{E} + \epsilon\frac{\partial E}{\partial \tau}$$

$$= \dot{I}\omega + \epsilon I\frac{d\omega}{d\tau},$$

and the new equations in (I, θ) variables become

$$\dot{I} = -\epsilon\frac{(d\omega/d\tau)}{\omega}I\cos 2\theta, \tag{1.2.21}$$

$$\dot{\theta} = \omega(I) + \frac{\epsilon}{2}\frac{(d\omega/d\tau)}{\omega}\sin 2\theta. \tag{1.2.22}$$

The right hand side of the action equation now averages to zero over a 2π interval, and during a time period when θ increases by π, I changes only by $O(\epsilon^2)$.

It is also instructive in this example to proceed formally, using a multi-scale asymptotic procedure, as described, for example in Kevorkian (1987), and more generally in Kevorkian and Cole (1996). We start with the equivalent slowly varying oscillator equation

$$\frac{d^2q}{dt^2} + \omega^2(\tau)q = 0 \tag{1.2.23}$$

$$q(0; \epsilon) = a, \qquad \frac{dq}{dt}(0; \epsilon) = b. \tag{1.2.24}$$

Suppose we make the transformation to a new, "fast" time variable

$$t^+ = \Omega(t; \epsilon), \qquad \Omega(0; \epsilon) = 0,$$

where the function Ω is as yet unspecified. Equation (1.2.23) in the new time variable then becomes

$$\frac{d^2q}{dt^{+2}} + \frac{(d^2\Omega/dt^2)}{(d\Omega/dt)}\frac{dq}{dt^+} + \frac{\omega^2(\tau)}{(d\Omega/dt)^2}\,q = 0$$

$$q(0; \epsilon) = a, \qquad \frac{dq}{dt^+}(0; \epsilon) = \frac{b}{\frac{d}{dt}(\Omega(0; \epsilon))}.$$

Our goal is to choose the function Ω so that oscillations in the q variable have constant frequency on a **fast time scale** t^+, hence we choose

$$\frac{d\Omega}{dt} \equiv \omega(\tau),$$

which when integrated gives

$$\Omega(t; \epsilon) \equiv \int_0^t w(\epsilon s) ds = \frac{1}{\epsilon} \int_0^\tau w(\tilde{s}) d\tilde{s}.$$

Our choice of time scales should then be

$$t^+ = \frac{1}{\epsilon} \int_0^\tau w(\tilde{s}) d\tilde{s}, \qquad \tau = \epsilon t.$$

Now expand the solution explicitly in these two times:

$$q(t; \epsilon) \equiv Q(t^+, \tau; \epsilon)$$
$$= Q_0(t^+, \tau) + \epsilon Q_1(t^+, \tau) + O(\epsilon^2)$$

giving

$$w^2 \mathcal{L} Q_0 + \epsilon \left[w^2 \mathcal{L} Q_1 + 2w \frac{\partial^2 Q_0}{\partial t^+ \partial \tau} + \frac{dw}{d\tau} \frac{\partial Q_0}{\partial t^+} \right] + O(\epsilon^2) = 0$$

$$Q_0(0,0) = a, \qquad w \frac{\partial Q_0}{\partial t^+}(0,0) = b,$$

where the linear operator $\mathcal{L}$ is defined as

$$\mathcal{L} \equiv \frac{\partial^2}{\partial t^{+2}} + 1.$$

This gives the sequence of linear equations that one gets by equating all terms of like powers of ϵ to zero:

$$w^2 \mathcal{L} Q_0 = 0$$
$$w^2 \mathcal{L} Q_1 = -2w \frac{\partial Q_0}{2\partial t^+ \partial \tau} - \frac{dw}{d\tau} \frac{\partial Q_0}{\partial t^+}$$
$$w^2 \mathcal{L} Q_n = F(Q_{n-1}, Q_{n-2}, \dots, Q_0) \qquad (n > 1).$$

Hence,

$$Q_0(t^+, \tau) = A(\tau) \cos t^+ + B(\tau) \sin t^+$$

with

$$\mathcal{L} Q_1 = \left[A \cdot \left(\frac{dw/d\tau}{w^2} \right) + \frac{2}{w} \frac{dA}{d\tau} \right] \sin t^+ - \left[B \cdot \left(\frac{dw/d\tau}{w^2} \right) + \frac{2}{w} \frac{dB}{d\tau} \right] \cos t^+.$$

This linear inhomogeneous equation has a right hand side that satisfies the corresponding homogeneous problem, hence solutions will be **secular** on the fast time scale (see Kevorkian and Cole (1996)), unless the right hand side is set to zero. This gives the solvability condition for the unknown slowly varying quantities $A(\tau)$ and $B(\tau)$,

$$A \cdot \left(\frac{d\omega/d\tau}{\omega^2} \right) + \frac{2}{\omega} \frac{dA}{d\tau} = 0$$

$$B \cdot \left(\frac{d\omega/d\tau}{\omega^2} \right) + \frac{2}{\omega} \frac{dB}{d\tau} = 0$$

whose solutions are

$$A(\tau) = A(0)\sqrt{\frac{\omega(0)}{\omega(\tau)}}, \qquad B(\tau) = B(0)\sqrt{\frac{\omega(0)}{\omega(\tau)}}.$$

The leading order solution to (1.2.23) can then be written as

$$Q_0(t^+, \tau) = \sqrt{\frac{\omega(0)}{\omega(\tau)}} \left[a \cos t^+ + \frac{b}{\omega(0)} \sin t^+ \right].$$

Using this, we can explicitly check the change in the action variable $I = E/\omega$ over a period using

$$I \equiv \frac{E}{\omega} = \frac{1}{2\omega} \left[(\frac{dq}{dt})^2 + \omega^2(\tau)q^2 \right].$$

One simply differentiates the action to get

$$\begin{aligned}
\frac{dI}{dt} &= \frac{1}{\omega} \frac{dE}{dt} - \frac{\epsilon}{\omega^2} E \frac{d\omega}{d\tau} \\
&= \frac{\epsilon}{2} \frac{d\omega}{d\tau} \left[q^2 - \frac{1}{\omega^2} (\frac{dq}{dt})^2 \right] \\
&= \epsilon \left[\omega(0) \left(a^2 - \frac{b^2}{\omega(0)^2} \right) \left(\frac{d\omega/d\tau}{2\omega} \right) \cos 2t^+ + ab \left(\frac{d\omega/d\tau}{\omega} \right) \sin 2t^+ \right] \\
&+ O(\epsilon^2),
\end{aligned}$$

which is purely oscillatory on the fast time scale and averages to zero over one period. By contrast, it can be verified directly that the change in energy is

$$\begin{aligned}
\frac{dE}{dt} &= \frac{\epsilon \omega(0)}{2} \frac{d\omega}{d\tau} \left[\left(a^2 + \frac{b^2}{\omega(0)^2} \right) + \frac{2ab}{\omega(0)} \sin 2t^+ + \left(a^2 - \frac{b^2}{\omega(0)^2} \right) \cos 2t^+ \right] \\
&+ O(\epsilon^2),
\end{aligned}$$

which has a nonoscillatory contribution from the first term. $\Diamond$

To formulate this more precisely we define the notion of an **adiabatic invariant**:

Definition 1.2.10. *A function $F(q(t), p(t); \tau(t))$ is an **adiabatic invariant** of order n if*

$$|F(t) - F(0)| \leq C_1 \epsilon^{n+1} |t| \quad \text{for } |t| \leq \epsilon^{-n}.$$

One can prove that for the perturbed action-angle system

$$\dot{I} = \epsilon f(\theta)$$
$$\dot{\theta} = \omega(I) \neq 0,$$

where $f(\theta)$ is a 2π-periodic function with mean

$$\langle f(\theta) \rangle = \frac{1}{2\pi} \int_0^{2\pi} f(\theta) d\theta,$$

the estimate $|I(t) - F(t)| < C_1 \cdot \epsilon$ for times $0 < t < \epsilon^{-1}$ holds, where C_1 is a constant and

$$F(t) = I(0) + \epsilon t \cdot \langle f(\theta) \rangle.$$

This means that when $\langle f(\theta) \rangle = 0$ as in (1.2.21), we have

$$|I(t) - I(0)| < C_1 \cdot \epsilon \quad \text{for } 0 < t < \epsilon^{-1},$$

showing that the action variable corresponding to the frozen system is an adiabatic invariant of order one. For general systems of $(M + N)$ equations of the form

$$\dot{p}_m = \epsilon f_m(p_i, q_i, \tilde{t}; \epsilon), \qquad\qquad m = 1, \dots, M$$
$$\dot{q}_n = \omega_n(p_i, \tilde{t}; \epsilon) + \epsilon g_n(p_i, q_i, \tilde{t}; \epsilon), \quad n = 1, \dots, N$$

one can define adiabatic invariants in the following way. Suppose f_m and g_n are 2π-periodic functions in the q_i variables. Following Kevorkian (1987), we can define an adiabatic invariant of order $O(\epsilon^K)$ as the function $\mathcal{A}(p_i, q_i; \tilde{t}; \epsilon)$, whose derivative must satisfy the following two conditions. Denote

$$\frac{d\mathcal{A}}{dt} = \phi(p_i, q_i, \tilde{t}; \epsilon).$$

(i) The function ϕ must have zero average to $O(\epsilon^{K+2})$ over all the q_i periods, i.e.,

$$\int_0^{2\pi} dq_1 \int_0^{2\pi} dq_2 \cdots \int_0^{2\pi} \phi(p_i, q_i, \tilde{t}; \epsilon) dq_N = O(\epsilon^{K+2}).$$

(ii) The function ϕ must have the property

$$\phi(p_i, q_i, \tilde{t}; \epsilon) = O(\epsilon^{K+1}) \quad \text{as} \quad \epsilon \to 0.$$

These conditions insure that the quantity $\mathcal{A}$ is constant to $O(\epsilon^K)$ uniformly in the interval $0 \leq t \leq T(\epsilon) = O(\epsilon^{-1})$. In his review article Kevorkian (1987) discusses these issues thoroughly.

Originally, the concept of adiabatic invariance was introduced by Ehrenfest (1916) in the context of the so-called "Old Quantum Theory," then by Born and Foch (1928) for the "New Quantum Theory." Through the 1950s and 60s the impetus for studies on adiabatic theory came primarily from applications in thermonuclear fusion, where the magnetic moment of a charged particle in a magnetic field was realized to be such an invariant. Important early papers include Kulsrud (1957), Lenard (1959), Gardner (1959), and Kruskal (1962). A comprehensive recent review of this interesting literature can be found in Henrard (1993) and a nice discussion, with references, can be also be found in Arnold (1988). We mention that there are detailed estimates on the change of an adiabatic invariant over both fixed and infinite time intervals. For example, it is known that the change in the adiabatic invariant over infinite time intervals is an exponentially small quantity

$$|I(\infty) - I(-\infty)| = C_1 \exp(-d/\epsilon)$$

(Dyhne (1960), Littlewood (1963), Meyer (1976), Fedorjuk (1976), Meyer (1980), Levi (1981), and Neistadt (1981)). See Holmes, Marsden and Scheurle (1988) for an interesting interpretation of the coefficient d in terms of the distance to the nearest pole in the complex t-plane. Alternatively, one can phrase the stability of the action variable as a "Nekhoroshev" type theorem. As long as the integrable part of the Hamiltonian $\mathcal{H}_0$ satisfies a certain so-called "steepness condition" (related to the convexity of the function), then

$$\|\mathbf{I}(t) - \mathbf{I}(0)\| < \epsilon^b$$

for all $t \in [0, T]$, where $T = \exp(C\epsilon^{-a})$ [3]. In a sense, Nekhoroshev's theorem (1971) and the KAM theorem complement each other in detailing the behavior of perturbed integrable systems. KAM is an existence result for invariant tori under perturbation, while Nekhoroshev's theorem quantifies the stability properties of the action variables, giving precise estimates on the stability timescale. See Dumas (1993) for further discussion on this and related literature. Additional recent research has focused on estimates for the change in the adiabatic invariant during a separatrix crossing event. See, for example, Timofeev (1978), Cary, Escande, Tennyson (1986), Bourland and Haberman (1990), Cary and Skodje (1989), Elskens and Escande (1991),

---

[3]For the general case of n frequencies, if the unperturbed Hamiltonian $\mathcal{H}_0$ is strictly convex, i.e., if the matrix $\partial^2 \mathcal{H}_0 / \partial I^2$ is positive definite, then one has the formulas $a = 2/(6n^2 - 3n + 14)$, $b = 3a/2$. See Arnold (1988) for more discussions of this and references.

Kaper and Wiggins (1992), Wiggins (1993), Neishtadt (1975), Neishtadt et al (1991), and Henrard (1993). In general, for a nice discussion of the more explicit and quantitative aspects of perturbed integrable systems, see Kevorkian (1987) and Kevorkian and Cole (1996).

We turn next to the change in the *angle variable* under slowly varying Hamiltonians, leading to the concept of the *geometric phase*. The recent literature on this topic was initiated by the papers of M. Berry (1984, 1985, 1988) in which he investigated the evolution of quantum systems whose Hamiltonian depends on external parameters that are slowly varied in a closed loop. The adiabatic theorem of quantum mechanics (Kato (1950)) says that for infinitely slow changes of the parameters the wave function whose evolution is described by the time-dependent Schrödinger equation is instantaneously in an eigenstate of the "frozen" Hamiltonian. At the end of a closed cycle, when the slowly varying parameters return to their original values, the wave function should return to the eigenstate in which it started, except for a possible phase change. Berry demonstrated that this phase change could be decomposed into two parts: a dynamic part due to the frozen system, and a "geometric" part due to the slow evolution of the system. The geometric phase factor has since been referred to as the "Berry phase," "Hannay–Berry phase," "Cartan–Hannay–Berry phase," or more generally the "geometric phase." [4] See Berry (1990) for historical developments of these ideas.

To briefly explain the presence of the geometric phase we follow the argument of Hannay (1985) in the classical context. Let $(I(\mathbf{R}(t)), \theta(\mathbf{R}(t)))$ represent the action-angle variables of a given integrable system, where $\mathbf{R}(t)$ is a vector of time varying external parameters. The Hamiltonian $\mathcal{H}(I(\mathbf{R}(t)), \mathbf{R}(t))$ depends on the action variables and external parameters. The frozen system $\mathbf{R} = const$ is governed by the equations

$$
\dot{I} = -\frac{\partial \mathcal{H}}{\partial \theta} \quad = 0
$$
$$
\dot{\theta} = \frac{\partial \mathcal{H}}{\partial I} \quad = \omega(I) \quad = const.
$$

By contrast, the time-dependent system is governed by

$$
\dot{I} = \dot{\mathbf{R}}(t) \cdot \frac{\partial I}{\partial \mathbf{R}}
$$
$$
\dot{\theta} = \omega(I) + \dot{\mathbf{R}}(t) \cdot \frac{\partial \theta}{\partial \mathbf{R}}.
$$

---

[4]Referred to by some as "the phase that launched a thousand scripts."

If **R** is small (i.e., **R** is slowly varying), then we can instead consider the averaged system [5]

$$\dot{I} = \dot{\mathbf{R}}(t) \cdot \left\langle \frac{\partial I}{\partial \mathbf{R}} \right\rangle$$

$$\dot{\theta} = \omega(I) + \dot{\mathbf{R}}(t) \cdot \left\langle \frac{\partial \theta}{\partial \mathbf{R}} \right\rangle,$$

where $\langle \ \rangle$ denotes the average around the level curves of the frozen Hamiltonian (i.e., one period of the frozen system). While the right hand side of the action equation is small due to the adiabatic invariance of the action, the right hand side of the angle equation is $O(1)$. Integrating with respect to time over the period T taken to traverse a closed level curve of the frozen Hamiltonian gives

$$\theta_T = \int_0^T \omega(I)dt + \int_0^T \dot{\mathbf{R}}(t) \cdot \left\langle \frac{\partial \theta}{\partial \mathbf{R}} \right\rangle dt$$
$$= \theta_d + \theta_g,$$

where θ_d is the **dynamic phase** due to the frozen system, while θ_g is the **geometric phase**. Its geometric nature can be seen by the simple formula

$$\theta_g = \int_0^T \dot{\mathbf{R}}(t) \cdot \left\langle \frac{\partial \theta}{\partial \mathbf{R}} \right\rangle dt$$
$$= \oint \left\langle \frac{\partial \theta}{\partial \mathbf{R}} \right\rangle \cdot d\mathbf{R},$$

where the contour integral is taken over the closed loop in parameter space, whose period is T. Notice that this formulation shows the term θ_g is independent of the time period T and depends only on geometric properties of

---

[5]This follows from the averaging theorem, which we summarize here. Consider the initial value problem for the vector function $x(t) \in \mathbb{R}^n$

$$\dot{x}(t) = \epsilon f(t, x) + \epsilon^2 g(t, x, \epsilon), \quad x(0) = x_0, \quad 0 < \epsilon \ll 1,$$

where $f(t, x)$ is T-periodic in t, with average $f^{(0)}(y)$ defined as

$$f^{(0)}(y) = \frac{1}{T} \int_0^T f(t, y)dt.$$

f and g can be any C^r functions with $r > 1$ and we assume that f, g, and $\partial f/\partial x$ are continuous and bounded by a constant M independent of ϵ for all t. The averaging theorem compares solutions $x(t)$ of the "exact" equations with solutions $y(t)$ of the "averaged" equation

$$\dot{y}(t) = \epsilon f^{(0)}(y), \quad y(0) = y_0.$$

The averaging theorem says that $|x(t) - y(t)| = O(\epsilon)$ on time scales $O(1/\epsilon)$. See Bogoliubov and Mitropolsky (1961), Volosov (1963), Hale (1980), Lochak and Meunier (1988), Arnold (1988), or Wiggins (1990) for more technical details.

the closed loop. In general, even in the limit $\epsilon \to 0$, this term will not vanish because of two compensating effects: On the one hand, $T \to \infty$, while on the other, $\dot{\mathbf{R}}(t) \to 0$. If the rates associated with these limits exactly balance, a residual term will remain when $\epsilon = 0$.

There are many physical settings in which this term appears, both adiabatically and nonadiabatically. The most familiar dynamical system exhibiting such a phase anholonomy is the Foucault pendulum (see Section 5.1). This well-known system shows a shift in the angle of the plane of the pendulum's swing when the Earth rotates through one complete circuit. The phase shift, equal to $2\pi \cos \alpha$ (see Section 5.1), where α is the colatitude, is derived by parallel-transporting an orthonormal frame through one revolution on the sphere along a fixed latitude. Another nice example for which many of the early ideas were developed is the "bead-on-hoop" problem. A frictionless bead is constrained to slide along a closed planar wire hoop, which encloses an area A and has length L. The hoop is oriented in a horizontal plane. As the bead slides around the closed hoop, the hoop is slowly rotated about a vertical axis through one full revolution. We are interested in a formula for the angular position of the bead after the hoop has rotated, compared with its position had the hoop been held fixed. The formula for this "geometric phase" $\Delta\theta$ is

$$\Delta\theta = -8\pi^2 \frac{A}{L^2}.$$

See Berry (1985), Hannay (1985), and Marsden and Ratiu (1999) for further discussion of these and other examples. The historical article by Berry (1990) is also quite useful in identifying some early contributions, particularly that of Rytov (1938) and Vladimirskii (1941), who anticipated some closely related ideas in their study of the circular polarization of light in an inhomogeneous medium. Another early contribution was made by Pancharatnam (1956) who developed some geometric phase concepts while investigating interference patterns produced by plates of anisotropic crystal.

Following Hannay's work, Berry obtained a semiclassical relationship between his phase and Hannay's angle (1985) . Aharonov and Anandan (1987) then showed that the geometric phase could be extracted from the total phase change not only for adiabatic but for any general cyclic evolution of a quantum system. Investigating the classical analogue, Berry and Hannay (1988) found a similar geometric phase for a general nonadiabatic cyclic change. Other contributions in the classical context include Golin (1988), Golin, Knauf, and Marmi (1989), and Golin and Marmi (1989, 1990).

The subject has a strong geometric flavor, starting with the work of Simon (1983). One can introduce a bundle of eigenstates of the slowly varying Hamiltonian as well as a natural *connection* on it — the Berry phase is the bundle holonomy associated with this connection. The curvature of the connection, when integrated over a closed two-dimensional surface in parameter space, gives the first Chern class characterizing the topological twisting of the bundle. To read more about this approach, one can begin with the papers of Gozzi and Thacker (1987a,b). Anandan and Stodolsky (1987) demonstrated that this interpretation can be extended to the holonomy in a vector bundle by considering all eigenspaces of the wave function. Marsden, Montgomery, and Ratiu (1989, 1990), Montgomery (1988) and Golin, Knauf, and Marmi (1989) develop these concepts further, especially in the classical case and in their extensions to nonintegrable systems, by showing that the geometric phase arises naturally in the reconstruction process — the opposite of reduction. An introduction to this point of view can be found in Marsden and Ratiu (1999).

The existence of a geometric phase and its importance have been established in a wide range of applications. Through examining the rotation of a free rigid body and the gravitational three-body problem, Montgomery (1996) illustrated how exact formulas for angle changes in these problems could be derived and related to the geometric phase, while Alber and Marsden (1992) showed how the phase shift formula for soliton interactions could be interpreted as a geometric phase. R. Newton (1994) derived the Berry phase formula associated with Schrödinger operators with continuous spectrum and showed its relation to the well-known S matrix from scattering theory. Shapere and Wilczek (1989) explained how a geometric phase arises in the context of self-propulsion of microorganisms at very low Reynolds numbers[6](see also Koiller, Ehlers, and Montgomery (1996)). This point of view is closely related to recent developments of the geometric phase in the context of control theory, discussed in Marsden et al (1991). In this context, the geometric phase idea has been used to explain how a falling cat manages to reorient itself from an upside down position, while maintaining zero angular momentum (see Montgomery (1993)). See Kane and Scher (1969) for a nice paper describing the phenomenon. These ideas have also been formulated more generally in terms of gauge theories (see Montgomery (1991) and the very thorough review article by Littlejohn and Reinsch (1997)). Marsden and Scheurle (1996) have applied the geometric phase idea for rotating mechanical systems with symmetries demonstrating how symmetric patterns can be brought out that would not otherwise be seen ("pattern evocation" — see Chapter 5). An interesting instance of this phenomenon for a point vortex system can be found in Kunin et al (1992). Newton (1994) and Shashikanth and Newton (1998) have demonstrated the presence of geometric phases in point vortex problems as shown below.

Example 1.12 Consider the motion of three planar point vortices of strength $(\Gamma_1, \Gamma_2, \Gamma_3)$ of the same sign in the configuration shown in Figure 1.6. We are interested in the phase angle $\theta(t)$ associated with vortex 2 during one full revolution of vortex 3, i.e., at the end of time $t = T$, where

---

[6]There is an interesting paper by Saffman (1967) describing a similar problem in the inviscid limit, although explicit use of a geometric phase concept is not used here.

$\varphi(t+T) = \varphi(t) + 2\pi$. In the adiabatic setting where $r/D \sim \epsilon << 1$, it was shown in Newton (1994) (for the restricted problem $\Gamma_1 = \Gamma_3 = \Gamma$, $\Gamma_2 = 0$) and Shashikanth and Newton (1998) (for the general case) that

$$\theta(T) = \theta_d + \theta_g,$$

where

$$\theta_g = \left(\frac{\Gamma_3}{\Gamma_1 + \Gamma_2 + \Gamma_3} \right) 2\pi \cos 2\theta(0),$$

$$\theta_d = \left(\frac{\Gamma_1 + \Gamma_2}{\Gamma_1 + \Gamma_2 + \Gamma_3} \right) (\frac{D}{r})^2.$$

The geometric quantity θ_g and the dynamic quantity θ_d can be understood in the following way. Since we are interested in the adiabatic limit, consider separately the 2-vortex configuration (Γ_1, Γ_2) in the absence of the third vortex Γ_3, then the 2-vortex problem $(\Gamma_1 + \Gamma_2, \Gamma_3)$ obtained by treating the two nearby vortices as one combined vortex of strength $\Gamma_1 + \Gamma_2$. The period of the first system is $T_{12} = r^2/(\Gamma_1 + \Gamma_2)$, while the period of the second is $T_{123} = D^2/(\Gamma_1 + \Gamma_2 + \Gamma_3)$. The vortex Γ_3 sweeps out a circle of radius $R = D(\Gamma_1 + \Gamma_2)/(\Gamma_1 + \Gamma_2 + \Gamma_3)$ and area $A = \pi R^2$. The dynamic phase term is then simply the ratio of the two periods

$$\theta_d = \frac{T_{123}}{T_{12}},$$

while the geometric phase can be written in terms of the limiting geometry

$$\theta_g = C \cdot \frac{A}{D^2},$$

where C is a constant function of the vortex strengths and initial phase angle. This is described in more detail in Chapter 5. $\Diamond$

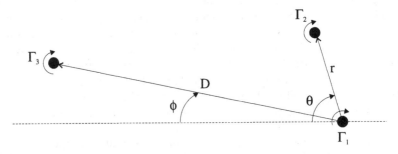

FIGURE 1.6. 3-vortex configuration exhibiting a geometric phase.

The collection of papers edited by Shapere and Wilczek (1989) offers a nice introduction to the important pre-1989 literature, while a general introduction to geometric phases can be found in Marsden and Ratiu (1999).

We end this section by mentioning that the modern subject of mechanics has a very geometrical literature associated with it, due mostly to the work of Arnold (1966), Smale (1970), Arnold (1978), and Abraham and Marsden (1978). A review of this approach lies beyond the scope of this book and we refer the reader to the recent book by Marsden (1992), Marsden and Ratiu (1999), as well as Arnold and Khesin (1998). The review article by Deift (1996) is also a useful starting point, particularly with regard to infinite-dimensional systems.

1.3   Summary of Basic Questions

We are now in a position to state the principal questions that will concern us in this book:

1. Under what conditions is an N-vortex system completely integrable? As we have seen, this question hinges on how many independent involutive conserved quantities exist in the $2n$-dimensional phase space. This in turn will depend on

   - the vortex strengths $(\Gamma_1, \dots, \Gamma_n) \in \mathbb{R}^n$, which we can think of as parameters in the problem;
   - the domain in question, i.e., planar, spherical, or with closed or open boundaries.

2. In the case where the system is completely integrable, do solutions exist for all time? Are they quasiperiodic or closed? Are there finite-time collisions of the vortices?

3. What kinds of fixed and relative equilibria exist and what are their stability properties?

4. Which vortex problems are nonintegrable and what are the mechanisms that break the integrability?

5. What kinds of instantaneous streamline patterns can exist for a given N-vortex configuration? In particular, what are the topological changes in the streamline structures that take place as the vortices evolve?

6. How can one make use of the discrete and continuous symmetries to construct special N-vortex configurations ?

7. Can one apply KAM theory to vortex systems, and, from it, what physical conclusions can one draw? What coordinates can one introduce to reduce the phase space dimension and apply KAM theory for a general vortex problem?

8. Which vortex configurations give rise to geometric phases? How does one compute them and why are they physically relevant?

9. What tools can one use to study the N-vortex problem for large N? What is meant by "two-dimensional turbulence?"

10. The problem of going from two dimensions to three is an interesting one: The two-dimensional problem can be thought of as studying the dynamics of perfectly straight vortex filaments perpendicular to the plane of motion. What can one say regarding the dynamics of these straight filaments under perturbation which, in effect, introduces a "weak" third dimension?

1.4 Exercises

1. Show that taking the divergence and curl of (1.1.1) gives the relations

$$\nabla \cdot \boldsymbol{\omega} = 0$$
$$\nabla \times \boldsymbol{\omega} = -\nabla^2 \mathbf{u},$$

and that the vector potential ψ satisfies the Poisson equation (1.1.5).

2. Derive the formula for the Biot–Savart kernel (1.1.9).

3. Take the curl of the Euler equations (1.1.12) to derive the vorticity evolution equation (1.1.13).

4. Using complex variable notation $z = x + iy$:

   (a) Show that

   $$\frac{\partial}{\partial z^*} \equiv \frac{1}{2}\left(\frac{\partial}{\partial x} + i\frac{\partial}{\partial y}\right).$$

   (b) Show that Hamilton's canonical equations can then be written

   $$\dot{z} = -2i\frac{\partial \mathcal{H}}{\partial z^*}.$$

   (c) For point vortex systems , show that

   $$\mathcal{H} = -\frac{1}{8\pi}\sum_{\alpha \neq \beta}^{N} \Gamma_\alpha \Gamma_\beta \log\left[(z_\alpha - z_\beta)(z_\alpha^* - z_\beta^*)\right].$$

5. Consider the Euler equations (1.1.12) with no body forces, i.e., $\mathbf{f} = 0$:

(a) Show that it can be written

$$\mathbb{J}\nabla(\psi_t) + \nabla(\frac{1}{2}\mid\nabla\psi\mid^2 + p)\Delta\psi\nabla\psi = 0.$$

(b) Use the complex variable notation

$$\nabla \equiv 2\partial_{z^*}$$
$$\mathbb{J} \equiv i$$
$$\Delta \equiv 4\partial_z\partial_{z^*}$$
$$\Delta\psi\nabla\psi \equiv 4\partial_z(\partial_{z^*}\psi)^2$$

to rewrite the Euler equation as

$$\partial_{z^*}g = 2\partial_z(\partial_{z^*}\psi)^2$$

for the function

$$g(z,t) \equiv 2\partial_z\psi\partial_{z^*}\psi + p - i\partial_t\psi.$$

(c) Rewrite the Euler equation as

$$d\left(gdz + 2(\partial_{z^*}\psi)^2 dz^*\right) = 0$$

and integrate this form over a domain Ω to obtain a weak form of Euler's equation

$$\int_{\partial\Omega} gdz + 2\int_{\partial\Omega}(\partial_{z^*}\psi)^2 dz^* = 0.$$

Discussions of this version of Euler's equation can be found in Flucher and Gustafsson (1997).

6. (a) Prove that the following quantities are conserved for the Euler equations:

$$E(t) = \frac{1}{2}\int_{\mathbb{R}^3}\|\mathbf{u}\|^2 d\mathbf{x}$$

$$W_1(t) = \int_{\mathbb{R}^3}\boldsymbol{\omega}(\mathbf{x})d\mathbf{x}$$

$$L(t) = \frac{1}{2}\int_{\mathbb{R}^3}(\mathbf{x}\times\boldsymbol{\omega})d\mathbf{x}$$

$$I(t) = -\frac{1}{2}\int_{\mathbb{R}^3}\|\mathbf{x}\|\boldsymbol{\omega}d\mathbf{x}$$

$$J(t) = \int_{\mathbb{R}^3}\mathbf{u}\cdot\boldsymbol{\omega}d\mathbf{x}.$$

(b) In two dimensions, prove that the moments of vorticity W_k are conserved for any integer k, where

$$W_k = \frac{1}{k} \int_{\mathbb{R}^2} \omega^k(\mathbf{x}).$$

(c) Derive the dynamical equation for W_2 in three dimensions, showing that it is not a conserved quantity.

7. For two-dimensional flows, derive the velocity and vorticity formulas in polar coordinates:

$$u_r = \frac{1}{r}\frac{\partial \psi}{\partial \theta}, \qquad u_\theta = -\frac{\partial \psi}{\partial r}$$

$$\omega = -\frac{1}{r}\frac{\partial}{\partial r}\left(r\frac{\partial \psi}{\partial r}\right) - \frac{1}{r^2}\frac{\partial^2 \psi}{\partial \theta^2}.$$

8. Take the equation of motion (1.1.22) for two corotating spiral vortices $\Gamma_1 = 1 + i = \Gamma_2$. Derive the solution for the vortex trajectories as well as the trajectory of a passive particle in the flowfield.

9. (a) Prove that the $\mathbb{J}$ matrix for rigid body dynamics satisfies the Jacobi condition (1.2.11).

(b) Prove that the angular momentum $\mathbf{u} \cdot \mathbf{u}$ is a Casimir.

10. Carry out the details from Example 1.6 for the harmonic oscillator.

11. For the 2-vortex problem described in Example 1.8 prove that D and C are conserved and derive the details of the solution in action-angle form.

12. For three equal strength vortices lying on an equilateral triangle prove that the configuration rotates rigidly and derive a solution formula for the rotation frequency.

13. For the forced pendulum system described in Example 1.10, derive the solution formula for the unperturbed homoclinic orbit, and evaluate the Melnikov amplitude integral using residues.

14. Consider the perturbed action-angle system

$$\dot{I} = \epsilon f(\theta)$$
$$\dot{\theta} = \omega(I) \neq 0,$$

where $f(\theta)$ is a 2π-periodic function with zero mean. Prove that the action variable $I(t)$ is an adiabatic invariant of order one.

2

N Vortices in the Plane

In this chapter we consider the equations for N-point vortices in the unbounded plane. First we give the general Hamiltonian formulation showing all conserved quantities. A by product of this discussion is the simple proof that the system is integrable for three or fewer vortices of any strength. The equations are then written in terms of their intervortical distances. In Section 2.2 we specialize to the case $N = 3$, characterizing all relative equilibria and collapsing states. Using the trilinear coordinates of Synge (1949) and Aref (1979), we then describe the phase plane formulation for more general states. In Section 2.3 we focus on the 4-vortex case. First we use the coordinates introduced in Khanin (1982) to reduce the Hamiltonian to one with two degrees of freedom. An extra feature of the reduction method is that it prepares the system for an application of the KAM theorem which we describe. We then use the Melnikov method to prove nonintegrability for a special 4-vortex configuration, following ideas of Ziglin (1980), Koiller and Carvalho (1985, 1989) and, more recently, Castilla, Moauro, Negrini, and Oliva (1993). Section 2.4 offers a brief summary of some classical and recent work that is particularly interesting and worth reading. Thus, taken together, we show in this chapter that the typical N-vortex problem contains regions of phase space that support periodic, quasiperiodic, and chaotic orbits, all coexisting.

2.1   General Formulation

The governing equations for a collection of vortices, each with strength $\Gamma_\alpha \in \mathbb{R}$, located at $z_\alpha = x_\alpha(t) + iy_\alpha(t)$, are given by (1.1.21), or equivalently, (1.1.22). The system is of Hamiltonian form

$$\Gamma_\alpha \dot{x}_\alpha = \frac{\partial \mathcal{H}}{\partial y_\alpha}, \qquad \Gamma_\alpha \dot{y}_\alpha = -\frac{\partial \mathcal{H}}{\partial x_\alpha}, \tag{2.1.1}$$

$\alpha = 1, \dots, N$, with

$$\mathcal{H} = -\frac{1}{4\pi} \sum_{\beta \neq \alpha}^{N} \Gamma_\alpha \Gamma_\beta \log(l_{\alpha\beta}), \tag{2.1.2}$$

$l_{\alpha\beta} = |z_\alpha - z_\beta|$. For convenience, throughout this chapter we will use the notation $\Gamma = \sum_\alpha^N \Gamma_\alpha$ to denote the total vorticity in the plane. The system (1.1.21) has the following four conserved quantities $(\mathcal{H}, Q, P, I)$

$$\mathcal{H} = -\frac{1}{4\pi} \sum_{\beta \neq \alpha}^{N} \Gamma_\alpha \Gamma_\beta \log(l_{\alpha\beta})$$

$$Q + iP = \sum_{\alpha=1}^{N} \Gamma_\alpha z_\alpha$$

$$I = \sum_{\alpha=1}^{N} \Gamma_\alpha |z_\alpha|^2.$$

The Hamiltonian $\mathcal{H}$ is the **interaction energy** for the vortex system — its conservation follows in the usual way from the canonical equations. The complex quantity $Q + iP$ is called the moment of vorticity. When divided by the total vorticity Γ, it is referred to as the **center of vorticity**, (or barycenter) of the system. Since the Hamiltonian only depends on the intervortical distances, it is invariant with respect to translations $z_\alpha \to z_\alpha + C$, and rotations $z_\alpha \to \exp(i\theta)z_\alpha$. Invariance to translations implies conservation of the components of the moment of vorticity (Q, P). These quantities are also called the components of **linear impulse** of the system and are the discrete analogues of the corresponding conserved quantity of the Euler equations given by (1.1.14). The conserved quantity I is called the angular impulse, also the discrete counterpart of the angular momentum of the Euler equations (1.1.15). Note also that the system (1.1.21) is invariant with respect to scale transformations $z \to \lambda \hat{z}$, $t \to \lambda^2 \hat{t}$, for any constant scaling factor λ which leads to the additional invariant

$$I_0 = \frac{1}{2i} \sum_{\alpha=1}^{N} \Gamma_\alpha(z_\alpha^* \dot{z}_\alpha - z_\alpha \dot{z}_\alpha^*)$$

$$\equiv \frac{1}{2} \sum_{\alpha \neq \beta} \Gamma_\alpha \Gamma_\beta$$

as discussed in more detail in Chapman (1978).

In addition to these continuous symmetries, the N-vortex system has the following discrete symmetries:

1. $t \to -t, \quad \Gamma_\beta \to -\Gamma_\beta$

2. $z \to -z$

3. $\Gamma_\beta \to -\Gamma_\beta, \quad z \to z^*$

4. Cyclic permutations of indices.

Remarks.

1. For the case $N = 3$ we assume $\Gamma_1 \geq \Gamma_2 > 0$, while $\Gamma_3 \in \mathbb{R}$. All other cases can be obtained using the discrete symmetry groups, hence there is no loss of generality in making this assumption.

2. These discrete symmetry groups have been used to generate special families of solutions (see, for example, Love (1894), Havelock (1931), Aref (1982), Koiller et al (1985), and Lewis and Ratiu (1996)).

3. It is sometimes useful to express the system (1.1.21) in polar coordinates using the canonical transformation

$$z_\beta = \sqrt{2R_\beta} \exp[i\theta_\beta].$$

Then (1.1.21) becomes

$$\dot{R}_\beta = \frac{1}{2\pi} \sum_{\alpha \neq \beta}^{N} \frac{\Gamma_\alpha \sqrt{R_\alpha R_\beta} \sin(\theta_\alpha - \theta_\beta)}{(R_\alpha + R_\beta) - 2\sqrt{R_\alpha R_\beta} \cos(\theta_\alpha - \theta_\beta)} \tag{2.1.3}$$

$$\dot{\theta}_\beta = \frac{1}{4\pi} \sum_{\alpha \neq \beta}^{N} \frac{\Gamma_\alpha(1 - \sqrt{R_\alpha/R_\beta} \cos(\theta_\alpha - \theta_\beta))}{((R_\alpha + R_\beta) - 2\sqrt{R_\alpha R_\beta} \cos(\theta_\alpha - \theta_\beta))}. \tag{2.1.4}$$

One can verify by direct calculation that the system is in canonical Hamiltonian form

$$\Gamma_\beta \dot{R}_\beta = \frac{\partial \mathcal{H}}{\partial \theta_\beta}, \quad \Gamma_\beta \dot{\theta}_\beta = -\frac{\partial \mathcal{H}}{\partial R_\beta}$$

for $\beta = 1, \ldots, N$, with governing Hamiltonian

$$\mathcal{H} = -\frac{1}{4\pi} \sum_{\alpha=1}^{N} \sum_{\beta \neq \alpha}^{N} \Gamma_\alpha \Gamma_\beta \log((R_\alpha + R_\beta) - 2\sqrt{R_\alpha R_\beta} \cos(\theta_\alpha - \theta_\beta)).$$

If one replaces (2.1.3) with a constraint that each vortex lies on a closed loop defined by

$$R_\beta = f_\beta(\theta_\beta),$$

where $R_\beta \geq 0$, $0 < \theta_\beta \leq 2\pi$, and $f_\beta(\theta_\beta)$ is required to be 2π-periodic,

$$f_\beta(\theta_\beta) = f_\beta(\theta_\beta + 2n\pi)$$

for any integer n, then one has a one-dimensional **constrained** vortex problem, described further in Newton (2001), where it is proven that if the loop is circular, the problem is integrable for all N.   ◊

With the canonical Poisson bracket (1.2.4) defined in Chapter 1, the bracket between two functions $f(z_\alpha)$, $g(z_\alpha)$ of the position variable z_α is

$$\{f, g\} = \sum_{\alpha=1}^{N} \frac{1}{\Gamma_\alpha} \left(\frac{\partial f}{\partial x_\alpha} \frac{\partial g}{\partial y_\alpha} - \frac{\partial f}{\partial y_\alpha} \frac{\partial g}{\partial x_\alpha} \right).$$

This gives the following three involutive quantities:

1. $\{\mathcal{H}, I\} = 0$

2. $\{\mathcal{H}, P^2 + Q^2\} = 0$

3. $\{P^2 + Q^2, I\} = 0.$

In addition to these, we have

1. $\{Q, P\} = \Gamma$

2. $\{Q, I\} = 2P$

3. $\{P, I\} = -2Q$

giving rise to the following fundamental result.

Theorem 2.1.1. *The N-vortex problem (1.1.21) for $N \leq 3$ is integrable for all values of Γ_α. If $\Gamma = 0$, the 4-vortex problem is integrable.*

Proof. For $N = 1$ and $N = 2$ the problem is trivially integrable as shown in the examples in Chapter 1. For $N = 3$ the three conserved quantities $\mathcal{H}, I, P^2 + Q^2$ are mutually involutive and functionally independent (see Novikov and Sedov (1978), Aref (1978), and Adams and Ratiu (1982)) for all Γ_α. If $\Gamma = 0$, then the four conserved quantities $\mathcal{H}, P, Q, I$ are mutually involutive since, without loss, we can shift the origin of the system so that $P = 0, Q = 0$. This is equivalent to adding a constant to P and Q so that their numerical value is zero — something which does not affect their Poisson bracket or the fact that they are conserved under the flow. $\square$

The integrability of the 3-vortex problem was known to Gröbli (1877), Poincaré (1893), and Synge (1949). The problem then lay dormant for 25 years before being reanalyzed [1] by Novikov (1975) for the case of equal Γ_α and Aref (1979) for general Γ_α. See Adams and Ratiu (1982) for further mathematical discussion on the integrability for $N = 3$. The integrable 4-vortex case is discussed in Aref and Pomphrey (1980, 1982), Aref (1982), and Eckhardt (1988).

Remark.   We know this result is sharp in the sense that nonintegrability for the 4-vortex problem for $\Gamma \neq 0$ has been proven, as discussed in detail in Section 2.3.3. $\Diamond$

It is useful to rewrite the N-vortex system in terms of their intervortex separations $l_{\alpha\beta}$. From the basic equations (1.1.21) one can derive the following closed system

$$\frac{d}{dt}(l_{\alpha\beta}^2) = \frac{2}{\pi} \sum_{\gamma \neq \alpha \neq \beta}^{N} \Gamma_\gamma \sigma_{\alpha\beta\gamma} A_{\alpha\beta\gamma} \left(\frac{1}{l_{\beta\gamma}^2} - \frac{1}{l_{\gamma\alpha}^2} \right) \qquad (2.1.5)$$

where $N \geq 3$. $\sigma_{\alpha\beta\gamma} = \pm 1$ denotes the arrangement of the three vortices $(\Gamma_\alpha, \Gamma_\beta, \Gamma_\gamma)$ spanning the triangle, defined as $+1$ if they appear counterclockwise (in Synge's terminology, this is called a "positive vortex circuit"), and -1 if they appear clockwise ("negative vortex circuit"). $A_{\alpha\beta\gamma}$ is the area of the vortex triangle expressed in terms of the lengths $l_{\alpha\beta}$ as

$$A_{\alpha\beta\gamma} = \frac{1}{4} \left[2(l_{\alpha\beta}^2 l_{\beta\gamma}^2 + l_{\beta\gamma}^2 l_{\gamma\alpha}^2 + l_{\gamma\alpha}^2 l_{\alpha\beta}^2) - l_{\alpha\beta}^4 - l_{\beta\gamma}^4 - l_{\gamma\alpha}^4 \right]^{1/2} \geq 0.$$

---

[1] Apparently Batchelor was unaware of the classic papers of Gröbli (1877) and Synge (1949) when he made the statement in his book (p. 531): "When $n = 3$ the details of the motion are not evident." This led Novikov (1975), who makes reference to Batchelor's statement, to study the problem, followed by Aref (1979). For more on the history of the problem, see Aref, Rott, and Thomann (1992). In general, the three papers of Synge, Novikov, and Aref complement each other and are well worth reading.

For a vortex circuit to change sign the three vortices must pass through a collinear state, which means $A_{\alpha\beta\gamma} = 0$. There are two coordinate independent conserved quantities that can be written directly in terms of the lengths

$$\mathcal{H} = -\frac{1}{4\pi} \sum_{\alpha \neq \beta}^{N} \Gamma_\alpha \Gamma_\beta \log(l_{\alpha\beta}) \tag{2.1.6}$$

$$M \equiv \sum_{\alpha \neq \beta}^{N} \Gamma_\alpha \Gamma_\beta l_{\alpha\beta}^2 = 2\Gamma \sum_\beta \Gamma_\beta (x_\beta^2 + y_\beta^2) - 2(\sum_\beta \Gamma_\beta x_\beta)^2 - 2(\sum_\beta \Gamma_\beta y_\beta)^2$$
$$= 2(\Gamma I - Q^2 - P^2). \tag{2.1.7}$$

Remarks.

1. The system (2.1.5) is a closed system of $N(N-1)/2$ variables, $(2N-3)$ of which are independent. For $N = 3$ there are three equations, which are independent.

2. If $N = 2$, the right side of (2.1.5) vanishes, since the distance between the vortices is constant, as was seen in Example 1.8 of Chapter 1. $N = 3$ is the smallest system capable of exciting new length scales since it has a nonzero right hand side. For any $N > 3$ the system is written in terms of all possible triad interactions. Since the problem $N = 3$ is integrable, the general (nonintegrable) system (for $N > 3$) can be viewed as a dynamical interaction of integrable triads, as remarked in Aref (1979) and Kraichnan (1980). From this point of view the 3-vortex'problem is fundamental for understanding the general case $N > 3$.

3. Cyclic permutations of indices do not change the sign of the vortex circuit. Anticyclic permutations, however, do cause the vortex circuit to change signs, which in general alters the dynamics. This distinction between positive and negative circuits is important when considering the collision process treated in detail in Section 2.2.3.   ◊

Definition 2.1.1. *When one of the vortex strengths of an N-vortex problem is taken to be zero, i.e., $\Gamma_\alpha = 0$ for some α, $1 \leq \alpha \leq N$, one refers to this as a **restricted N-vortex problem**.*

This terminology is borrowed from the celestial mechanics literature (see Szebehely (1967)) where one refers to a "restricted" N-body problem when one of the masses is taken to be zero. See Neufeld and Tél (1997) and Whipple and Szebehely (1984) for some discussion. The equations of motion for a particle located at position z is then coupled to the equations governing the remaining $N - 1$ vortices,

$$\dot{z}^* = -\frac{i}{2\pi} \sum_{\beta=1}^{N-1} \frac{\Gamma_\beta}{(z - z_\beta)}$$

$$\dot{z}_\alpha^* = -\frac{i}{2\pi} \sum_{\beta \neq \alpha}^{N-1} \frac{\Gamma_\beta}{(z_\alpha - z_\beta)}.$$

In his classic paper of 1949 Synge formulated the following theorem for a general configuration of N vortices.

Theorem 2.1.2 (Synge).

1. *Given a vortex configuration, if the strengths of all vortices are reversed, the system will retrace the sequence of configurations through which it came.*

2. *If a configuration is collinear at time $t = t_0$, then the configuration at times $t = t_0 \pm \tau$ are reflections of one another for all values of τ.*

3. *A system cannot pass through more than two distinct collinear configurations. The times required to pass from one collinear configuration to the other are all the same.*

4. *Suppose there are two systems S_1 and S_2 of vortices, each consisting of the same number of vortices, with the strength of each vortex in S_2 being λ^2 times the strength of each in S_1. Suppose further that initially the configurations are similar, without reflection, with the lengths in S_2 being L times those in S_1, where L is a constant. Then the subsequent configuration of S_2 after time t_2 is similar, without reflection, to the configuration of S_1 after time t_1, where $t_2 = t_1(L^2/\lambda^2)$.*

5. *If the strengths of all the vortices have the same sign, their mutual distances are bounded above and below for all time.*

Proof. Parts 1 and 4 follow directly from the scaling invariants mentioned earlier, while part 5 follows directly from the conservation of $\mathcal{H}$. We leave the detailed proof of the statements as an exercise. □

We finish this subsection by defining a special class of vortex motions called *self-similar* evolutions.

Definition 2.1.2. *A system of vortices evolves **self-similarly** if $z_\alpha = \lambda_\alpha f(t)$ for $\alpha = 1, \ldots, N$, where $\lambda_\alpha \in \mathbb{C}$ are complex-valued scale factors and $f(t) \equiv r(t) \exp(i\theta(t))$ is a complex-valued function of time.*

There is no loss of generality in taking $r(0) = 1$, since the λ_α can have arbitrary modulus. Consider the planar vortex system written as

$$\dot{z}_\alpha = \frac{i}{2\pi} \sum_{\substack{\beta \neq \alpha}}^{N} \frac{\Gamma_\beta}{(z_\alpha^* - z_\beta^*)},$$

for $\alpha = 1, \ldots, N$. Assuming the system evolves self-similarly, then we have

$$\lambda_\alpha \dot{f} f^* = \frac{i}{2\pi} \sum_{\substack{\beta \neq \alpha}}^{N} \frac{\Gamma_\beta}{(\lambda_\alpha^* - \lambda_\beta^*)},$$

which yields the dynamical equation for $f(t)$

$$\dot{f} f^* = C \equiv A + iB, \tag{2.1.8}$$

along with the algebraic system of N complex equations

$$\lambda_\alpha C = \frac{i}{2\pi} \sum_{\substack{\beta \neq \alpha}}^{N} \frac{\Gamma_\beta}{(\lambda_\alpha^* - \lambda_\beta^*)}. \tag{2.1.9}$$

To solve the system (2.1.8) let $f(t) = r(t) \exp(i\theta(t))$, which gives the coupled amplitude-phase system

$$\frac{1}{2} \frac{d(r^2)}{dt} = A, \quad r(0) = 1,$$
$$\frac{d\theta'}{dt} = \frac{B}{r^2},$$

whose solution is

$$r(t) = \sqrt{2At + 1}$$
$$\theta(t) = \begin{cases} \frac{B}{2A} \log(2At + 1) & (A \neq 0) \\ Bt + \theta(0) & (A = 0). \end{cases}$$

The vortices collide self-similarly (at their center of vorticity) if $r(t^*) = 0$ for some $t^* > 0$. Hence, if $A < 0$, then $t^* = -1/2A > 0$. Solving the algebraic system (2.1.9) for the scaling constants λ_α for general N is complicated and a systematic treatment has yet to be given. However, special cases are not difficult to construct.

Example 2.1 (Kimura (1988)) Take as a special case $(\Gamma_1, \Gamma_2, \Gamma_3) = (2, 2, -1)$ with initial conditions

$$z_1(0) = x_1(0) + iy_1(0) = \left(\frac{3 + \sqrt{3}\cos\theta}{6} + i\frac{\sqrt{3}\sin\theta}{6} \right) l_{12}$$

$$z_2(0) = x_2(0) + iy_2(0) = \left(\frac{-3 + \sqrt{3}\cos\theta}{6} + i\frac{\sqrt{3}\sin\theta}{6} \right) l_{12}$$

$$z_3(0) = x_3(0) + iy_3(0) = \left(\frac{2\sqrt{3}\cos\theta}{3} + i\frac{2\sqrt{3}\sin\theta}{3} \right) l_{12},$$

where $l_{12} = \|x_1 - x_2\|$ and the θ are arbitrary. The center of vorticity of this configuration is at the origin where the vortices collide. Solving for A and B gives

$$A = -\frac{6\sin 2\theta}{(5 - 3\cos 2\theta)l_{12}^2}$$

$$B = \frac{18 - 6\cos 2\theta}{(5 - 3\cos 2\theta)l_{12}^2}$$

with collision time t^* given by

$$t^* = \frac{5 - 3\cos 2\theta}{12\sin 2\theta} l_{12}^2.$$

Necessary and sufficient conditions for collision for $N = 3$ will be described in the next subsection. Figure 2.1 shows a typical collapsing spiral for $N = 3$. $\qquad\qquad\diamond$

2.2 $N = 3$

In this section, we describe in more detail the motion of three vortices. We will make the assumption throughout that $\Gamma_1 \geq \Gamma_2 > 0$ while $\Gamma_3 \in \mathbb{R}$. First we characterize all fixed and relative equilibrium states, then we describe the process of self-similar collapse. After that we follow in some detail the papers of Synge (1949) and Aref (1979), making use of the trilinear coordinates to describe the general phase plane dynamics.

We start with an example showing that, in contrast to the solution of the restricted three body problem of celestial mechanics (Szebehely (1967)), solving the restricted 3-vortex problem is relatively straightforward.

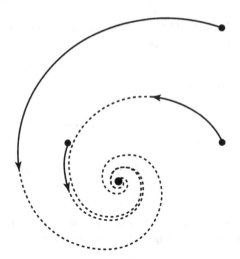

FIGURE 2.1. Self-similar collapse of three vortices.

Example 2.2   (Koiller, Carvalho (1985)) The equations governing the motion of two vortices are

$$\dot{z}_1 = \frac{i\Gamma_2}{2\pi} \cdot \frac{(z_1 - z_2)}{|z_1 - z_2|^2}$$

$$\dot{z}_2 = \frac{i\Gamma_1}{2\pi} \cdot \frac{(z_2 - z_1)}{|z_1 - z_2|^2}.$$

We will consider the case $\Gamma_1 + \Gamma_2 \neq 0$, so the two primaries move in circular orbits around the center of vorticity

$$C_0 = \frac{(\Gamma_1 z_1 + \Gamma_2 z_2)}{(\Gamma_1 + \Gamma_2)}$$

with frequency

$$\omega = \frac{(\Gamma_1 + \Gamma_2)}{2\pi D^2}$$

and D the distance between the two vortices. We then consider the motion of a third vortex of negligible strength, $\Gamma_3 \approx 0$. Its governing equation of motion, assuming it does not affect the motion of Γ_1 and Γ_2, is

$$\dot{z}_3 = \frac{i\Gamma_1}{2\pi} \cdot \frac{(z_3 - z_1)}{|z_3 - z_1|^2} + \frac{i\Gamma_2}{2\pi} \cdot \frac{(z_3 - z_2)}{|z_3 - z_2|^2},$$

where we have fixed the coordinates so that the center of vorticity is at the origin. We now transform to a rotating reference frame in which (Γ_1, Γ_2) are at rest, hence

$$\xi_j = \exp(-i\omega t) z_j, \quad j = 1, 2, 3,$$

giving an equation with periodic coefficients

$$\dot{\xi}_3 = -iw\xi_3 + \frac{i\Gamma_1}{2\pi} \cdot \frac{(\xi_3 - \xi_1)}{|\xi_3 - \xi_1|^2} + \frac{i\Gamma_2}{2\pi} \cdot \frac{(\xi_3 - \xi_2)}{|\xi_3 - \xi_2|^2},$$

with $\Gamma_1\xi_1 + \Gamma_2\xi_2 = 0$. As in the restricted three-body problem (Szebehely (1967)), it is convenient to normalize the problem by choosing units of length and time so that

$$\Gamma_1 + \Gamma_2 = 2\pi$$
$$|\xi_1 - \xi_2| = 1,$$

and we choose the coordinates ξ_1, ξ_2 to lie on the real axis. The system can be conveniently expressed by letting

$$\xi_1 = -\lambda$$
$$\xi_2 = 1 - \lambda$$
$$\Gamma_2 = 2\pi\lambda$$

for $\lambda \in (-\infty, \infty)$. The equation for $\xi_3 \equiv u + iv$ is a Hamiltonian system, with Hamiltonian given by

$$\mathcal{H}(u, v) = -\frac{1}{2}(u^2 + v^2) + (1 - \lambda)\log(\sqrt{(u + \lambda)^2 + v^2})$$
$$+ \lambda\log(\sqrt{(u + \lambda - 1)^2 + v^2}).$$

The level curves of the Hamiltonian are shown in Figure 2.6 for the symmetric case of equal primaries $\lambda = 1/2$. The equilibria correspond to two equilateral triangle configurations at the fixed points $\xi = \pm i\sqrt{3}/2$, and three collinear states at $\xi = 0, \pm\sqrt{5}/2$. For the general case $0 < \lambda < 1$, the same five possible equilibria are present, their coordinates, of course, depend on the specific value of λ. These are the analogues of the Lagrange points for the restricted three-body problem (Szebehely (1967)). The stability of the equilibria can be determined by evaluating the Hessian matrix at each critical point, giving

$$\mathcal{H}''(0, 0) = \begin{pmatrix} -5 & 0 \\ 0 & 3 \end{pmatrix} \quad \text{(saddle point)}$$

$$\mathcal{H}''(\pm\sqrt{5}/2, 0) = \begin{pmatrix} -5/2 & 0 \\ 0 & 1/2 \end{pmatrix} \quad \text{(saddle point)}$$

$$\mathcal{H}''(0, \pm\sqrt{3}/2) = \begin{pmatrix} -1/2 & 0 \\ 0 & -3/2 \end{pmatrix} \quad \text{(stable equilibrium)}$$

Hence, the equilateral triangle configurations are nonlinearly stable, while the collinear states are unstable.   ◊

For the general 3-vortex problem the equations written in terms of the intervortical distances $l_{\alpha\beta}$ are

$$(l_{12}^2)' = \frac{2}{\pi}\Gamma_3\sigma A\left(\frac{1}{l_{23}^2} - \frac{1}{l_{31}^2}\right) \tag{2.2.1}$$

$$(l_{23}^2)' = \frac{2}{\pi}\Gamma_1\sigma A\left(\frac{1}{l_{31}^2} - \frac{1}{l_{12}^2}\right) \tag{2.2.2}$$

$$(l_{31}^2)' = \frac{2}{\pi}\Gamma_2\sigma A\left(\frac{1}{l_{12}^2} - \frac{1}{l_{23}^2}\right). \tag{2.2.3}$$

Remarks.

1. Synge (1949) uses the notation $(R_1, R_2, R_3) \equiv (l_{23}, l_{31}, l_{12})$ to denote the triangle sides, with corresponding vertex angles denoted $(\theta_1, \theta_2, \theta_3)$, as shown in Figure 2.2. It follows that

$$\theta_1 + \theta_2 + \theta_3 = \pi,$$

   and by the triangle inequality we have

$$l_{23} \leq l_{31} + l_{12}$$
$$l_{31} \leq l_{23} + l_{12}$$
$$l_{12} \leq l_{23} + l_{31},$$

   with equality holding in each case only if the vortices are collinear.

2. By geometric considerations, one can derive the area formulas (Heron's formulas) [2]

$$A = \frac{1}{2}l_{31}l_{12}\sin\theta_1 = \frac{1}{2}l_{12}l_{23}\sin\theta_2 = \frac{1}{2}l_{23}l_{31}\sin\theta_3$$
$$= [r(r - l_{23})(r - l_{31})(r - l_{12})]^{1/2}$$
$$\equiv f_1(l_{12}, l_{23}, l_{31})$$

   with $r = \frac{1}{2}(l_{12} + l_{23} + l_{31})$.

---

[2] We list here for convenience several other classical triangle formulas:

Law of Sines: $\frac{R_1}{\sin\theta_1} = \frac{R_2}{\sin\theta_2} = \frac{R_3}{\sin\theta_3}$

Law of Cosines: $R_3^2 = R_1^2 + R_2^2 - 2R_1 R_2\cos\theta_3$

Mollweide's Equations:

$$\frac{R_1 + R_2}{R_3} = \frac{\cos\frac{1}{2}(\theta_1 - \theta_2)}{\sin(\frac{1}{2}\theta_3)}$$

$$\frac{R_1 - R_2}{R_3} = \frac{\sin\frac{1}{2}(\theta_1 - \theta_2)}{\cos(\frac{1}{2}\theta_3)}$$

3. By differentiating the area formula one obtains the system

$$\dot{A} = f_2(l_{12}, l_{23}, l_{31})$$

with right hand side defined as

$$f_2(l_{12}, l_{23}, l_{31}) = \frac{1}{4\pi}[\sum \Gamma_1 l_{23}^{-1}(l_{31}^{-2} - l_{12}^{-2})][(r - l_{23})(r - l_{31})(r - l_{12})$$
$$+ r\sum(r - l_{31})(r - l_{12})]$$
$$- \frac{r}{2\pi}\sum \Gamma_1 l_{23}^{-1}(l_{31}^{-2} - l_{12}^{-2})(r - l_{31})(r - l_{12}).$$

4. From here on we adopt the notation that $\sum$ denotes summation over cyclic permutations of indices — hence, for example, the first term in brackets reads

$$\sum \Gamma_1 l_{23}^{-1}(l_{31}^{-2} - l_{12}^{-2}) \equiv \Gamma_1 l_{23}^{-1}(l_{31}^{-2} - l_{12}^{-2}) + \Gamma_2 l_{31}^{-1}(l_{12}^{-2} - l_{23}^{-2})$$
$$+ \Gamma_3 l_{12}^{-1}(l_{23}^{-2} - l_{31}^{-2}). \qquad \Diamond$$

Since not all collinear configurations simultaneously satisfy the condition

$$f_2(l_{12}, l_{23}, l_{31}) = 0,$$

not all collinear initial conditions remain collinear. For a collinear configuration to remain collinear for all time it is necessary and sufficient that the triangle sides satisfy $A = \dot{A} = 0$, i.e., they satisfy the algebraic system

$$f_1(l_{12}, l_{23}, l_{31}) = 0$$
$$f_2(l_{12}, l_{23}, l_{31}) = 0.$$

2.2.1 Equilibria

We first state the following theorem concerning fixed and relative equilibria.

Theorem 2.2.1 (Equilibria).

1. *(Fixed equilibria, $N = 3$) Necessary and sufficient conditions for fixed equilibria in the plane are:*

   (a) *The vortices are collinear with* $(\mathbf{x}_2 - \mathbf{x}_1) = \frac{\Gamma_2}{\Gamma_3}(\mathbf{x}_1 - \mathbf{x}_3)$.

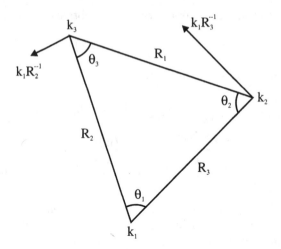

FIGURE 2.2. 3-vortex triangle.

(b) $\sum \Gamma_1 \Gamma_2 = 0.$[3]

2. *(Relative equilibria, $N = 3$) The only relative equilibria are collinear states or equilateral triangles.*

    (a) *All equilateral triangles form relative equilibria that rotate rigidly about their center of vorticity with frequency $\omega = \Gamma/2\pi s^2$, where s is the triangle side. In the limiting case where $\Gamma = 0$, the center of vorticity is at infinity and the vortices translate in parallel with velocity $v = [\frac{1}{2}(\Gamma_1^2 + \Gamma_2^2 + \Gamma_3^2)]^{1/2}/2\pi s$.*

    (b) *Collinear states can form relative equilibria if and only if $A = 0$, $\dot{A} = 0$, where A is the triangle area; equivalently $f_1(l_{12}, l_{23}, l_{31}) = f_2(l_{12}, l_{23}, l_{31}) = 0$.*

Proof. We start with the planar equations written in Cartesian form

$$\dot{\mathbf{x}}_\alpha = \sum_{\beta \neq \alpha} \frac{\Gamma_\beta}{2\pi} \frac{(\mathbf{x}_\beta - \mathbf{x}_\alpha) \times \mathbf{e}_z}{l_{\alpha\beta}^2} \qquad (\alpha = 1, 2, 3).$$

To prove part 1, start with the necessary condition $\dot{\mathbf{x}}_\alpha = 0$, which gives

$$\sum_{\beta \neq \alpha} \frac{\Gamma_\beta}{2\pi} \frac{(\mathbf{x}_\beta - \mathbf{x}_\alpha) \times \mathbf{e}_z}{l_{\alpha\beta}^2} = 0.$$

For $\alpha = 1$

$$\frac{\Gamma_2(\mathbf{x}_2 - \mathbf{x}_1)}{l_{12}^2} = \frac{\Gamma_3(\mathbf{x}_1 - \mathbf{x}_3)}{l_{31}^2},$$

---

[3]This condition can also be written $(\sum \Gamma_i)^2 - \sum \Gamma_i^2 = 0$.

which shows that $(\mathbf{x}_2 - \mathbf{x}_1)$ and $(\mathbf{x}_1 - \mathbf{x}_3)$ are scalar multiples of each other, hence the three vortices are collinear. Then taking the dot product on the left with $(\Gamma_2/l_{12}^2)(\mathbf{x}_2 - \mathbf{x}_1)$ and on the right with $(\Gamma_3/l_{31}^2)(\mathbf{x}_1 - \mathbf{x}_3)$ gives

$$\frac{\Gamma_2^2}{l_{12}^2} = \frac{\Gamma_3^2}{l_{31}^2}$$

and the result that

$$(\mathbf{x}_2 - \mathbf{x}_1) = \frac{\Gamma_2}{\Gamma_3}(\mathbf{x}_1 - \mathbf{x}_3).$$

For condition (b), start with the three equations

$$0 = \frac{\Gamma_2(\mathbf{x}_2 - \mathbf{x}_1)}{l_{12}^2} + \frac{\Gamma_3(\mathbf{x}_3 - \mathbf{x}_1)}{l_{31}^2} \tag{2.2.4}$$

$$0 = \frac{\Gamma_1(\mathbf{x}_1 - \mathbf{x}_2)}{l_{12}^2} + \frac{\Gamma_3(\mathbf{x}_3 - \mathbf{x}_2)}{l_{23}^2} \tag{2.2.5}$$

$$0 = \frac{\Gamma_1(\mathbf{x}_1 - \mathbf{x}_3)}{l_{31}^2} + \frac{\Gamma_2(\mathbf{x}_2 - \mathbf{x}_3)}{l_{23}^2}. \tag{2.2.6}$$

Now take $\Gamma_1\mathbf{x}_1 \cdot (2.2.4) + \Gamma_2\mathbf{x}_2 \cdot (2.2.5) + \Gamma_3\mathbf{x}_3 \cdot (2.2.6)$ to get

$$\frac{\Gamma_1\Gamma_2}{l_{12}^2}\mathbf{x}_1 \cdot (\mathbf{x}_2 - \mathbf{x}_1) + \frac{\Gamma_1\Gamma_3}{l_{31}^2}\mathbf{x}_1 \cdot (\mathbf{x}_3 - \mathbf{x}_1) + \frac{\Gamma_1\Gamma_2}{l_{12}^2}\mathbf{x}_2 \cdot (\mathbf{x}_1 - \mathbf{x}_2) +$$

$$\frac{\Gamma_2\Gamma_3}{l_{23}^2}\mathbf{x}_2 \cdot (\mathbf{x}_3 - \mathbf{x}_2)\frac{\Gamma_1\Gamma_3}{l_{31}^2}\mathbf{x}_3 \cdot (\mathbf{x}_1 - \mathbf{x}_3) + \frac{\Gamma_3\Gamma_2}{l_{23}^2}\mathbf{x}_3 \cdot (\mathbf{x}_2 - \mathbf{x}_3) = 0.$$

Then use the fact that $\mathbf{x}_\alpha \cdot (\mathbf{x}_\beta - \mathbf{x}_\alpha) = l_{\alpha\beta}^2/2$ giving

$$\frac{\Gamma_1\Gamma_2}{l_{12}^2}\frac{l_{12}^2}{2} + \frac{\Gamma_1\Gamma_3}{l_{31}^2}\frac{l_{31}^2}{2} + \frac{\Gamma_1\Gamma_2}{l_{12}^2}\frac{l_{12}^2}{2} + \frac{\Gamma_2\Gamma_3}{l_{23}^2}\frac{l_{23}^2}{2} + \frac{\Gamma_3\Gamma_1}{l_{31}^2}\frac{l_{31}^2}{2} + \frac{\Gamma_2\Gamma_3}{l_{23}^2}\frac{l_{23}^2}{2} = 0$$

$$\Rightarrow \sum_{\alpha \neq \beta}^{3} \Gamma_\alpha\Gamma_\beta = 0.$$

To prove that the conditions are sufficient start with $\sum \Gamma_1\Gamma_2 = 0$, hence

$$\Gamma_3 = -\frac{\Gamma_1\Gamma_2}{(\Gamma_1 + \Gamma_2)}.$$

Then

$$\dot{\mathbf{x}}_1 = \frac{1}{2\pi}\left[\frac{\Gamma_2(\mathbf{x}_2 - \mathbf{x}_1) \times \mathbf{e}_z}{l_{12}^2} + \frac{\Gamma_3(\mathbf{x}_3 - \mathbf{x}_1) \times \mathbf{e}_z}{l_{31}^2}\right].$$

Since

$$\frac{\Gamma_2(\mathbf{x}_2 - \mathbf{x}_1)}{l_{12}^2} = \frac{\Gamma_3(\mathbf{x}_1 - \mathbf{x}_3)}{l_{31}^2},$$

the right hand side is zero, hence $\dot{x}_1 = 0$. Similarly, we can show that $\dot{x}_2 = \dot{x}_3 = 0$, which completes the proof of part 1.

To prove part 2 of the theorem start with the condition that the system is in relative equilibrium

$$(l_{12}^2)' = \frac{2}{\pi}\Gamma_3 \sigma A \left(\frac{1}{l_{23}^2} - \frac{1}{l_{31}^2} \right) = 0$$

$$(l_{23}^2)' = \frac{2}{\pi}\Gamma_1 \sigma A \left(\frac{1}{l_{31}^2} - \frac{1}{l_{12}^2} \right) = 0$$

$$(l_{31}^2)' = \frac{2}{\pi}\Gamma_2 \sigma A \left(\frac{1}{l_{12}^2} - \frac{1}{l_{23}^2} \right) = 0.$$

This shows that either $A = 0$ (collinear), or $l_{12}^2 = l_{23}^2 = l_{31}^2 \equiv s^2$ (equilateral). The derivation of the rotational frequencies are left as an exercise. $\square$

Remarks.

1. For a fixed collinear configuration, without loss we can take $(z_1, z_2, z_3) = (0, 1, -\Gamma_3/\Gamma_2)$.

2. Notice the similarity of the equilateral triangle case for $N = 3$ with the formulas given in Example 1.8 for $N = 2$. $\lozenge$

Example 2.3   Consider three vortices each of unit strength placed in an equilateral triangle configuration on the unit circle. Since each side must have length $s = \sqrt{3}$, the triangle rotates rigidly with frequency $\omega = 1/2\pi$. To verify this directly write the vortex coordinates as

$$z_j(t) = \exp(i\theta_j(t)) \quad j = (1, 2, 3)$$

where $\theta_j(0) = 2\pi j/3$. The governing equation for z_j is

$$\dot{z}_j^* = \frac{i}{2\pi}\left(\frac{1}{z_k - z_j} + \frac{1}{z_l - z_j} \right) \quad (j \neq k \neq l),$$

which gives the angle equation

$$\theta_j(t) = \frac{t}{2\pi} + \frac{2\pi}{3}j \quad (j = 1, 2, 3).$$

The equation for a particle located at $z(t)$ in the flowfield is given by

$$\dot{z}^* = \frac{i}{2\pi}\left(\frac{1}{z_1 - z} + \frac{1}{z_2 - z} + \frac{1}{z_3 - z} \right).$$

Since the equation has periodic coefficients, it is advantageous to move into a rotating frame of reference $z \mapsto \zeta \exp(it/2\pi)$, giving the system

$$\dot{\zeta}^* = -\frac{i}{2\pi}\left(\frac{3\zeta^2}{\zeta^3 - 1} - \zeta^*\right). \tag{2.2.7}$$

By letting $\zeta = \xi + i\eta$, one can write this equation in Hamiltonian form

$$\dot{\xi} = \frac{d\mathcal{H}_0}{d\eta}, \qquad \dot{\eta} = -\frac{d\mathcal{H}_0}{d\xi}$$

where

$$\mathcal{H}_0 = -\frac{1}{2}\sum_{j=1}^{3}\log|\zeta - \exp(\frac{2\pi i}{3}j)|^2, \tag{2.2.8}$$

whose phase portrait is shown in Figure 1.5(a). This is the starting point for the nonintegrability proof of Ziglin (1980).   ◇

2.2.2 Trilinear Formulation

As was recognized by Synge (1949) and Aref (1979), based on the invariant M (2.1.7), it is useful to introduce the (dimensionless) trilinear variables[4] (b_1, b_2, b_3) defined as

$$b_1 \equiv \frac{l_{23}^2}{\Gamma_1 \tilde{M}}, \qquad b_2 \equiv \frac{l_{31}^2}{\Gamma_2 \tilde{M}}, \qquad b_3 \equiv \frac{l_{12}^2}{\Gamma_3 \tilde{M}} \tag{2.2.9}$$

where $\tilde{M} = M/(3\Gamma_1\Gamma_2\Gamma_3)$. If we assume, for the moment, $\tilde{M} \neq 0$, it follows that

$$b_1 + b_2 + b_3 = 3$$

and we can view the vortex motion in the "trilinear coordinate plane" shown in Figure 2.3 (reprinted with permission from Aref (1979)).

Since we make the assumption that $\Gamma_1 \geq \Gamma_2 > 0$, only the regions marked I, II, III are physically allowable, and because of the triangle inequalities, only subregions of these domains — called the "physical regions" by Aref (1979) — are accessible. Using the trilinear coordinates (2.2.9) in the triangle inequalities, one can show that the physical regions are bounded by conic sections (Salmon (1854))

$$(\Gamma_1 b_1)^2 + (\Gamma_2 b_2)^2 + (\Gamma_3 b_3)^2 \leq 2(\Gamma_1\Gamma_2 b_1 b_2 + \Gamma_2\Gamma_3 b_2 b_3 + \Gamma_3\Gamma_1 b_3 b_1) \tag{2.2.10}$$

---

[4]For a general discussion of the use of trilinear coordinates in fluid mechanics see the article by Aref (1992), or more generally, Ferrers (1866).

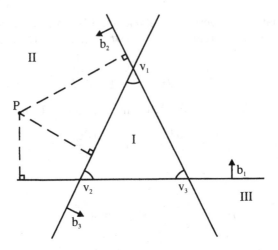

FIGURE 2.3. Trilinear coordinate plane.

with equality corresponding to the boundary. There are then three distinct cases to consider based on the sign of the coefficients. The conic section is either an

(i) ellipse, if $\Gamma_3 > 0$, $\Gamma_1 + \Gamma_2 + \Gamma_3 > 0$, or $\Gamma_3 < 0$, $\Gamma_1 + \Gamma_2 + \Gamma_3 < 0$, or

(ii) parabola, if $\Gamma_1 + \Gamma_2 + \Gamma_3 = 0$, or

(iii) hyperbola, if $\Gamma_3 < 0$, $\Gamma_1 + \Gamma_2 + \Gamma_3 > 0$.

We show as an example case (i) in Figure 2.4 from Aref (1979) where the vortex strengths are equal ($\Gamma_1 = \Gamma_2 = \Gamma_3 = 1$), hence the conic section is circular and the phase diagram has a threefold symmetry. In this case it is possible to derive exact solutions, as derived in Appendix A of Kuznetsov and Zaslavsky (1998).

To further classify the orbits, note that using the trilinear coordinates (2.2.9) in the conserved Hamiltonian requires a given phase trajectory in the trilinear plane be a level curve of the function $f(b_1, b_2, b_3)$, namely,

$$f(b_1, b_2, b_3) = |b_1|^{1/\Gamma_1} |b_2|^{1/\Gamma_2} |b_3|^{1/\Gamma_3} \equiv 1/\theta$$

where

$$\theta = (\Gamma_1^{1/\Gamma_1})(\Gamma_2^{1/\Gamma_2})|\Gamma_3|^{1/\Gamma_3} |\tilde{M}|^{3h} \exp(4\pi\mathcal{H}/g^3).$$

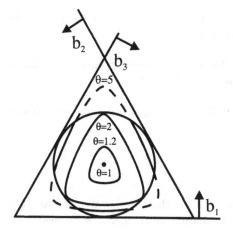

FIGURE 2.4. Phase plane diagram for the case of equal vortices.

Here, h and g are the harmonic mean and geometric mean of the vortex strengths, defined by

$$h = \frac{1}{3}\left(\frac{1}{\Gamma_1} + \frac{1}{\Gamma_2} + \frac{1}{\Gamma_3}\right) = \frac{1}{3}\frac{\sum \Gamma_1 \Gamma_2}{\Gamma_1 \Gamma_2 \Gamma_3},$$
$$g = (\Gamma_1 \Gamma_2 \Gamma_3)^{1/3}.$$

To better understand the phase trajectories within the physical regions one must understand the level curve structure of $f(b_1, b_2, b_3)$. The first important observation is that f has a stationary point at P^*, where

$$P^* \equiv (b_1^*, b_2^*, b_3^*) = \frac{1}{h}\left(\frac{1}{\Gamma_1}, \frac{1}{\Gamma_2}, \frac{1}{\Gamma_3}\right),$$

and we assume that $h \neq 0$. This stationary point corresponds to an equilateral triangle state, as seen by substituting the coordinates P^* into the equations (2.2.1), (2.2.2), (2.2.3) giving

$$l_{12}^2 = l_{23}^2 = l_{31}^2 = \frac{\tilde{M}}{h}.$$

As described in Section 2.2.1, these states are relative equilibria. Their linear stability can be summarized (Synge (1949)) in the following theorem.

Theorem 2.2.2 (Linear Stability of Equilateral Triangles). [5]

The linearization of the system around the fixed point P^* yields eigenvalues $\lambda^2 = -(\sum \Gamma_2 \Gamma_3)$. There are therefore three cases to consider:

---

[5]Lord Kelvin (1867, 1910) was the first to state the result that three identical vortices forming an equilateral triangle is stable.

(i) $\sum \Gamma_2\Gamma_3 > 0$: *The eigenvalues are pure imaginary, hence the equilateral triangle state is neutrally stable.*

(ii) $\sum \Gamma_2\Gamma_3 < 0$: *The eigenvalues are real, one positive, one negative, hence the equilateral triangle is unstable.*

(iii) $\sum \Gamma_2\Gamma_3 = 0$: *The eigenvalues are both zero, hence the equilateral triangle is neutrally stable.*

Remarks.

1. Aref (1979) classifies the possible dynamical states according to the sign of the harmonic mean h. Since he also makes the assumption that $\Gamma_1 \geq \Gamma_2 > 0$, his three cases are:

   (i) $\Gamma_3 > 0, h > 0$, which coincides with case (i) above.

   (ii) $\Gamma_3 < 0, h < 0$, also case (i) above.

   (iii) $\Gamma_3 < 0, h > 0$, corresponding to case (ii) above. Case (iii) above is treated in a separate discussion.

2. The nonlinear stability can be inferred from the level curves in the neighborhood of the fixed point, as discussed in Synge (1949) and Aref (1979). We show the relevant phase diagrams. Figure 2.5a (reprinted with permission from Synge (1949)) demonstrates that the curves are closed around the neutrally stable fixed point in case (i), hence the equilateral triangle state is nonlinearly stable. Figure 2.5b shows the saddle point structure for the unstable case (ii). Case (iii) is unstable, but more delicate, and we refer the reader to the discussion in Tavantzis and Ting (1988). In general, the paper by Tavantzis and Ting (1988) contains a wealth of detailed stability information. One of their results is that the self-similar expanding triangle states are stable, while the partner collapsing states are unstable.   ◊

There is much more detailed dynamical information that one can extract from the trilinear coordinates, and we refer the reader to Synge (1949), Novikov (1975), and Aref (1979) for more.

2.2.3 Collapse

We next derive necessary and sufficient conditions for three vortices to collapse self-similarly. It is simplest to work directly with the equations

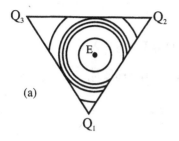
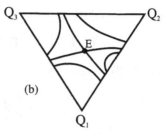

FIGURE 2.5. Level curves of $f(b_1, b_2, b_3)$. (a) Case (i) — nonlinearly stable, (b) Case (ii) — unstable.

governing the intervortical distances (l_{12}, l_{23}, l_{31}). We first derive the necessary conditions for collapse by looking explicitly for solutions that retain the ratios λ_1, λ_2, where

$$\lambda_1 = \left(\frac{l_{12}(t)}{l_{31}(t)} \right)^2$$

$$\lambda_2 = \left(\frac{l_{23}(t)}{l_{31}(t)} \right)^2.$$

Consider the two conserved quantities (2.1.6), (2.1.7). The first can more usefully be written

$$\tilde{\mathcal{H}} = \exp(-4\pi \mathcal{H}/\Gamma_1\Gamma_2\Gamma_3) = (l_{12}(t))^{1/\Gamma_3}(l_{23}(t))^{1/\Gamma_1}(l_{31}(t))^{1/\Gamma_2}.$$

Then, using the above constrained ratios gives

$$\tilde{\mathcal{H}} = \lambda_1^{1/\Gamma_3}\lambda_2^{1/\Gamma_1}(l_{31}(t))^{3h}.$$

In order for the right side to remain constant, we obtain the first necessary condition for collapse.

Condition 1 $h = 0$.
This implies that the three strengths cannot be of the same sign, hence we take $\Gamma_3 < 0$. Since, at collision time $t = t^*$, we know all triangle sides have length zero, the second conserved quantity M must be zero, the second important necessary condition.

Condition 2 $M = 0$.
Using the scaled intervortical distances in (2.2.1), (2.2.2), (2.2.3) yields

$$\lambda_1 \frac{d}{dt}(l_{31}^2) = \frac{2}{\pi}\Gamma_3\sigma A \left(\frac{1}{\lambda_2 l_{31}^2} - \frac{1}{l_{31}^2} \right) \qquad (2.2.11)$$

$$\lambda_2 \frac{d}{dt}(l_{31}^2) = \frac{2}{\pi}\Gamma_1\sigma A \left(\frac{1}{l_{31}^2} - \frac{1}{\lambda_1 l_{31}^2} \right) \qquad (2.2.12)$$

$$\frac{d}{dt}(l_{31}^2) = \frac{2}{\pi}\Gamma_2\sigma A \left(\frac{1}{\lambda_1 l_{31}^2} - \frac{1}{\lambda_2 l_{31}^2} \right). \qquad (2.2.13)$$

These give the relations

$$\Gamma_3(1 - \lambda_2) = \Gamma_1(\lambda_1 - 1) = \Gamma_2(\lambda_2 - \lambda_1).$$

Using (2.2.13) gives the scalar equation for the side l_{31}^2

$$\frac{d}{dt}(l_{31}^2) = -\sigma\omega,$$

whose solution is

$$l_{31}(t) = \sqrt{-\sigma\omega t + l_{31}^2(0)}$$

where

$$\omega = \frac{\Gamma_2}{2\pi}\left(\frac{\lambda_1 - \lambda_2}{\lambda_1\lambda_2}\right)\gamma^{1/2}$$
$$\gamma = \left[2(\lambda_1 + \lambda_2) - (\lambda_1 - \lambda_2)^2 - 1\right].$$

Therefore, if we take $\omega > 0$, $\gamma > 0$, blow-up occurs at time t^* if the vortex circuit is positive ($\sigma = 1$), where

$$t^* = \frac{l_{31}^2(0)}{\omega}. \tag{2.2.14}$$

Before stating the general theorem, notice that conditions 1 and 2 do not rule out the possibility that the vortices are in relative equilibrium, which clearly precludes their collapse. We can now summarize with the following theorem.

Theorem 2.2.3 ($N = 3$ Collapse). *Necessary and sufficient conditions for the self-similar collapse of three vortices are:*

1. $h = 0$.

2. $M = 0$.

3. *The initial configuration is not an equilibrium.*

4. $\sigma > 0$.

To understand why these conditions are also sufficient we need to examine the phase diagrams more carefully for the case $\tilde{M} = 0$, $\Gamma_3 < 0$. Use as trilinear variables

$$b_1 = \frac{l_{23}^2}{\Gamma_1}, \quad b_2 = \frac{l_{31}^2}{\Gamma_2}, \quad b_3 = \frac{l_{12}^2}{\Gamma_3},$$

which gives

$$b_1 + b_2 + b_3 = 0.$$

Then (b_1, b_2) can be viewed as Cartesian variables, with $b_3 = -(b_1 + b_2)$. If we use this in the equation defining the physical region boundaries (2.2.10), we get a standard equation for a degenerate conic section (Salmon (1854)). The phase trajectories in this case are all level curves of the function $\tilde{f}(b_1, b_2)$ where

$$\tilde{f}(b_1, b_2) = b_1^{1/\Gamma_1} \cdot b_2^{1/\Gamma_2} \cdot (b_1 + b_2)^{1/\Gamma_3} \equiv 1/\theta.$$

When we choose $\tilde{M} = 0$, $h = 0$, each phase curve is a straight line through the origin

$$b_2 = \lambda b_1, \quad \lambda > 0$$

$$\theta = (1 + \lambda)^{1/\Gamma_1} \cdot \left(1 + \frac{1}{\lambda}\right)^{1/\Gamma_2}$$

as shown in Figure 2.7. Each point on the curve marked $\lambda = \Gamma_1/\Gamma_2$ is a rigidly rotating equilateral triangle state, while those on the line segments bounding the wedge-shaped region are collinear states that rigidly rotate. These are all the relative equilibria, and by assumption, the initial conditions do not lie on these exceptional curves. All other line segments go through the origin, and depending on the sign of the vortex circuit, represent collapsing states ($b_1 \to 0, b_2 \to 0$) or expanding states ($b_1 \to \infty, b_2 \to \infty$). To see this notice that

$$b_3 = -(1 + \lambda)b_1,$$

hence l_{13} and l_{12} are proportional to l_{23}, and the right hand sides of (2.2.11), (2.2.12), (2.2.13) are constant. If we assume $\sigma\omega > 0$, we have already shown explicitly that the configuration collapses, hence the proof of sufficiency is complete.

Remarks.

1. One can learn much more about the details of the 3-vortex problem by reading the papers of Synge (1949), Novikov (1975), and Aref (1979). In particular, Aref (1979) describes a method for explicitly constructing the initial conditions leading to collapse.

2. The self-similar states expand if the vortex circuit is negative, i.e., $\sigma = -1$. Each collapsing state can be related to an expanding state by simply reversing the sign of the vortex circuit, which amounts to switching two of the vortex initial positions in an anticyclic manner. We call such collapsing-expanding configurations **partner states**.

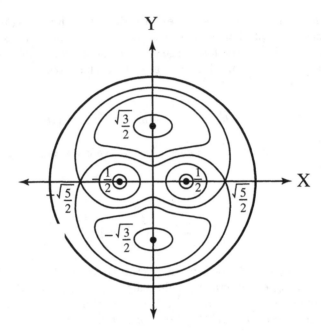

FIGURE 2.6. Phase curves for the restricted 3-vortex problem.

3. One can derive the following explicit formulas for the collapse process

$$l_{\alpha\beta}(t) = l_{\alpha\beta}(0)(1 - t/t^*)^{1/2} \qquad (2.2.15)$$

$$\rho_\alpha(t) = r_\alpha(0)(1 - t/t^*)^{1/2}$$

$$\dot{\phi}_\alpha(t) = \omega = \frac{(\Gamma_1 + \Gamma_2)^3 + \Gamma_1^3(1 + 1/\lambda) + \Gamma_2^3(1 + \lambda)}{4\pi(\Gamma_1 + \Gamma_2)^2 l_{12}^2(0)(1 - t/t^*)}$$

where the collapse time t^* is given by (2.2.14). The polar coordinates $(\rho_\alpha(t), \phi_\alpha(t))$ measure the vortex position $z_\alpha(t)$ with respect to the center of vorticity, or collapse point. ◊

2.3 $N = 4$

In this subsection, we consider the 4-vortex problem. We start by showing a special example of "leapfrogging" vortex pairs, which is a two-dimensional version of axisymmetric coaxial rings. Then, using a special coordinate system introduced by Khanin (1982), we show how to explicitly reduce the problem to one with two degrees of freedom. The method can be generalized to the case of N vortices where one can reduce the degrees of freedom

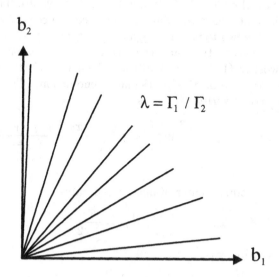

FIGURE 2.7. Phase curves for collapsing and expanding states.

from N to $N - 2$. This is achieved, in general, by decomposing the vortex positions into a hierarchy of groups so that the distances between the various groups are much larger than the distances between the vortices in each group. By separating out the spatial and temporal scales in this way, each vortex group acts effectively as a single vortex whose strength is the sum of all the vortex strengths in its group. The overall vortex motion is a hierarchical superposition of the motion of each group and the more detailed motions of vortices within a given group. This hierarchical decomposition is analogous to similar procedures used in the celestial mechanics literature (see Pollard (1962)). There are other reduction coordinates that have been used for the vortex problem; we mention the papers by Aref and Pomphrey (1980, 1982) and Lim (1990, 1991 a,b). To prove nonintegrability for the 4-vortex problem we then follow the paper of Castilla, Moauro, Negrini, and Oliva (1993).

Example 2.4 (Leapfrogging vortex pairs) Consider a 4-vortex configuration made up of two pairs of vortex dipoles arranged so that the x-axis perpendicularly bisects the two line segments that join each vortex with its partner. Thus, at time $t = 0$ we place one pair of strength $(\Gamma, -\Gamma)$ at position $(x_1, \pm y_1)$ and the other pair of the same strength at position $(x_2, \pm y_2)$. The positive circulation (hence counterclockwise) vortices are placed in the upper half-plane, while the negative strength ones are placed in the lower half-plane. We also assume that initially $x_2 > x_1$, $y_2 > y_1$, so that the larger pair is in front of the smaller pair. The geometry is shown in Figure 1 of Péntek, Tél, and Toroczkai (1995a) where this example is treated in some detail. See also Péntek, Tél, and Toroczkai (1995b). The configuration can be thought of as a two-dimensional slice of a coaxial pair of vortex rings, which are known to pass through one another, hence "leapfrog" their way forward if the conditions are set up carefully (see Yamada and Matsui (1978), Maxworthy (1977), or Shariff and Leonard (1992) and the references therein for more on this particular phenomenon). The Hamiltonian for this symmetric 4-vortex system is

$$\mathcal{H}(x_1, x_2; y_1, y_2) = \frac{\Gamma^2}{2\pi} \log \left[4y_1 y_2 \frac{(x_1 - x_2)^2 + (y_1 + y_2)^2}{(x_1 - x_2)^2 + (y_1 - y_2)^2} \right]$$
$$\equiv E. \tag{2.3.1}$$

It is useful to introduce center-of-mass coordinates

$$x_0 = \frac{x_1 + x_2}{2} \qquad y_0 = \frac{y_1 + y_2}{2},$$

and relative coordinates

$$x_r = x_2 - x_1, \qquad y_r = y_2 - y_1.$$

In these variables the Hamiltonian becomes

$$\mathcal{H}(x_0, x_r; 2y_0, y_r/2) = \frac{1}{2} \log \left[\frac{(x_r^2 + (2y_0)^2)((2y_0)^2 - 4(y_r/2)^2)}{x_r^2 + 4(y_r/2)^2} \right]. \tag{2.3.2}$$

Since the variable x_0 is cyclic, the conjugate variable y_0 is conserved,

$$2y_0 = (y_1 + y_2) = const.$$

Then, in terms of relative coordinates (x_r, y_r), we have

$$\frac{1}{1 - y_r^2} - \frac{1}{1 + x_r^2} \equiv \exp(-E). \tag{2.3.3}$$

Notice from this expression that the trajectories are symmetric with respect to the axes $x_r = 0$, $y_r = 0$. The equations of motion in these variables are

$$\dot{x}_r = -2y_r \frac{(2y_0)^2 + x_r^2}{((2y_0)^2 - y_r^2)(x_r^2 + y_r^2)} \tag{2.3.4}$$

$$\dot{y}_r = 2x_r \frac{(2y_0)^2 - y_r^2}{((2y_0)^2 + x_r^2)(x_r^2 + y_r^2)} \tag{2.3.5}$$

$$\dot{x}_0 = 2y_0 \frac{x_r^2 - y_r^2 + 2(2y_0)^2}{((y_0)^2 - y_r^2)((2y_0)^2 + x_r^2)} \tag{2.3.6}$$

$$\dot{y}_0 = 0. \tag{2.3.7}$$

Using (2.3.3) in (2.3.4) gives the relation

$$dt = -\left(\frac{\exp(-E)}{2}\right)\frac{(1 - y_r^2(x_r))^2}{y_r(x_r)}dx_r$$

with

$$y_r = \gamma^{-1}\left(\frac{\gamma^2 - x_r^2}{\beta^2 + x_r^2}\right)^{1/2}$$

$$\gamma \equiv \sqrt{\frac{\exp(-E)}{(1 - \exp(-E))}}$$

$$\beta \equiv \sqrt{\frac{(1 + \exp(-E))}{\exp(-E)}}.$$

As x_r increases from zero to its maximum value of $x_r = \gamma$, the leapfrogging pair periodically oscillates through a quarter period, hence

$$T = 2\exp(-E)\int_0^\gamma \frac{(1 - y_r^2(x_r))^2}{y_r(x_r)}dx_r,$$

which, as shown in Péntek, Tél, and Toroczkai (1995), can be expressed in terms of elliptic integrals in the following way

$$T = 4\exp(E)\left[\frac{1}{1 - \exp(-2E)}F(\exp(-E)) - K(\exp(-E))\right],$$

where F and K are complete elliptic integrals of the first and second type. In the high energy limit $E \gg 1$ the expression can be expanded to yield

$$T \sim \pi\exp(-E) + \frac{9\pi}{8}\exp(-3E) + \dots$$

If we define the mean translational velocity of the pair over one period as

$$<v> \equiv \frac{1}{T}\int_0^T \dot{x}_0(t)dt,$$

then the exact expression can be found

$$<v> = \frac{\exp(-2E)F(\exp(-E))}{F(\exp(-E)) - (1 - \exp(-2E))K(\exp(-E))},$$

whose high energy expansion expansion is given by

$$<v> \sim 2 - \frac{3}{4}\exp(-2E) + \dots$$

The details of this example are contained in one of the exercises.   ◇

2.3.1   Reduction

We now explicitly show how to make use of the conserved quantities and thereby reduce the number of degrees of freedom of the system. This is the first step in applying KAM-type results for proving the persistence of quasiperiodic vortex trajectories. We start by making the assumption that all vortex circulations are of the same sign (without loss, we take them to be positive), hence by Theorem 2.1.2, the vortices remain in a bounded region of the plane for all time and cannot collide. The main idea behind the choice of coordinates is to group the four vortices into two subgroups, (Γ_1, Γ_2) forming one subgroup with center of vorticity C_{12}, and (Γ_3, Γ_4) forming the other with center of vorticity C_{34}, as shown in Figure 2.8. The centers of vorticity of each subgroup are given by

$$C_{12} = (x^{(1)}, y^{(1)}) = \left(\frac{\Gamma_1 x_1 + \Gamma_2 x_2}{\Gamma_1 + \Gamma_2}, \frac{\Gamma_1 y_1 + \Gamma_2 y_2}{\Gamma_1 + \Gamma_2} \right)$$

$$C_{34} = (x^{(2)}, y^{(2)}) = \left(\frac{\Gamma_3 x_1 + \Gamma_4 x_2}{\Gamma_3 + \Gamma_4}, \frac{\Gamma_3 y_1 + \Gamma_4 y_2}{\Gamma_3 + \Gamma_4} \right).$$

We denote the conserved center of vorticity of the 4-vortex system

$$(x^{(0)}, y^{(0)}) = \left(\frac{\sum \Gamma_i x_i}{\Gamma}, \frac{\sum \Gamma_i y_i}{\Gamma} \right).$$

The coordinates of Khanin (1982) can best be described by a sequence of transformations.

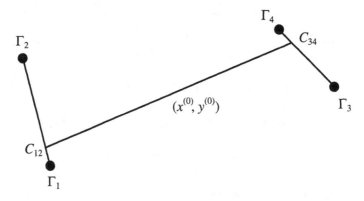

FIGURE 2.8. Vortex subgroupings.

Step 1 First, we transform variables to new ones that make use of the center of vorticity of the full system, the relative distances between vortices in each subgroup, and the relative distances between the center of vorticities of each of the subgroups.

$$z \equiv (x_1, x_2, x_3, x_4; y_1, y_2, y_3, y_4)^T \mapsto Z \equiv (x^{(0)}, X_2, X_3, X_4; y^{(0)}, Y_2, Y_3, Y_4)^T$$

where

$$(X_2, Y_2) = (x_1 - x_2, y_1 - y_2)$$
$$(X_3, Y_3) = (x_3 - x_4, y_3 - y_4)$$
$$(X_4, Y_4) = (x^{(1)} - x^{(2)}, y^{(1)} - y^{(2)})$$

In matrix form the transformation is given as

$$Z = Mz$$

where M is the (8×8) block diagonal matrix

$$M = \begin{bmatrix} M_1 & 0 \\ 0 & M_1 \end{bmatrix}$$

$$M_1 = \begin{bmatrix} \frac{\Gamma_1}{\Gamma} & \frac{\Gamma_2}{\Gamma} & \frac{\Gamma_3}{\Gamma} & \frac{\Gamma_4}{\Gamma} \\ 1 & -1 & 0 & 0 \\ 0 & 0 & 1 & -1 \\ \frac{\Gamma_1}{\Gamma_1 + \Gamma_2} & \frac{\Gamma_2}{\Gamma_1 + \Gamma_2} & -\frac{\Gamma_3}{\Gamma_3 + \Gamma_4} & -\frac{\Gamma_4}{\Gamma_3 + \Gamma_4} \end{bmatrix}.$$

A similar transformation is used in Lim (1990) for the general N-vortex problem.

Step 2 Having used the center of vorticity invariant, we next introduce a polar coordinate transformation

$$(X_2, X_3, X_4; Y_2, Y_3, Y_4) \mapsto (\tilde{p}_1, \tilde{p}_2, \tilde{p}_3; \tilde{\theta}_1, \tilde{\theta}_2, \tilde{\theta}_3)$$

where

$$\left(\frac{2(\Gamma_1 + \Gamma_2)}{\Gamma_1 \Gamma_2} \tilde{p}_1 \right)^{1/2} = \sqrt{X_2^2 + Y_2^2}$$

$$\tilde{\theta}_1 = \tan^{-1} \left(\frac{Y_2}{X_2} \right),$$

$$\left(\frac{2(\Gamma_3 + \Gamma_4)}{\Gamma_3 \Gamma_4} \tilde{p}_2 \right)^{1/2} = \sqrt{X_3^2 + Y_3^2}$$

$$\tilde{\theta}_2 = \tan^{-1} \left(\frac{Y_3}{X_3} \right)$$

$$\left(\frac{2\Gamma}{(\Gamma_1 + \Gamma_2)(\Gamma_3 + \Gamma_4)} \tilde{p}_3 \right)^{1/2} = \sqrt{X_4^2 + Y_4^2}$$

$$\tilde{\theta}_3 = \tan^{-1} \left(\frac{Y_4}{X_4} \right).$$

The direct transformation which combines steps 1 and 2 is then

$$x_1 = \eta \tilde{x} + \beta \sqrt{\tilde{p}_3} \cos \tilde{\theta}_3 + \gamma \sqrt{\tilde{p}_1} \cos \tilde{\theta}_1$$

$$y_1 = \eta \tilde{y} + \beta \sqrt{\tilde{p}_3} \sin \tilde{\theta}_3 + \gamma \sqrt{\tilde{p}_1} \sin \tilde{\theta}_1$$

$$x_2 = \eta \tilde{x} + \beta \sqrt{\tilde{p}_3} \cos \tilde{\theta}_3 - \gamma \frac{\Gamma_1}{\Gamma_2} \sqrt{\tilde{p}_1} \cos \tilde{\theta}_1$$

$$y_2 = \eta \tilde{y} + \beta \sqrt{\tilde{p}_3} \sin \tilde{\theta}_3 - \gamma \frac{\Gamma_1}{\Gamma_2} \sqrt{\tilde{p}_1} \sin \tilde{\theta}_1$$

$$x_3 = \eta \tilde{x} - \beta \frac{\Gamma_1 + \Gamma_2}{\Gamma_3 + \Gamma_4} \sqrt{\tilde{p}_3} \cos \tilde{\theta}_3 + \delta \sqrt{\tilde{p}_2} \cos \tilde{\theta}_2$$

$$y_3 = \eta \tilde{y} - \beta \frac{\Gamma_1 + \Gamma_2}{\Gamma_3 + \Gamma_4} \sqrt{\tilde{p}_3} \sin \tilde{\theta}_3 + \delta \sqrt{\tilde{p}_2} \sin \tilde{\theta}_2$$

$$x_4 = \eta \tilde{x} - \beta \frac{\Gamma_1 + \Gamma_2}{\Gamma_3 + \Gamma_4} \sqrt{\tilde{p}_3} \cos \tilde{\theta}_3 - \delta \frac{\Gamma_3}{\Gamma_4} \sqrt{\tilde{p}_2} \cos \tilde{\theta}_2$$

$$y_4 = \eta \tilde{y} - \beta \frac{\Gamma_1 + \Gamma_2}{\Gamma_3 + \Gamma_4} \sqrt{\tilde{p}_3} \sin \tilde{\theta}_3 - \delta \frac{\Gamma_3}{\Gamma_4} \sqrt{\tilde{p}_2} \sin \tilde{\theta}_2$$

where the constants $\eta, \beta, \gamma, \delta$ are chosen to make the transformation canonical

$$\eta = \frac{1}{\sqrt{\Gamma}} \qquad \beta = \eta \sqrt{\frac{2(\Gamma_3 + \Gamma_4)}{\Gamma_1 + \Gamma_2}}$$

$$\gamma = \sqrt{\frac{2\Gamma_2}{\Gamma_1(\Gamma_1 + \Gamma_2)}} \qquad \delta = \sqrt{\frac{2\Gamma_4}{\Gamma_3(\Gamma_3 + \Gamma_4)}}.$$

The equations of motion are then

$$\dot{\tilde{p}}_i = \frac{\partial \mathcal{H}}{\partial \tilde{\theta}_i}, \qquad \dot{\tilde{\theta}}_i = -\frac{\partial \mathcal{H}}{\partial \tilde{p}_i} \tag{2.3.8}$$

where $\mathcal{H}$ is the Hamiltonian (2.1.6) written in the new variables $(\tilde{p}_1, \tilde{p}_2, \tilde{p}_3; \tilde{\theta}_1, \tilde{\theta}_2, \tilde{\theta}_3)$. In terms of these coordinates, the system is a canonical three-degree-of-freedom system which depends only on the angle differences, i.e., $\mathcal{H}(\tilde{p}_1, \tilde{p}_2, \tilde{p}_3; \tilde{\theta}_1 - \tilde{\theta}_2, \tilde{\theta}_1 - \tilde{\theta}_3, \tilde{\theta}_2 - \tilde{\theta}_3)$. From this it is straightforward to prove that

$$\frac{\partial \mathcal{H}}{\partial \tilde{\theta}_1} + \frac{\partial \mathcal{H}}{\partial \tilde{\theta}_2} + \frac{\partial \mathcal{H}}{\partial \tilde{\theta}_3} = 0,$$

from which it follows that $\tilde{p}_1 + \tilde{p}_2 + \tilde{p}_3$ is a conserved quantity. In terms of the original conserved quantities of the system,

$$\tilde{p}_1 + \tilde{p}_2 + \tilde{p}_3 = \frac{1}{2\Gamma}(\Gamma I - Q^2 - P^2) = const.$$

Step 3

The final transformation is given by $(\tilde{p}_1 \; \tilde{p}_2, \tilde{p}_3; \tilde{\theta}_1, \tilde{\theta}_2, \tilde{\theta}_3) \mapsto (p_1, p_2, p_3; \theta_1, \theta_2, \theta_3)$ where

$$
\begin{aligned}
p_1 &= \tilde{p}_1 & q_1 &= \tilde{\theta}_1 - \tilde{\theta}_3 \\
p_2 &= \tilde{p}_2 & q_2 &= \tilde{\theta}_2 - \tilde{\theta}_3 \\
p_3 &= \tilde{p}_1 + \tilde{p}_2 + \tilde{p}_3 = const & q_3 &= \tilde{\theta}_3.
\end{aligned}
$$

Fixing p_3 then gives the reduced two-degree-of-freedom system

$$
\hat{\mathcal{H}} = \mathcal{H}_0(p_1, p_2) + \mathcal{H}'(p_1, p_2, \theta_1, \theta_2),
$$

where

$$
\mathcal{H}_0 = -\frac{1}{4\pi} \left[\Gamma_1 \Gamma_2 \log(p_1) + \Gamma_3 \Gamma_4 \log(p_2) + (\Gamma_1 + \Gamma_2)(\Gamma_3 + \Gamma_4) \log(p_3) \right]
$$

$$
\mathcal{H}' = -\frac{1}{4\pi} \sum_{\alpha=1,2;\beta=3,4} \Gamma_\alpha \Gamma_\beta \log \left[1 + \frac{1}{p_3}(\sum_{i=1,2} a_{\alpha\beta}^{(i)} p_i \right.
$$
$$
\left. + b_{\alpha\beta}^{(i)} \sqrt{p_i} \sqrt{p_3 - p_1 - p_2} \cos\theta_i + b_{\alpha\beta}^{(3)} \sqrt{p_1 p_2} \cos(\theta_1 - \theta_2)) \right].
$$

2.3.2 Application of KAM Theory

Based on the above Hamiltonian decomposition, Khanin (1982) is able to prove, by use of KAM arguments, that the 4-vortex problem, and more generally the N-vortex problem, has positive measure regions in the phase space made up of quasiperiodic vortex trajectories. The quasiperiodic solutions physically correspond to regions of positive measure where vorticity tends to cluster and persist. Using the fact that, for large enough values of p_3, the Hamiltonian $\hat{\mathcal{H}}$ is an analytic function in some domain F, where $\mathcal{H}' \to 0$ uniformly as $p_3 \to \infty$, Khanin first proves the following result having to do with adiabaticity.

Theorem 2.3.1 (Khanin). *Consider the region defined by*

$$
F_R = \{p_i, \theta_i : 0 < B \leq p_1, \; p_2 \leq A, \; p_3 \geq R \; (0 \leq \theta_i < 2\pi; i = 1, 2, 3)\}.
$$

For any $\epsilon > 0$ there exists an $\tilde{R}(\epsilon, A, B, \Gamma_\alpha)$ such that, for $R \geq \tilde{R}$, any solution of (2.3.8) with initial conditions in the set F_R satisfies the inequality

$$
\mid \tilde{p}_i(t) - \tilde{p}_i(0) \mid < \epsilon, \quad -\infty < t < \infty.
$$

This is a statement that the variables p_1 and p_2 are adiabatic invariants. Khanin then states the following KAM-type result.

Theorem 2.3.2 (Khanin). *Consider the region in the reduced 4-vortex phase space* $(p_1, p_2, p_3; \theta_1, \theta_2, \theta_3)$ *defined by*

$$F_R = \{p_i, \theta_i : 0 \leq B \leq p_1, \ p_2 \leq A, \ R \leq p_3 \leq 2R; \ 0 \leq \theta_i < 2\pi; \ i = 1, 2, 3\}.$$

For any $\epsilon > 0$ *there exists an* $\tilde{R}$ *such that, for all* $R \geq \tilde{R}$*:*

1. *The set* F_R *is the union of two sets* $F_R^{(1)}$ *and* $F_R^{(2)}$*, hence* $F_R = F_R^{(1)} + F_R^{(2)}$ *where* $F_R^{(1)}$ *is invariant under the dynamics and* $F_R^{(2)}$ *has small Lebesgue measure.*

2. $F_R^{(1)}$ *is foliated by invariant three-dimensional analytical tori* I_ω*:*

$$p_i = P_\omega^{(i)} + f_\omega^{(i)}(Q_1, Q_2), \quad i = 1, 2$$
$$p_3 = P_\omega^{(3)}$$
$$q_i = Q_i + g_\omega^{(i)}(Q_1, Q_2), \quad i = 1, 2, 3$$

   where $f_\omega^{(i)}(Q_1, Q_2)$ *and* $g_\omega^{(i)}(Q_1, Q_2)$ *are* 2π*-periodic analytic functions of* Q_1 *and* Q_2 *and* $\omega = (\omega_1, \omega_2, \omega_3)$ *is a parameter that labels the tori* I_ω *.*

3. $f_\omega^{(i)}(Q_1, Q_2)$ *and* $g_\omega^{(i)}(Q_1, Q_2)$ *are smaller than* ϵ*.*

4. *The motion on the torus* I_ω *is quasiperiodic in the phase space with frequencies* $\omega_1, \omega_2, \omega_3$*, i.e.,*

$$\dot{Q}_i = \omega_i, \quad i = 1, 2, 3.$$

Remarks.

1. The paper of Celletti and Falcolini (1988) uses Khanin's result to derive an estimate on the size of the perturbation necessary to break the KAM torus with a given rotation number. This result is interesting and quite explicit, showing that the typical perturbation sizes for which KAM theorems are proven are far smaller than the actual size at which breakdown occurs.

2. The works of Lim (1990, 1991a,b, 1993a) are the most thorough with regard to applications of KAM arguments to vortex systems. In particular, they are more general than that of Khanin (1982) in several respects:

   (a) A general reduction method making use of the Jacobi variables is carried out for the general N-vortex problem and for a wide class of N-body problems (Lim (1991a,b)).

(b) KAM theory is applied to vortex lattice systems and desingularized vortex models (Lim (1990)), making the results applicable to a wide range of vortex numerical methods as well as physical models of vortex sheets and vortex trails (Lim (1993b)).

(c) The paper of Lim (1993a) introduces a combinatorial perturbation method applicable to a general class of N-body problems, where the existence of Arnold's "whiskered tori" is rigorously proven.

$\Diamond$

2.3.3   Nonintegrable Cases

The original proof of nonintegrability is Ziglin's (1980), followed by that of Koiller and Carvalho (1985, 1989) for a different configuration. A third, more recent and detailed proof is given in Castilla, Moauro, Negrini, and Oliva (1993) which is the one we will follow in most detail (see also Oliva (1991)). All three proofs are based on the Melnikov method, though each uses a different vortex configuration.

The proof of Castilla et al (1993) is based on the motion of three vortices located at $z_1 = x_1 + iy_1$, $z_2 = x_2 + iy_2$, $z_3 = x_3 + iy_3$, each with unit circulation, perturbed by a fourth, located at $z_4 = x_4 + iy_4$ with circulation ϵ. Let

$$C_\epsilon \equiv \frac{(z_1 + z_2 + z_3) + \epsilon z_4}{3 + \epsilon}$$

be the center of vorticity of the 4-vortex system, while $C_0 \equiv \frac{1}{3}(z_1 + z_2 + z_3)$ is the center of vorticity of the 3-unit vortex subsystem. One can derive the relationships between the centers of vorticity and the vortex positions,

$$z_1 - C_\epsilon = \frac{1}{2}[(z_1 - z_2) + (C_0 - z_3)] + \frac{\epsilon}{3 + \epsilon}(C_0 - z_4)$$

$$z_2 - C_\epsilon = -\frac{1}{2}[(z_1 - z_2) - (C_0 - z_3)] + \frac{\epsilon}{3 + \epsilon}(C_0 - z_4)$$

$$z_3 - C_\epsilon = -(C_0 - z_3) + \frac{\epsilon}{3 + \epsilon}(C_0 - z_4)$$

$$z_4 - C_\epsilon = -\frac{3}{3 + \epsilon}(C_0 - z_4)$$

$$(C_0 - C_\epsilon) = -\frac{\epsilon}{3 + \epsilon}(z_4 - C_0).$$

In the limiting case $\epsilon \to 0$, the centers of vorticity coincide, i.e., $C_\epsilon = C_0$. The first step in the proof is to introduce appropriate new coordinates based on the center of vorticity and polar variables $(x_0, y_0, \tilde{p}_1, \theta_1, \tilde{p}_2, \theta_2, \tilde{p}_3, \theta_3)$, where

$$C_\epsilon = (\eta x_0, \eta y_0)$$
$$z_1 - z_2 = \alpha\sqrt{\tilde{p}_1}\exp(i\theta_1)$$
$$C_0 - z_3 = \beta\sqrt{\tilde{p}_2}\exp(i\theta_2)$$
$$C_0 - z_4 = \gamma\sqrt{\tilde{p}_3}\exp(i\theta_3).$$

We choose the constants $\eta, \alpha, \beta, \gamma$ so that the transformation

$$(z_1, z_2, z_3, \sqrt{\epsilon}z_4) \mapsto (x_0, y_0, \tilde{p}_1, \theta_1, \tilde{p}_2, \theta_2, \tilde{p}_3, \theta_3)$$

is canonical:

$$\eta = \sqrt{\frac{1}{3+\epsilon}}$$
$$\alpha = 2$$
$$\beta = \frac{2\sqrt{3}}{3}$$
$$\gamma = \sqrt{\frac{2(3+\epsilon)}{3\epsilon}}.$$

The full transformation is given by

$$x_1 = \frac{1}{\sqrt{3+\epsilon}}x_0 + [\sqrt{\tilde{p}_1}\cos\theta_1 + \frac{\sqrt{3}}{3}\sqrt{\tilde{p}_2}\cos\theta_2] + \sqrt{\frac{2}{3}}\sqrt{\frac{\epsilon}{3+\epsilon}}\sqrt{\tilde{p}_3}\cos\theta_3$$

$$y_1 = \frac{1}{\sqrt{3+\epsilon}}y_0 + [\sqrt{\tilde{p}_1}\sin\theta_1 + \frac{\sqrt{3}}{3}\sqrt{\tilde{p}_2}\sin\theta_2] + \sqrt{\frac{2}{3}}\sqrt{\frac{\epsilon}{3+\epsilon}}\sqrt{\tilde{p}_3}\sin\theta_3$$

$$x_2 = \frac{1}{\sqrt{3+\epsilon}}x_0 - [\sqrt{\tilde{p}_1}\cos\theta_1 - \frac{\sqrt{3}}{3}\sqrt{\tilde{p}_2}\cos\theta_2] + \sqrt{\frac{2}{3}}\sqrt{\frac{\epsilon}{3+\epsilon}}\sqrt{\tilde{p}_3}\cos\theta_3$$

$$y_2 = \frac{1}{\sqrt{3+\epsilon}}y_0 - [\sqrt{\tilde{p}_1}\sin\theta_1 - \frac{\sqrt{3}}{3}\sqrt{\tilde{p}_2}\sin\theta_2] + \sqrt{\frac{2}{3}}\sqrt{\frac{\epsilon}{3+\epsilon}}\sqrt{\tilde{p}_3}\sin\theta_3$$

$$x_3 = \frac{1}{\sqrt{3+\epsilon}}x_0 - \frac{2\sqrt{3}}{3}\sqrt{\tilde{p}_2}\cos\theta_2 + \sqrt{\frac{2}{3}}\sqrt{\frac{\epsilon}{3+\epsilon}}\sqrt{\tilde{p}_3}\cos\theta_3$$

$$y_3 = \frac{1}{\sqrt{3+\epsilon}}y_0 - \frac{2\sqrt{3}}{3}\sqrt{\tilde{p}_2}\sin\theta_2 + \sqrt{\frac{2}{3}}\sqrt{\frac{\epsilon}{3+\epsilon}}\sqrt{\tilde{p}_3}\sin\theta_3$$

$$\bar{x}_4 \equiv \sqrt{\epsilon}x_4 = \sqrt{\frac{\epsilon}{3+\epsilon}}x_0 - \sqrt{2}\sqrt{\frac{3}{3+\epsilon}}\sqrt{\tilde{p}_3}\cos\theta_3$$

$$\bar{y}_4 \equiv \sqrt{\epsilon}y_4 = \sqrt{\frac{\epsilon}{3+\epsilon}}y_0 - \sqrt{2}\sqrt{\frac{3}{3+\epsilon}}\sqrt{\tilde{p}_3}\sin\theta_3.$$

Notice that when $\epsilon = 0$ the transformation canonically relates the coordinates of the three unit vortices, (z_1, z_2, z_3), to C_0, $\tilde{p}_1, \theta_1, \tilde{p}_2, \theta_2$, and $\bar{x}_4 = -\sqrt{2\tilde{p}_3}\cos\theta_3$, $\bar{y}_4 = -\sqrt{2\tilde{p}_3}\sin\theta_3$.

The governing Hamiltonian is given by

$$\mathcal{H} = \mathcal{H}_0 + \epsilon \mathcal{H}_1$$
$$= -\frac{1}{4\pi}[(\log r_{12}^2 + \log r_{13}^2 + \log r_{23}^2) + \epsilon(\log r_{14}^2 + \log r_{24}^2 + \log r_{34}^2)]$$
$$= -\frac{1}{4\pi}[\log(r_{12}^2 \cdot r_{13}^2 \cdot r_{23}^2) + \epsilon \log(r_{14}^2 \cdot r_{24}^2 \cdot r_{34}^2)],$$

where $r_{ij} = |z_i - z_j|$. In terms of the new coordinates, we have

$$r_{12}^2 = 4\tilde{p}_1$$
$$r_{13}^2 = \tilde{p}_1 + 3\tilde{p}_2 + 2\sqrt{3}\sqrt{\tilde{p}_1 \tilde{p}_2} \cos(\theta_1 - \theta_2)$$
$$r_{23}^2 = \tilde{p}_1 + 3\tilde{p}_2 - 2\sqrt{3}\sqrt{\tilde{p}_1 \tilde{p}_2} \cos(\theta_1 - \theta_2),$$

and

$$r_{14}^2 \cdot r_{24}^2 \cdot r_{34}^2 = \frac{1}{\epsilon^3}\left(8\tilde{p}_3^3 + \epsilon\phi_1 + \epsilon^{3/2}\phi_2\right)$$

where

$$\phi_1 = -16\tilde{p}_2\tilde{p}_3^2 \cos^2(\theta_3 - \theta_2) + 8\tilde{p}_3^3 + 8\tilde{p}_2\tilde{p}_3^2 + 8\tilde{p}_1\tilde{p}_3^2 - 16\tilde{p}_1\tilde{p}_3^2 \cos(\theta_1 - \theta_3)$$

$$\phi_2 = \sqrt{\epsilon}\left[-\frac{64}{9}\tilde{p}_3^2\tilde{p}_2 \cos^2(\theta_3 - \theta_2) + \frac{4A}{3}\tilde{p}_2 + \frac{2A}{3}\tilde{p}_3 + 2B\tilde{p}_3\right.$$
$$\left. -8C(3+\epsilon)\sqrt{\tilde{p}_3\tilde{p}_2}\cos(\theta_3 - \theta_2)\right]$$
$$+ \epsilon\left[-\sqrt{2(3+\epsilon)}B\sqrt{\tilde{p}_3\tilde{p}_2}\cos(\theta_3 - \theta_2) + \frac{4C}{3}\sqrt{2(3+\epsilon)}\tilde{p}_2\right.$$
$$\left. -8C(3+\epsilon)\sqrt{\tilde{p}_2\tilde{p}_3}\cos(\theta_3 - \theta_2)\right]$$
$$+ \frac{32}{9}\sqrt{2(3+\epsilon)}(\tilde{p}_2\tilde{p}_3)^{3/2}\cos(\theta_3 - \theta_2) + \frac{16}{9}\sqrt{2(3+\epsilon)}\tilde{p}_3^2\sqrt{\tilde{p}_3\tilde{p}_2}\cos(\theta_3 - \theta_2)$$
$$- \frac{4A}{3}\sqrt{2(3+\epsilon)}\sqrt{\tilde{p}_3\tilde{p}_2}\cos(\theta_3 - \theta_2) + 2C\sqrt{2(3+\epsilon)}\tilde{p}_3$$

with

$$A = \frac{8}{3}\tilde{p}_3^2 + 4\tilde{p}_1\tilde{p}_2 + \frac{4}{3}\tilde{p}_2\tilde{p}_3 + \frac{8}{3}\tilde{p}_2\tilde{p}_3 \cos^2(\theta_3 - \theta_2) - 8\tilde{p}_1\tilde{p}_3 \cos^2(\theta_3 - \theta_1)$$

$$B = \tilde{p}_1^2 + \frac{1}{9}\tilde{p}_2^2 + \frac{4}{9}\tilde{p}_3^2 + \frac{4}{3}\tilde{p}_1\tilde{p}_3 + \frac{2}{3}\tilde{p}_1\tilde{p}_2 + \frac{4}{9}\tilde{p}_2\tilde{p}_3 + \frac{8}{9}\tilde{p}_2\tilde{p}_3 \cos^2(\theta_3 - \theta_2)$$
$$- \frac{4}{3}\tilde{p}_1\tilde{p}_2 \cos^2(\theta_1 - \theta_2) - \frac{8}{3}\tilde{p}_1\tilde{p}_3 \cos^2(\theta_3 - \theta_1)$$

$$C = \sqrt{\tilde{p}_2\tilde{p}_3}\left[\cos(\theta_3 - \theta_2)\left(\frac{4}{3}\tilde{p}_1 + \frac{8}{9}\tilde{p}_3 + \frac{4}{9}\tilde{p}_2\right) - \frac{8}{3}\tilde{p}_1 \cos(\theta_1 - \theta_3)\cos(\theta_1 - \theta_2)\right].$$

In the new variables, after some algebra, we have the following expressions for the leading order and perturbed Hamiltonian terms:

$$\mathcal{H}_0 = -\frac{1}{4\pi} \log \left[\tilde{p}_1 (\tilde{p}_1 + 3\tilde{p}_2)^2 - 12\tilde{p}_1^2 \tilde{p}_2 \cos^2(\theta_1 - \theta_2) \right] \tag{2.3.9}$$

$$\mathcal{H}_1 = -\frac{1}{4\pi} \log(r_{14}^2 \cdot r_{24}^2 \cdot r_{34}^2) \tag{2.3.10}$$

(we have dropped all constants added to $\mathcal{H}_0$). Notice that $\mathcal{H}_0$ is independent of (x_0, y_0) and only depends on the angle differences $\theta_1 - \theta_2$, hence (x_0, y_0) and $\tilde{p}_1 + \tilde{p}_2$ are constants of the motion. As a final transformation, we make the canonical variable change

$$p_1 = \tilde{p}_1$$
$$q_1 = \theta_1 - \theta_2$$
$$p_2 = \tilde{p}_1 + \tilde{p}_2$$
$$q_2 = \theta_2 - \theta_3$$
$$p_3 = \tilde{p}_1 + \tilde{p}_2 + \tilde{p}_3$$
$$q_3 = \theta_3,$$

which gives the unperturbed Hamiltonian

$$\mathcal{H}_0 = -\frac{1}{4\pi} \log[p_1(p_1 + 3(p_2 - p_1))^2 - 12p_1^2(p_2 - p_1) \cos^2 q_1]$$

with equations of motion for the three unit vortices

$$\dot{p}_1 = \frac{\partial \mathcal{H}_0}{\partial q_1}, \quad \dot{q}_1 = -\frac{\partial \mathcal{H}_0}{\partial p_1}$$

$$\dot{p}_2 = 0, \quad \dot{q}_2 = -\frac{\partial \mathcal{H}_0}{\partial p_2}.$$

Analysis of Unperturbed System

We first analyze in detail the unperturbed system corresponding to $\epsilon = 0$. Let

$$V = -[p_1(p_1 + 3(p_2 - p_1))^2 - 12p_1^2(p_2 - p_1) \cos^2 q_1]$$

with $0 < p_1 < p_2$. Then rescale time

$$\tau = \frac{1}{4\pi} \exp(4\pi \mathcal{H}_0)t,$$

so that we can write the equations

$$\frac{dp_1}{d\tau} = \frac{\partial V}{\partial q_1}, \quad \frac{dq_1}{d\tau} = -\frac{\partial V}{\partial p_1}$$

$$\frac{dp_2}{\partial \tau} = 0, \quad \frac{dq_2}{d\tau} = -\frac{\partial V}{\partial p_2}.$$

Since p_2 is a constant, it can be treated as a parameter in the equations, calling $p_2 = \mu$. Then one has the planar system

$$\frac{dp_1}{d\tau} = -24p_1^2(\mu - p_1) \cos q_1 \sin q_1 \qquad (2.3.11)$$

$$\frac{dq_1}{d\tau} = \cos^2 q_1 [36p_1^2 - 24\mu p_1] + (3\mu - 2p_1)(3\mu - 6p_1) \qquad (2.3.12)$$

where $0 < p_1 < \mu$, with the following fixed point structure:

1. Equilateral triangles: $p_1 = \frac{1}{2}\mu$, $\cos q_1 = 0$.

2. Collisions ($z_2 = z_3$ or $z_1 = z_3$): $p_1 = \frac{3}{4}\mu$, $\sin q_1 = 0$.

3. Collinear states $((z_1 z_2 z_3)$ or $(z_3 z_1 z_2))$: $p_1 = \frac{1}{4}\mu$, $\sin q_1 = 0$.

It is straightforward to show that the fixed points corresponding to equilateral triangles and collisions are centers, while the collinear states are saddles. The value of V at the saddle point $p_1 = \frac{1}{4}\mu$, $\sin q_1 = 0$, is $-\mu^3$, hence the saddle connection lies on the energy level

$$V = -\mu^3.$$

We can then factor V as

$$V(p_1, q_1; \mu) + \mu^3 = (p_1 - \mu)\left(p_1 - \frac{\mu}{2(2 + \sqrt{3}\sin q_1)}\right)\left(p_1 - \frac{\mu}{2(2 - \sqrt{3}\sin q_1)}\right)$$

to get the fact that the curve defined by

$$p_1 = \frac{\mu}{2(2 + \sqrt{3}\sin q_1)}, \qquad 0 < q_1 < \pi, \ \ p_2 = \mu, \qquad (2.3.13)$$

is a saddle connection (separatrix connecting the two saddle points $q_1 = 0$, $q_1 = \pi$). The phase portrait is shown in Figure 2.9.

We next consider in detail the solutions corresponding to a fixed level curve

$$V(p_1, q_1, p_2) = -\mu^3 < 0.$$

For fixed μ this equation can be solved for p_2 to give

$$p_2 = \frac{2}{3}p_1(1 + \cos^2 q_1) \pm \frac{1}{3}\sqrt{\frac{\mu^3 - p_1^3 \sin^2 2q_1}{p_1}} \equiv h_0(p_1, q_1, \mu).$$

The + branch is the one corresponding to (2.3.13), so we choose it. We can now reparametrize the system with q_2, since in the neighborhood of the

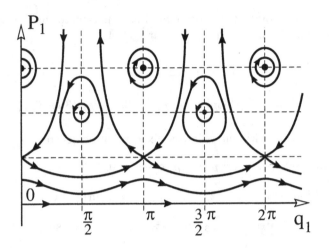

FIGURE 2.9. Phase portrait for reduced system.

separatrix (2.3.13), $\partial V / \partial p_2 \neq 0$. The unperturbed system in its cleanest form then reads

$$\frac{dp_1}{dq_2} = \frac{\partial h_0}{\partial q_1}(p_1, q_1, \mu)$$

$$\frac{dq_1}{dq_2} = -\frac{\partial h_0}{\partial p_1}(p_1, q_1, \mu).$$

The second equation can be written

$$\frac{dq_1}{dq_2} = \frac{2 \sin q_1 (\sqrt{3} + 2 \sin q_1)(2 + \sqrt{3} \sin q_1)}{4 + \sin^2 q_1 + 3\sqrt{3} \sin q_1}$$

which, when integrated, gives

$$q_2(q_1) - q_2^0 = \int_{q_1^0}^{q_1} \frac{4 + \sin^2 t + 3\sqrt{3} \sin t}{2 \sin t(\sqrt{3} + 2 \sin t)(2 + \sqrt{3} \sin t)} dt$$

$$= \frac{1}{2\sqrt{3}} \log \left[\frac{\sqrt{3}\tan^2(q_1/2) + \frac{1}{\sqrt{3}}\tan(q_1/2) + \tan(q_1^0/2)}{\sqrt{3}\tan^2(q_1^0/2) + \frac{1}{\sqrt{3}}\tan(q_1^0/2) + \tan(q_1/2)} \right]$$

$$- \tan^{-1}(\sqrt{3} + 2\tan\frac{q_1}{2}) + \tan^{-1}(\sqrt{3} + 2\tan\frac{q_1^0}{2}).$$

Following the notation of Castilla et al (1993), let

$$q_1 = x + \frac{\pi}{2}, \qquad -\frac{\pi}{2} < x < \frac{\pi}{2}$$

$$q_1^0 = x_0 + \frac{\pi}{2}, \qquad -\frac{\pi}{2} < x_0 < \frac{\pi}{2},$$

giving

$$q_2(x, x_0, q_2^0) = q_2(x + \frac{\pi}{2}) = q_2^0 + \frac{1}{2\sqrt{3}} \log \left(\frac{F(x)}{F(x_0)} \right)$$

$$- \tan^{-1} \left[\frac{(2 + \sqrt{3}) + (2 - \sqrt{3}) \tan(x/2)}{1 - \tan(x/2)} \right]$$

$$+ \tan^{-1} \left[\frac{(2 + \sqrt{3}) + (2 - \sqrt{3}) \tan(x_0/2)}{1 - \tan(x_0/2)} \right]$$

$$= q_2^0 + \frac{1}{2\sqrt{3}} \log \left(\frac{F(x)}{F(x_0)} \right) \tan^{-1} \left[(2 - \sqrt{3}) \tan \left(\frac{x}{2} \right) \right]$$

$$+ \tan^{-1} \left[(2 - \sqrt{3}) \tan \left(\frac{x_0}{2} \right) \right]$$

$$= q_2^0 + s(x) - s(x_0)$$

where

$$F(x) = \frac{(1 + \tan(x/2))^2}{(1 - \tan(x/2))^2} \frac{1 + \sqrt{3} - (\sqrt{3} - 1) \tan(x/2)}{1 + \sqrt{3} + (\sqrt{3} - 1) \tan(x/2)}$$

$$s(x) = \frac{1}{2\sqrt{3}} \log F(x) - \tan^{-1} \left[(2 - \sqrt{3}) \tan \left(\frac{x}{2} \right) \right].$$

Analysis of Perturbed System

The last step in the analysis is to consider the perturbed system $\epsilon > 0$. In the new coordinates the full Hamiltonian can be written

$$\mathcal{H} = -\frac{1}{4\pi} \log[-W_\epsilon]$$

where

$$W_\epsilon(p_1, q_1, p_2, q_2, p_3) = V(p_1, q_1, p_2)[8(p_3 - p_2)^3 + \epsilon\phi_1 + \epsilon^{3/2}\phi_2]^\epsilon$$

$$= V(p_1, q_1, p_2)\{1 + 3\epsilon \log[2(p_3 - p_2)]$$

$$+ \frac{\epsilon^2}{2}[9 \log^2(2(p_3 - p_2)) + \frac{\phi_1}{4(p_2 - p_3)^3}]\} + O(\epsilon^2).$$

Note that W_ϵ is independent of q_3, 2π-periodic in q_2, and is defined for $p_3 > p_2$. The perturbed equations of motion are

$$\frac{dp_1}{d\tau} = \frac{\partial W_\epsilon}{\partial q_1}, \quad \frac{dq_1}{d\tau} = -\frac{\partial W_\epsilon}{\partial p_1}$$

$$\frac{dp_2}{d\tau} = \frac{\partial W_\epsilon}{\partial q_2}, \quad \frac{dq_2}{d\tau} = -\frac{\partial W_\epsilon}{\partial p_2}$$

$$\frac{dp_3}{d\tau} = 0, \quad \frac{dq_3}{d\tau} = -\frac{\partial W_\epsilon}{\partial p_3}.$$

In these variables the system has the two first integrals

$$p_3 = const$$
$$W_\epsilon(p_1, q_1, p_2, q_2, p_3; \epsilon) = const$$

As before, along the separatrix curve we want to solve

$$W_\epsilon(p_1, q_1, p_2, q_2, \mu + \alpha, \epsilon) = -\mu^3$$

for p_2, with $\alpha > 0$, $0 < \epsilon \ll 1$. The solution can be expanded in powers of ϵ

$$p_2(p_1, q_1, q_2, \alpha, \epsilon, \mu) = h_0(p_1, q_1, \mu) + \epsilon h_1 + \epsilon^2 h_2 + O(\epsilon^3)$$

and the perturbed equations then become

$$\frac{dp_1}{dq_2} = \frac{\partial h_0}{\partial q_1} + \epsilon \left[\frac{\partial h_1}{\partial q_1} + \epsilon \frac{\partial h_2}{\partial q_1} + O(\epsilon^2) \right]$$
$$\frac{dq_1}{dq_2} = -\frac{\partial h_0}{\partial p_1} - \epsilon \left[\frac{\partial h_1}{\partial p_1} + \epsilon \frac{\partial h_2}{\partial p_1} + O(\epsilon^2) \right].$$

These are the perturbed phase plane equations, which we analyze in the separatrix neighborhood — we refer the motivated reader to the original paper of Castilla et al (1993) for the detailed forms of h_1, h_2.

The system is now prepared for a Melnikov analysis. Hence, we compute the Melnikov integral

$$M(q_2^0) = -\int_{-\infty}^{\infty} \left[\frac{\partial h_0}{\partial p_1} \frac{\partial h_1}{\partial q_1} - \frac{\partial h_0}{\partial q_1} \frac{\partial h_1}{\partial p_1} \right] dq_2$$

which ultimately reduces to

$$M(q_2^0) = \frac{2}{3} \mu^2 \epsilon \left[I_1 \sin(2q_2^0) + I_2 \cos(2q_2^0) \right] + O(\epsilon^2). \qquad (2.3.14)$$

Hence, one must ascertain that

$$I_1 \sin(2q_2^0) + I_2 \cos(2q_2^0)$$

has a simple zero. It is enough to check that $I_1 \neq 0$, where

$$I_1 = -2\sqrt{3} \int_0^{\pi/2} \left[\frac{\cos^2 x \cos(2s(x))}{(\sqrt{3} + 2\cos x)(2 + \sqrt{3}\cos x)} - \frac{\sqrt{3}\sin x \sin(2s(x))}{3(\sqrt{3} + 2\cos x)} \right] dx$$
$$\approx 0.2621,$$

thus completing the proof.

The proof of Koiller and Carvalho

Koiller and Carvalho's (1989) proof is based on a different unperturbed configuration. We briefly describe the three main steps in the proof:

Step 1 The flowfield is made up of two "primary" vortices $\Gamma_1 = -\Gamma_2$ and two satellites $\Gamma_3 = \Gamma_4 = \epsilon$, $0 << \epsilon << 1$. Since the satellites are sufficiently weak, their effect on the primaries, as a first step, is ignored. However, the mutual influence of each satellite on the other is taken into account.

Step 2 In a frame of reference moving at the average velocity of the primaries, the unperturbed ($\epsilon = 0$) problem gives the streamfunction structure shown in Figure 2.10 [6]. Imagine that for $\epsilon = 0$, Γ_3 is chosen to move along the unperturbed separatrix, while Γ_4 is chosen to move along one of the closed orbits inside the separatrix. As the parameter ϵ is turned on, the vortex Γ_3 undergoes chaotic oscillations in the neighborhood of the unperturbed separatrix, which is proven by a Melnikov calculation.

Step 3 Rigorously justify that the mechanism inducing chaos in step 2 is not affected by the approximations in step 1, i.e., carefully estimate the errors in the approximations.

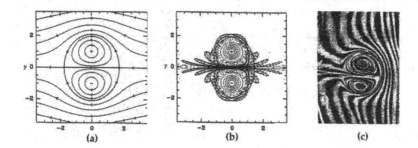

(a)   (b)   (c)

FIGURE 2.10. Three views of the streamfunction structure for a vortex couple: (a) Translating vortex couple in frame moving with the vortices. (b) Poincaré map of the perturbed flow. (c) Experimental realization of the flow.

The vortex couple shown in Figure 2.10 has been used as one of the canonical configurations to develop Melnikov techniques — in particular, the work of Rom-Kedar, Leonard, and Wiggins (1990) focuses on this flowfield under the influence of a time-varying background strain field. For the purposes of applying the Melnikov method we start with their problem because it is simpler than the one studied by Koiller and Carvalho (1989).

---

[6]This configuration has been studied experimentally by Couder and Basdevant (1986) who call it a "Batchelor couple," a double pun which pays homage to G. Batchelor while also referring to the fact that the configuration is stable.

Example 2.5 (Rom-Kedar, Leonard, Wiggins (1990)) Consider the pair of equal and opposite vortices of strength $\pm\Gamma$ moving in the presence of an oscillating external strain field whose streamfunction is given by

$$\psi_2 = \epsilon xy \sin \omega t.$$

The total streamfunction for the flow moving in a frame of reference with the average velocity of the vortices is

$$\psi = -\frac{\Gamma}{4\pi} \log \left[\frac{(x-x_v)^2 + (y-y_v)^2}{(x+x_v)^2 + (y+y_v)^2} \right] - V_v y + \epsilon xy \sin \omega t,$$

where the vortex positions are denoted $(x_v(t), \pm y_v(t))$, V_v is the average velocity of the pair, ϵ is the strain rate of the external field, which we take to be weak, $0 < \epsilon << 1$. We can nondimensionalize the system by using the variables

$$\frac{x}{d} \to x, \quad \frac{y}{d} \to y, \quad \frac{\Gamma t}{2\pi d^2} \to t,$$

$$\frac{\epsilon}{\omega} \to \epsilon, \quad \frac{2\pi V_v}{\Gamma} \to \nu_v, \quad \frac{\Gamma}{2\pi \omega d^2} \to \gamma, \quad \frac{2\pi\psi}{\Gamma} \to \psi$$

giving

$$\dot{x} = - \left[\frac{(y-y_v)}{(x-x_v)^2 + (y-y_v)^2} - \frac{(y+y_v)}{(x-x_v)^2 + (y+y_v)^2} \right] - \nu_v + \epsilon \frac{x}{\gamma} \sin\left(\frac{t}{\gamma}\right)$$

$$\dot{y} = (x - x_v) \left[\frac{1}{(x-x_v)^2 + (y-y_v)^2} - \frac{1}{(x-x_v)^2 + (y+y_v)^2} \right] - \epsilon \frac{y}{\gamma} \sin\left(\frac{t}{\gamma}\right).$$

The equations can then be expanded in powers of ϵ

$$\dot{x} = f_1(x, y) + \epsilon g_1(x, y, t/\gamma; \gamma) + O(\epsilon^2)$$
$$\dot{y} = f_2(x, y) + \epsilon g_2(x, y, t/\gamma; \gamma) + O(\epsilon^2)$$

where

$$f_1 = -\frac{y-1}{I_-} + \frac{y+1}{I_+} - \frac{1}{2}$$

$$f_2 = x \left[\frac{1}{I_-} - \frac{1}{I_+} \right]$$

$$g_1 = [\cos(\frac{t}{\gamma}) - 1] \left[\frac{1}{I_-} + \frac{1}{I_+} - \frac{2(y-1)^2}{I_-^2} - \frac{2(y+1)^2}{I_+^2} \right]$$
$$+ \frac{x}{\gamma} \sin(\frac{t}{\gamma}) \left[\gamma^2 \left(\frac{y-1}{I_-^2} - \frac{y+1}{I_+^2} \right) + 1 \right] - \frac{1}{2}$$

$$g_2 = 2x[\cos(\frac{t}{\gamma}) - 1] \left[\frac{y-1}{I_-^2} + \frac{y+1}{I_+^2} \right]$$
$$= \frac{1}{\gamma} \sin(\frac{t}{\gamma}) \left[\frac{\gamma^2}{2} \left(\frac{1}{I_-} - \frac{1}{I_+} \right) - x^2\gamma^2 \left(\frac{1}{I_-^2} - \frac{1}{I_+^2} \right) - y \right]$$

$$I_\pm \equiv x^2 + (y \pm 1)^2.$$

The vortex locations are governed by

$$\dot{x}_v = \frac{1}{2y_v} - \nu_v + \frac{\epsilon x_v}{\gamma} \sin(\frac{t}{\gamma})$$

$$\dot{y}_v = -\frac{\epsilon y_v}{\gamma} \sin(\frac{t}{\gamma})$$

which, with $x_v(0) = 0, y_v(0) = 1$, can be integrated to give

$$x_v(t) = \frac{1}{2}\gamma \exp[-\epsilon(\cos(\frac{t}{\gamma}) - 1)] \int_0^{\frac{t}{\gamma}} \{1 - 2\nu_v \exp[\epsilon(\cos s - 1)]\} \, ds$$

$$y_v(t) = \exp[\epsilon(\cos(\frac{t}{\gamma}) - 1)].$$

The mean velocity is given by

$$\nu_v = \frac{\exp(\epsilon)}{2I_0(\epsilon)}$$

where I_0 is the modified Bessel function of zero order. The details of this calculation will be carried out in an exercise.

As described in Rom-Kedar et al (1990), the Melnikov function for the system is

$$M(t_0) = \int_{-\infty}^{\infty} [f_1(\overline{q}(t))g_2(\overline{q}(t), t + t_0) - f_2(\overline{q}(t))g_1(\overline{q}(t), t + t_0)]dt$$

$$= \frac{F(\gamma)}{\gamma} \sin(\frac{t_0}{\gamma})$$

where $\overline{q}(t)$ represents the heteroclinic orbit shown in Figure 2.10. Since a simple closed form expression for the heteroclinic orbit is not available, the best one can hope for is a numerical solution of the Melnikov amplitude $F(\gamma)/\gamma$, which is given in Rom-Kedar et al (1990) and shown to be nonzero.

◊

The nonintegrability proof of Koiller and Carvalho (1989) is based on the same unperturbed flow configuration. However, the perturbation terms are more complicated since they involve equations that couple them to the satellites — the time dependence is only implicit. As a first step, one can show that a particle located at $z = u + iv$ in a coordinate system where the primaries are at rest at $z_1 = -z_2 = i$ is governed by the Hamiltonian system

$$\dot{u} = \frac{\partial F}{\partial v}, \quad \dot{v} = -\frac{\partial F}{\partial u}$$

$$F(u, v) = \frac{v}{4} + \frac{1}{2} \log(\frac{r_1}{r_2})$$

$$r_1 = |z_1 - i|$$

$$r_2 = |z_2 + i|.$$

The Hamiltonian that includes the effects of the satellites is

$$\mathcal{H}_\epsilon(p_1, q_1, p_2, q_2) = F(p_1, q_1) + F(p_2, q_2) + \epsilon\mathcal{H}_1(p_1, q_1, p_2, q_2)$$

where $z_3 = p_1 + iq_1$, $z_4 = p_2 + iq_2$ and

$$\mathcal{H}_1 = \log|z_3 - z_4|.$$

The system now exactly corresponds to the "F-G" system analyzed in Holmes and Marsden (1982a,b), who consider an unperturbed integrable Hamiltonian $\mathcal{H}_0$ of the form

$$\mathcal{H}_0(q, p; x, y) = F(q, p) + G(x, y).$$

They then make the following additional assumptions:

1. $G(x, y)$ can be transformed to action-angle variables (I, θ) so that the equations of motion can be written

$$\dot{q} = \frac{\partial F}{\partial p}, \quad \dot{p} = -\frac{\partial F}{\partial q} \tag{2.3.15}$$

$$\dot{\theta} = \Omega(I), \quad \dot{I} = 0 \tag{2.3.16}$$

   where $G(0) = 0$, $\Omega(I) = G'(I) > 0$ for $I > 0$. The more general case, where $\mathbf{q}, \mathbf{p} \in \mathbb{R}^n$ for $n > 1$, is treated in Holmes and Marsden (1982a, b), but is not needed for our purposes.

2. Assume that the F system (2.3.15) contains a homoclinic orbit

$$(\bar{q}(t - t_0), \bar{p}(t - t_0)),$$

   which joins the saddle point (q_0, p_0) to itself, while the G system contains a family of 2π-periodic orbits

$$\theta(t) = \Omega(I_0)t + \theta_0, \quad I(t) = I_0.$$

The unperturbed F-G system can be thought of as made up of "products of homoclinic orbits and periodic orbits," shown in Figure 2.11 (reprinted with permission from Holmes and Marsden (1982a)). Equivalently, one could take the case where F has a heteroclinic orbit connecting two different saddle points.

We then consider the perturbed system

$$\mathcal{H}_\epsilon(q, p; \theta, I) = F(q, p) + G(I) + \epsilon\mathcal{H}_1(q, p; \theta, I) + O(\epsilon^2)$$

where $\mathcal{H}_1$ is smooth (say C^∞) and 2π-periodic in θ. Then the perturbed equations are

$$\dot{q} = \frac{\partial F}{\partial p} + \epsilon\frac{\partial\mathcal{H}_1}{\partial p}, \quad \dot{p} = -\frac{\partial F}{\partial q} - \epsilon\frac{\partial\mathcal{H}_1}{\partial q} \tag{2.3.17}$$

$$\dot{\theta} = \Omega(I) + \epsilon\frac{\partial\mathcal{H}_1}{\partial I}, \quad \dot{I} = -\epsilon\frac{\partial\mathcal{H}_1}{\partial\theta}. \tag{2.3.18}$$

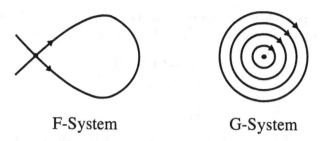

F-System   G-System

FIGURE 2.11. F-G system.

The principal result from Holmes and Marsden (1982a,b) is that if the Melnikov integral

$$M(t_0) = \int_{-\infty}^{\infty} \{F, \mathcal{H}_1\}(t - t_0)dt \qquad (2.3.19)$$

is nonzero, the Poincaré map associated with $\mathcal{H}_\epsilon$ contains Smale horseshoes on each fixed energy surface, hence via the Smale–Birkhoff homoclinic theorem (see Wiggins (1988)), there is no analytic second integral and the system is nonintegrable. See Holmes and Marsden (1982a,b) for a precise statement of the theorem along with some simple examples.

To apply this to the 4-vortex problem for $\epsilon = 0$ take Γ_4 moving along a closed orbit of the G system, specified by setting $G = h$. Take Γ_3 to move along the separatrix $(\bar{p}_1(t), \bar{q}_1(t))$ of F given by

$$\bar{q}_1(t) = 0$$
$$\dot{\bar{p}}_1 = \frac{1}{4} \frac{(\bar{p}_1^2 - 3)}{(1 + \bar{p}_1^2)}.$$

Solving for $\bar{p}_1$

$$\int \frac{(1 + \bar{p}_1^2)}{(\bar{p}_1^2 - 3)} d\bar{p}_1 = \int \left(\frac{2/\sqrt{3}}{\bar{p}_1 - \sqrt{3}} + \frac{\bar{p}_1 + 1/\sqrt{3}}{\bar{p}_1 + \sqrt{3}} \right) d\bar{p}_1$$
$$= \frac{t}{4} + const$$

which gives

$$t = 4\bar{p}_1 + \frac{8}{\sqrt{3}} \log \left(\frac{\sqrt{3} - \bar{p}_1}{\sqrt{3} + \bar{p}_1} \right).$$

As ϵ is turned on the motion of Γ_4 is only slightly perturbed, while that of Γ_3 is more complicated. Each time Γ_3 approaches one of the saddle points, it essentially "flips a coin" to decide which branch to choose. The details of the Melnikov calculation are carried out in the region $z_1 \sim i$, $z_2 \sim -i$, where it is possible to expand the periodic orbit in powers of the radius r representing the deviation from $z_1 = i$, $z_2 = -i$

$$p_2 \sim r \sin \omega t + O(r^2)$$
$$q_2 \sim 1 + r \cos \omega t + O(r^2)$$
$$\omega \sim \frac{1}{2r^2}.$$

In this limit, Koiller and Carvalho (1989) show that the Melnikov integral $M(t_0, r)$ has the asymptotic expansion

$$M(t_0, r) = \sum_{j=1}^{n} r^j B_j(t_0, \omega)$$

with lead term given by

$$B_1(t_0) = \pi \exp[-(1/2r^2)(8\pi/3\sqrt{3} - 4)] \sin t_0.$$

The higher order terms B_n ($n > 1$) decrease exponentially in ω. Hence, to leading order, the zeros of the Melnikov integral are $t_0 = n\pi$ and the separatrices intersect transversally. The remaining details of this calculation are carried out in a series of exercises at the end of the chapter.

Remarks.

1. In general, one can extract more information from the Melnikov calculation. Typically, one is interested in the details of the so-called "homoclinic tangle dynamics" associated with a given system as this has a direct bearing on transport processes associated with the efficiency of fluid mixing in the perturbed flow. The mixing, of course, is enhanced by the transverse intersection of separatrices, which form distinct lobe structures in the Poincaré section. In Rom-Kedar et al (1990), it is shown that the area of the lobes, μ is proportional to the product of the perturbation size ϵ and the Melnikov amplitude $F(\gamma)$,

$$\mu \sim 2\epsilon \|F(\gamma)\| + O(\epsilon^2).$$

This quantifies the size of the mixing region due to homoclinic chaos. For general discussion of these ideas, see Beigie, Leonard, and Wiggins (1994) and Ottino (1989). These types of detailed calculations for pure vortex flows (i.e., flows where all of the time dependence is generated by the mutual interaction of the vortices) have not been carried out and would certainly be worth pursuing.

2. As mentioned earlier, strictly speaking, since the unperturbed configurations of Koiller and Carvalho (1989) and Rom-Kedar et al (1990)

involves a *heteroclinic* structure, one needs an additional argument to show a recurrence mechanism necessary for producing the horseshoes. Neither paper does this, although based on numerics there is no doubt in this case that the horseshoes are present.   ◊

The proof of Ziglin

Ziglin's proof (1980) relies on a different configuration, but is also based on a Melnikov calculation. In Example 2.3 the equation for a particle moving in the field of a rigidly rotating triangular configuration of vortices was written as (2.2.7). The (unperturbed) coordinate of each vortex is

$$z_j^{(0)}(t) = \exp[i(\frac{t}{2\pi} + \frac{2\pi}{3}j)] \;\; j = 1, 2, 3.$$

Ziglin (1980) then considers how each vortex evolves in the neighborhood of this equilateral state, demonstrating that one can write the perturbed vortex coordinates as

$$z_j(t) \sim \exp[i(\Omega(\epsilon^2)p(\epsilon^2)t + \frac{2\pi}{3}j)] \cdot [1 + \epsilon\gamma_j \exp(-ip(\epsilon^2)t) + O(\epsilon^2)],$$

where ϵ measures the deviation from equilibrium, $\gamma_j = const$, and

$$p(\epsilon^2) \sim \frac{1}{2\pi} + O(\epsilon^2)$$
$$\Omega(\epsilon^2) \sim 1 + O(\epsilon^2).$$

Then, as in Example 2.3, one can write the particle equations in this perturbed flow as a Hamiltonian system,

$$\mathcal{H} = \mathcal{H}_0(\xi, \eta) + \epsilon\mathcal{H}_1(\xi, \eta; t) + O(\epsilon^2),$$

where $\mathcal{H}_0(\xi, \eta)$ is the unperturbed Hamiltonian given by (2.2.8), while

$$\mathcal{H}_1(\xi, \eta; t) \equiv C \cdot \text{Re}\left[\left(\frac{\zeta}{\zeta^3 - 1}\right)\exp(-i\frac{t}{2\pi})\right] \qquad (2.3.20)$$

with $\zeta = \xi + i\eta$ denoting the position of a passive particle. The Poincaré map for this flow is shown in Figure 1.5(b), while the Melnikov integral is given by

$$M(t_0) = \int_{-\infty}^{\infty} \{\mathcal{H}_0, \mathcal{H}_1\}(\overline{\zeta}(t - t_0), t)dt,$$

which is numerically integrated and shown to be nonzero. The orbit $\overline{\zeta}$ is the unperturbed heteroclinic orbit. Many details in the paper by Ziglin (1980) not provided here are contained in a series of exercises at the end of the chapter.

2.4   Bibliographic Notes

We mention briefly several other promising lines of work. A large literature focuses on constructing vortex equilibria of the N-vortex problem based on the discrete symmetry groups. This originated with the classical paper by Lord Kelvin (1867) concerning his idea of the "vortex atom," in which he considered vortex polygons that are regular N-gons with identical point vortices placed at each corner. He was the first to derive the linear stability result that for $N > 7$ the configuration is unstable, while for $N \leq 7$ it is stable. Other results along these lines can be found in Havelock (1931). More generally, one can look for concentric ring configurations of vortices, say of m rings, with n evenly spaced vortices on each ring, where $N = nm$. Configurations of this type have been considered by Havelock (1931), Morikawa and Swenson (1971), Khazin (1976), Mertz (1978), Campbell and Ziff (1978, 1979), Aref (1982), Koiller et al (1985), and Lewis and Ratiu (1996). The work of Aref (1982), followed by that of Koiller et al (1985) and Lewis and Ratiu (1996), makes the further assumption that there is a center of symmetry associated with the configuration. There have been several attempts at deriving general results for all possible equilibrium configurations in the plane, a difficult problem for large N. We mention in this context the work of Campbell and Ziff (1978, 1979), as well as Palmore (1982), who uses ideas from Morse theory to derive estimates on the number of equilibria as a function of N. This was followed by the work of O'Neal (1987).

A new technique for finding *asymmetric* equilibrium configurations was recently introduced by Aref and Vainchtein (1998). From a relative equilibrium configuration of N equal strength vortices, they first identify the stagnation points in the flow where zero-strength vortices stay fixed relative to the N-vortex configuration. Then, by a numerical continuation method, an $N + 1$-vortex equilibrium is "grown" by increasing the circulation of the zero-strength vortex, while allowing the full configuration to deform. In certain situations this results in asymmetric $N + 1$-vortex equilibria, with N vortices of equal strength and an additional vortex of a different strength. We mention also some recent experimental results of Fine, Cass, Flynn, and Driscoll (1995) using plasmas, in which certain classes of initial conditions relax to equilibrium states, dubbed "vortex crystal" states. A nice review of this and related work can be found in O'Neil (1999).

The special case where $\Gamma = 0$ was considered in some detail by Rott (1989), Aref (1989), and most recently Aref and Stremler (1999). It allows for a particularly elegant treatment for $N = 3$. Aref (1989) shows that the solutions for this case can be understood by studying an equivalent restricted 4-vortex problem for a passive particle in the flowfield of three vortices placed in an equilibrium collinear configuration — the level curves of this time-independent Hamiltonian give the solutions of the general 3-vortex problem. For $N = 4$, as proven in Theorem 2.1.1, the condition

$\Gamma = 0$ makes the problem integrable, a case that was extensively studied in Eckhardt (1988) where appropriate coordinates are introduced, which reduce the problem to a phase plane analysis.

Recent work of Aref and Bröns (1997) discusses the number and location of stagnation points for the N-vortex problem and identifies all possible streamline topologies for the 3-vortex problem in the plane. This work is part of an ongoing attempt to understand and classify all possible flow topologies for much more general three-dimensional turbulent flows, and in this context we mention the work of Perry and Farlie (1974), Perry and Chong (1986), Dallman (1988), and the review article of Cantwell (1981). We will describe recent work of this type on the sphere in Chapter 4.

There is a growing body of work that is primarily numerical, whose goal is to quantify and understand the statistical properties of passive tracer particles in flows populated by point vortices. Thus, Kuznetsov and Zaslavsky (1998, 2000) consider the evolution of a particle under the influence of three identical vortices, while Boatto and Pierrehumbert (1999) consider the evolution of a tracer particle in the far-field of four identical vortices. By contrast, Babiano, Boffetta, Provenzale, and Vulpiani (1994) study the dynamics of passive tracers in point vortex flows near the point vortices in the case where the vortices undergo chaotic motion. The work of Péntek, Tél, and Toroczkai (1995, 1996) considers the evolution of tracers in the flow composed of two pairs of point vortices undergoing a leapfrogging cycle. The paper of Min, Mezić, and Leonard (1996) describes both numerically and analytically the probablility density functions (pdf's) of the velocity fields induced by a population of vortices and, in particular, they show that the pdf's for the velocity differences follow a Cauchy distribution. These papers and the references therein offer a glimpse into much of the recent literature concerned with particle transport and mixing in vortex-dominated flows, including the use of statistical concepts to analyze the particle distribution functions associated with tracers. This work is one of the promising avenues for analysis in the much broader context of turbulent diffusion, as reviewed recently by Majda and Kramer (1999).

There is beginning to be a substantial body of work on N-layer problems, where vortices are placed in each of the fluid layers and interact with each other within their own layer as well as across layers. These models are used in geophysical settings where the different layers model vertical density stratification. The interaction potential is typically of Bessel function type, making analysis more difficult than that done with logarithmic potentials. Papers dealing directly with point vortices include the work of Gryanik (1983), Hogg and Stommel (1985a,b), Young (1985), Pedlosky(1985), Lim and Majda (2000), and the recent thesis of Jamaloodeen (2000). In particular, it is shown in the thesis of Jamaloodeen that the 3-vortex, two-layer model admits a type of finite-time collision that is not self-similar. This is the only non-self-similar collision result that we know of.

2.5   Exercises

1. For the general N-vortex problem prove that the quantities $\mathcal{H}$, Q, P, and I are conserved.

2. Prove each of the five separate parts of Synge's Theorem 2.1.2.

3. Work out the details of the collapsing 3-vortex configuration described in Example 2.1. What happens to a particle in such a flow-field?

4. Prove that a necessary and sufficient condition for a 3-vortex collinear configuration to remain collinear is that

$$A = \dot{A} = 0,$$

   where A is the area of the vortex triangle.

5. (a) Derive the equations of motion for N vortices moving in a rotating frame of reference with angular velocity Ω centered at the center of vorticity of the system.

   (b) Derive the equations for a particle in the flowfield of N vortices in a rotating frame of reference as in (a).

6. Work out the details of the nonintegrability proof of Castilla et al (1993). In particular:

   (a) Derive the relationships between the centers of vorticity and the vortex positions, i.e., the formulas for $z_i - C_\epsilon$ ($i = 1, \ldots, 4$) and $C_0 - C_\epsilon$.

   (b) Derive the equations for $\mathcal{H}_0$, $\mathcal{H}_1$ as shown in (2.3.9), (2.3.10).

   (c) Derive the equations for the unperturbed planar system (2.3.11), (2.3.12). Show that the fixed points corresponding to equilateral triangles and collisions are centers, while collinear states are saddles.

   (d) Derive the perturbed phase plane equations of motion and compute the Melnikov integral (2.3.14).

7. Work out the details of Example 2.4. In particular:

   (a) Derive the Hamiltonian (2.3.1) and equations of motion for the leapfrogging pairs in the original variables $(x_1, x_2; y_1, y_2)$.

   (b) Derive the Hamiltonian (2.3.2) and equations of motion for the leapfrogging pairs in the new variables $(x_r, x_0; y_r, y_0)$.

   (c) Show that the transformation from $(x_1, x_2; y_1, y_2) \mapsto (x_r, x_0; y_r, y_0)$ is canonical.

(d) Carry out the details of the formulas for the period T and average translational velocity $<v>$, both the exact formulas and the high energy expansions.

(e) Obtain the streamfunction and equations of motion for a passive particle both in the fixed frame and a comoving frame which moves with the vortex pairs.

(f) Plot level curves of the streamlines through one full period in the comoving frame.

8. Work out the details of Example 2.5. In particular:

   (a) Derive the formula for the total streamfunction in Example 2.5.

   (b) Derive solution formulas for the vortex locations $(x_\nu(t), y_\nu(t))$ and the mean velocity ν_v.

   (c) Derive the formula for the Melnikov function $M(t_0)$.

9. Work out the details of the Koiller and Carvalho (1989) proof of nonintegrability.

10. Work out the details of Ziglin's (1980) proof of nonintegrability.

3
Domains with Boundaries

In this chapter we consider the dynamics of N vortices in planar domains with solid boundaries. The domains could be closed, simply connected regions in the plane (e.g., interior of a unit circle), closed, multiply connected regions (e.g., interior of an annular region), or they could be unbounded but containing a solid boundary (e.g. upper half-plane or corner geometries). This problem differs considerably from that described in the previous chapter for two important reasons:

- On the solid boundary $\partial \mathcal{D}$ the condition of no fluid penetration must be imposed:

$$\mathbf{u} \cdot \mathbf{n} = 0 \, |_{\partial \mathcal{D}} \, . \tag{3.0.1}$$

- The presence of boundaries can break symmetries that would otherwise exist and thus reduce the number of conserved quantities.

We know in the unbounded plane, for general Γ_α, $N = 3$ is the maximum number of vortices allowing integrable motion. Since we also know by Noether's theorem that symmetries give rise to conservation laws, we would expect that the maximum number of vortices allowing integrable motion should decrease with the presence of boundaries that break symmetries. Hence, for example, in the upper half-plane with the x-axis as boundary rotational symmetry is lost, and $N = 2$ should be the maximum number of vortices allowing integrable motion. In a perturbed circular domain, where both rotational and translational symmetries are lost, we would expect the maximum number to be $N = 1$. Thus, on general principles we would expect that the motion of a particle in the flowfield of a single vortex in a perturbed circular domain should be chaotic — results of this type are described in more detail in Section 3.4.

Since we consider inviscid flows, the "no-slip" condition on the boundary is not imposed, hence there is no tangential constraint on the velocity at the boundary. In two dimensions, the boundary condition (3.0.1) can be written either in terms of the streamfunction or the velocity potential. Since

$$\mathbf{u} \equiv \nabla^{\perp} \psi$$
$$\mathbf{n} \equiv \nabla \psi,$$

(3.0.1) is automatically satisfied on any constant streamline, and typically we choose the constant to be zero, i.e., $\psi = 0$. In terms of the velocity potential, since $\mathbf{u} = \nabla \phi$, (3.0.1) becomes a Neumann condition

$$\mathbf{n} \cdot \nabla \phi \equiv \frac{\partial \phi}{\partial n} = 0 \mid_{\partial \mathcal{D}} .$$

In the first section we review some classical results from potential theory regarding the construction of a Green's function in a domain with boundaries — called a Green's function of the first or second kind. We show how to find the Hamiltonian for the N-vortex problem associated with a given domain in terms of the Green's function of the first kind. In section Section 3.2 we describe the method of images used to construct the Green's function for domains with special symmetries. Section 3.3 outlines how to make use of conformal mapping results to construct the Hamiltonian in more general domains. In section Section 3.4 we describe the recent results of Zannetti and Franzese (1993, 1994), who prove nonintegrability for the restricted 2-vortex problem in a closed domain without symmetries. The proof, which we briefly sketch, makes use of the Melnikov method as described in Chapter 1.

3.1   Green's Function of the First Kind

We start by reviewing some classical results from potential theory regarding the construction of Green's functions in closed planar regions. Consider the problem of constructing a **Green's function of the first kind** G_I in a closed, simply or multiply connected domain $\mathcal{D}$ in the plane, with boundary denoted $\partial \mathcal{D}$

$$\nabla^2 G_I(\mathbf{x}; \mathbf{x}_\alpha) + \delta \|\mathbf{x} - \mathbf{x}_\alpha\| = 0, \quad \mathbf{x} \in \mathcal{D} \qquad (3.1.1)$$
$$G_I(\mathbf{x}; \mathbf{x}_\alpha) = 0, \quad \mathbf{x} \in \partial \mathcal{D}. \qquad (3.1.2)$$

The delta function represents a source located at $\mathbf{x}_\alpha \in \mathcal{D}$, where we are using the notation $\mathbf{x} \equiv (x, y)$, $\mathbf{x}_\alpha \equiv (x_\alpha, y_\alpha)$. The solution $G_I(\mathbf{x}; \mathbf{x}_\alpha)$ is also called the Dirichlet function since it can be used to solve the Dirichlet problem

$$\nabla^2 u = 0 \in \mathcal{D}$$
$$u = f \in \partial\mathcal{D}$$

via the generalized Poisson formula

$$u = -\int_{\partial\mathcal{D}} f \frac{\partial G_I}{\partial n} \, dS \qquad (3.1.3)$$

where $\partial G_I/\partial n \equiv \mathbf{n}\cdot\nabla G_I$ is the normal derivative of G_I on $\partial\mathcal{D}$, $\mathbf{n}$ being the outward unit normal. The basic set-up is shown in Figure 3.1. The construc-

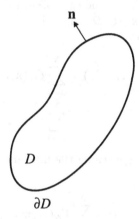

n

D

∂D

FIGURE 3.1. Dirichlet problem in region $\mathcal{D}$ of the plane.

tion of G_I is intimately connected to the details of the boundary shape, and in general, unless the boundary has certain symmetry properties, it is not easy to satisfy condition (3.1.2).

As a first step, we can decompose G_I into the sum of two distinct parts,

$$G_I = G + G_H^{(\alpha)},$$

where G is the Green's function in the unbounded plane (see (1.1.7), (1.1.8))

$$G(\mathbf{x}; \mathbf{x}_\alpha) = -\frac{1}{2\pi} \log \|\mathbf{x} - \mathbf{x}_\alpha\|.$$

Since G does not satisfy the boundary condition (3.1.2), we need to add to it the harmonic function $G_H^{(\alpha)}$,

$$\nabla^2 G_H^{(\alpha)} = 0,$$

and to satisfy the boundary condition (3.1.2), we must enforce

$$G_H^{(\alpha)} = -G = \frac{1}{2\pi} \log \|\mathbf{x} - \mathbf{x}_\alpha\| \quad \mathbf{x} \in \partial \mathcal{D}.$$

With this choice of G and $G_H^{(\alpha)}$ (3.1.1) and (3.1.2) are both satisfied.

Remarks.

1. One can prove that $G_I(\mathbf{x}; \mathbf{x}_\alpha)$ satisfies the reciprocity property

$$G_I(\mathbf{x}; \mathbf{x}_\alpha) = G_I(\mathbf{x}_\alpha; \mathbf{x}),$$

   by standard application of Green's formulas.

2. The Green's function for the case of an arbitrary number of distinct sources located at $\mathbf{x}_\alpha \in \mathcal{D}$, $\alpha = 1, \ldots, N$, can be constructed by linear superposition

$$G_I^{(N)}(\mathbf{x}; \mathbf{x}_1, \ldots, \mathbf{x}_N) = \sum_{\alpha=1}^{N} G_I(\mathbf{x}; \mathbf{x}_\alpha)$$

$$= \sum_{\alpha=1}^{N} G(\mathbf{x}; \mathbf{x}_\alpha) + \sum_{\alpha=1}^{N} G_H^{(\alpha)}(\mathbf{x}).$$

The streamfunction governing the fluid motion at an arbitrary point $\mathbf{x}$ is given by

$$\psi \equiv \sum_{\alpha=1}^{N} \Gamma_\alpha G_I(\mathbf{x}; \mathbf{x}_\alpha).$$

Then, to compute the motion of a vortex at $\mathbf{x} = \mathbf{x}_\beta$ one must subtract the singular contribution

$$\Psi_\beta \equiv \left(\psi + \frac{\Gamma_\beta}{2\pi} \log \|\mathbf{x} - \mathbf{x}_\beta\| \right)_{\mathbf{x}=\mathbf{x}_\beta}$$

and get the velocity field in the usual way from the relations

$$(u, v)_\beta \equiv (\dot{x}_\beta, \dot{y}_\beta) = \left(\frac{\partial \Psi_\beta}{\partial y_\beta}, -\frac{\partial \Psi_\beta}{\partial x_\beta} \right).$$

Equivalently, one can derive the equations of motion from the Hamiltonian which is constructed by forming the sum

$$\mathcal{H}(\mathbf{x}_1, \ldots, \mathbf{x}_N) = \frac{1}{2} \sum_{\alpha \neq \beta}^{N} \Gamma_\alpha \Gamma_\beta G(\mathbf{x}_\alpha, \mathbf{x}_\beta) + \frac{1}{2} \sum_{\alpha=1}^{N} \Gamma_\alpha^2 G_H^{(\alpha)}(\mathbf{x}_\alpha).$$

$$(3.1.4)$$

The first summation represents the energy due to vortex-vortex interactions, while the second is due to vortex-boundary interactions. This gives rise to the canonical equations

$$\Gamma_\alpha \dot{x}_\alpha = \frac{\partial \mathcal{H}}{\partial y_\alpha}$$

$$\Gamma_\alpha \dot{y}_\alpha = -\frac{\partial \mathcal{H}}{\partial x_\alpha}.$$

See Lin (1941a) for more details.

3. In the case of a single vortex, the first summation in (3.1.4) is absent and it is more conventional to write the Hamiltonian as

$$\mathcal{H}(x_1, y_1) = \frac{\Gamma_1}{2} G_H^{(1)}(\mathbf{x}_1)$$

with equations of motion

$$\dot{x}_1 = \frac{\partial \mathcal{H}}{\partial y_1}$$

$$\dot{y}_1 = -\frac{\partial \mathcal{H}}{\partial x_1}.$$

The literature, however, is somewhat inconsistent on the multiplicative constant that appears in the definition of the Hamiltonian. In the classical literature, the Hamiltonian is referred to as the Kirchhoff–Routh path function, and the vortex trajectories are given by $\mathcal{H} = const$. The problem was first solved by Masotti (1931). Flucher and Gustafsson (1997) refer to it as the **Robin function**.

4. It is a classical result that G_I for a closed, simply or multiply connected domain is unique. This is easily proven by noting that the difference between any two solutions $G_I^{(1)}$, $G_I^{(2)}$ must satisfy the homogeneous Dirichlet problem. Let $w = G_I^{(1)} - G_I^{(2)}$; then

$$\nabla^2 w = 0 \in \mathcal{D}$$

$$w = 0 \in \partial\mathcal{D},$$

and by standard methods such as those found in Courant and Hilbert (1953), one can conclude that $w = const \in \mathcal{D}$. However, since its value is zero on the boundary, we can conclude that $w = 0 \in D$, hence

$$G_I^{(1)} \equiv G_I^{(2)} \in \mathcal{D}.$$

5. The classical Neumann problem

$$\nabla^2 u = 0 \in \mathcal{D}$$

$$\frac{\partial u}{\partial n} = f \in \partial\mathcal{D}$$

is solved via the formula

$$u = \int_{\partial D} fG_{II}dS. \tag{3.1.5}$$

G_{II} is called the **Green's function of the second kind** (or Neumann function) and is a solution of

$$\nabla^2 G_{II}(\mathbf{x}; \mathbf{x}_\alpha) + \delta\|\mathbf{x} - \mathbf{x}_\alpha\| = 0 \in D$$
$$\frac{\partial G_{II}}{\partial n} = const \in \partial D.$$

The Neumann function is used for constructing the velocity potential instead of the streamfunction — for our purposes we will be more concerned with the construction of G_I, since the streamfunction gives the Hamiltonian. In general, G_{II} is unique only up to an additive constant.

6. The uniqueness result for G_I still holds in the case where the boundary values, specified by f, are not continuous. However, if the region is unbounded, like the upper half-plane, we no longer have uniqueness unless an additional condition at infinity (such as requiring that the solution be bounded) is imposed. This can easily be seen by considering the functions

$$u_1 = y$$
$$u_2 = xy.$$

Both are harmonic and have the same value at the boundary $y = 0$, but are not equal.   ◊

3.2   Method of Images

A standard method for constructing a Green's function of the first kind for a domain that has special symmetries is the method of images. For a given vortex located inside the domain, one places "image" vortices outside the domain at strategic locations and with various strengths so that the boundary condition (3.0.1) is satisfied. The method is classical and we present it through a series of examples.

Example 3.1   (Vortex in upper half-plane) Consider a point vortex of strength Γ located at $\mathbf{x}_1 = (x_1, y_1)$ in the upper half-plane D, where

$$D = \{\mathbf{x} = (x, y)|y > 0\}$$
$$\partial D = \{\mathbf{x} = (x, y)|y = 0\}.$$

G_I decomposes into the sum of two terms:

$$G_I = G + G_H^{(1)}$$
$$= -\frac{1}{2\pi} \log \|\mathbf{x} - \mathbf{x}_1\| + \frac{1}{2\pi} \log \|\mathbf{x} - \mathbf{x}_1^*\|,$$

where $\mathbf{x}_1 = (x_1, y_1)$, $\mathbf{x}_1^* = (x_1, -y_1)$. The first term is the Green's function in the unbounded plane, while the second term is constructed so that it is harmonic in $\mathcal{D}$ and cancels the value of G on $\partial\mathcal{D}$. From this one constructs the streamfunction corresponding to a point vortex of strength Γ located at $\mathbf{x}_1 = (x_1, y_1) \in \mathcal{D}$:

$$\psi = -\frac{\Gamma}{2\pi} \log \|\mathbf{x} - \mathbf{x}_1\| + \frac{\Gamma}{2\pi} \log \|\mathbf{x} - \mathbf{x}_1^*\|$$
$$= \psi_1 + \psi_2.$$

Thus, ψ_1 corresponds to a streamfunction associated with a vortex of strength Γ located at $\mathbf{x} = \mathbf{x}_1$, while ψ_2 represents a streamfunction for an opposite strength vortex reflected across the boundary, hence located at the image point $\mathbf{x} = \mathbf{x}_1^*$. In this way, it is clear that on the x-axis, $\mathbf{x} = (x, 0)$, and $\psi = 0$, hence (3.0.1) is satisfied. To compute the velocity at the point $\mathbf{x} = \mathbf{x}_1$ in order to determine the vortex motion, one must subtract off the singularity at $\mathbf{x} = \mathbf{x}_1$, hence

$$\left(\psi + \frac{\Gamma}{2\pi} \log \|\mathbf{x} - \mathbf{x}_1\| \right)_{x=x_1} = \psi_2|_{x=x_1} = \frac{\Gamma}{2\pi} \log \|\mathbf{x}_1 - \mathbf{x}_1^*\| = G_H^{(1)}(\mathbf{x}_1).$$

Then the Hamiltonian is given by

$$\mathcal{H} = \frac{\Gamma}{2} G_H^{(1)}(\mathbf{x}_1)$$
$$= \frac{\Gamma}{4\pi} \log \|\mathbf{x}_1 - \mathbf{x}_1^*\|$$
$$= \frac{\Gamma}{4\pi} \log |2y_1|,$$

from which the velocity can be computed as

$$(u, v) = \left(\frac{\partial \mathcal{H}}{\partial y_1}, -\frac{\partial \mathcal{H}}{\partial x_1} \right)$$
$$= \left(\frac{\Gamma}{4\pi y_1}, 0 \right),$$

showing that the vortex moves in a straight line parallel to the wall, with velocity inversely proportional to the distance from the wall. ◊

Example 3.2 (Evenly spaced array of point vortices) Consider a row of point vortices, each having strength Γ, evenly spaced along the x-axis, separated by distance a. The streamfunction is given by

$$\psi = -\frac{\Gamma}{4\pi} \log[\cosh k(y - y_0) - \cos k(x - x_0)]$$

where $k = 2\pi/a$ and (x_0, y_0) denotes the position of one of the vortices. By symmetry the vortices remain in a fixed equilibrium position, while the fluid velocity is given by

$$u = -\frac{\Gamma}{2a} \cdot \frac{\sinh ky}{\cosh ky - \cos kx} \qquad (3.2.1)$$

$$v = \frac{\Gamma}{2a} \cdot \frac{\sin kx}{\cosh ky - \cos kx}. \qquad (3.2.2)$$

One arrives at this formula by linearly superposing the contribution of each vortex, then summing up the resulting infinite series. For details of the derivation of these formulas, see Lamb (1932) or Pozrikidis (1997), as well as the recent article by Aref (1995) on the stability properties of a vortex row, along with some historical context.   ◊

Example 3.3   (Vortex between parallel planes) It is interesting to compare Example 3.1 with the problem of a point vortex of strength Γ moving between two parallel plane walls, parallel to the x-axis, separated by distance h. The image system is made up of two periodic arrays of point vortices distributed along the y direction with wavelength $2h$. The first array, made up of vortices each of strength Γ, contains the original point vortex. The second array, made up of vortices each of strength $-\Gamma$, contains the image of the original vortex with respect to either wall. Using the result from Example 3.2, the streamfunction ψ_p due to a periodic array of point vortices (periodic along the x-axis) separated by distance $2h$ is given by

$$\psi_p = -\frac{\Gamma}{4\pi} \log\left[\cosh k(y - y_0) - \cos k(x - x_0)\right],$$

where $k = \pi/h$ and $z_0 = x_0 + iy_0$ is the location of one of the vortices in the array. Linearly superposing two rows, one constructs the streamfunction for the double array as

$$\psi(z) = -\frac{\Gamma}{4\pi} \log\left[\frac{\cosh k(x - x_0) - \cos k(y - y_0)}{\cosh k(x - x_0) - \cos k(y - y_0^{im})}\right]$$

where y_0^{im} is the y-coordinate of any image point vortex. From this, one can compute the velocity of the vortex as

$$(u, v) = \left(\frac{\partial \psi}{\partial y}, -\frac{\partial \psi}{\partial x}\right)$$

$$= \left(\frac{\Gamma}{4h} \cdot \frac{\sin 2kb}{1 - \cos 2kb}, 0\right),$$

where b is the distance from the point vortex to the lower wall. Hence, the vortex moves in a straight line parallel to the wall. In the limit $kb \to 0$ one recovers the velocity formula associated with a vortex above a wall in the semi-infinite plane, as derived in the Example 3.1.   ◊

Example 3.4 (Vortex in upper right quadrant) Consider a vortex of strength Γ located at position $\mathbf{x}_1$ in the upper right quadrant, the coordinate axes being solid boundaries. If one places image vortices of strength $-\Gamma$ at $(x_1, -y_1)$, $-\Gamma$ at $(-x_1, y_1)$, Γ at $(-x_1, -y_1)$, then the boundary condition (3.0.1) is satisfied. G_I is given by

$$
\begin{aligned}
G_I &\equiv G + G_H^{(1)} \\
&= -\frac{1}{2\pi} \log \|\mathbf{x} - \mathbf{x}_1\| \\
&\quad + \left[\frac{1}{2\pi} \log \|\mathbf{x} - \mathbf{x}_1^*\| + \frac{1}{2\pi} \log \|\mathbf{x} + \mathbf{x}_1^*\| - \frac{1}{2\pi} \log \|\mathbf{x} + \mathbf{x}_1\| \right],
\end{aligned}
$$

while the streamfunction is

$$
\begin{aligned}
\psi &= -\frac{\Gamma}{2\pi} \log \|\mathbf{x} - \mathbf{x}_1\| + \frac{\Gamma}{2\pi} \log \|\mathbf{x} - \mathbf{x}_1^*\| \\
&\quad + \frac{\Gamma}{2\pi} \log \|\mathbf{x} + \mathbf{x}_1^*\| - \frac{\Gamma}{2\pi} \log \|\mathbf{x} + \mathbf{x}_1\| \\
&= \psi_1 + \psi_2 + \psi_3 + \psi_4.
\end{aligned}
$$

On the y-axis, $\mathbf{x} = (x, 0)$, $\psi = 0$, since the first two terms cancel and the third and fourth cancel, while on the x-axis $\mathbf{x} = (0, y)$, $\psi = 0$, since the first and third cancel and the second and fourth cancel. The computation of the vortex trajectory is left as an exercise.   ◊

Example 3.5 (Vortex in upper right quadrant with background corner flow) We now subject the point vortex of the previous example to a background stagnation point flow of the form $\psi = -Kxy$. As shown in Suh (1993), the vortex will undergo periodic motion around an elliptic fixed point. The streamfunction for the combined vortex-stagnation point flow, after simplification, is given by

$$
\psi(x, y) = -Kxy + \frac{\Gamma}{4\pi} \log xy - \frac{\Gamma}{8\pi} \log(x^2 + y^2).
$$

Hence, the vortex motion is governed by

$$
\begin{aligned}
\dot{x} &= \frac{\partial \psi}{\partial y} = -Kx + \frac{\Gamma}{4\pi} \cdot \frac{1}{y} - \frac{\Gamma}{4\pi} \cdot \frac{y}{x^2 + y^2} \\
\dot{y} &= -\frac{\partial \psi}{\partial x} = Ky - \frac{\Gamma}{4\pi} \cdot \frac{1}{x} + \frac{\Gamma}{4\pi} \cdot \frac{x}{x^2 + y^2}.
\end{aligned}
$$

We can show that the point

$$(x, y) = \left(\sqrt{\frac{\Gamma}{8\pi K}}, \sqrt{\frac{\Gamma}{8\pi K}} \right)$$

is an elliptic fixed point for the point vortex and that in the neighborhood of this fixed point the vortex oscillates with period π/K. In more recent work of Noack, Banaszuk, and Mezić (2000) this configuration was studied by subjecting the flow to a time-dependent forcing, with the control objective of maximizing flux across the separatrix, hence in some quantitative sense maximizing mixing in the flow. See also the paper of Conlisk and Rockwell (1981) for discussions of related models.   ◊

Example 3.6   (Vortex inside or outside circular domain) Consider a vortex of strength Γ located at position $\mathbf{x}_1$ inside a circular cylinder of radius a centered at $\mathbf{x}_c$. If one places an image vortex of strength $-\Gamma$ located at the inverse point $\mathbf{x}_1^{im}$, where

$$\mathbf{x}_1^{im} = \mathbf{x}_c + \frac{(\mathbf{x}_1 - \mathbf{x}_c)a^2}{\|\mathbf{x}_1 - \mathbf{x}_c\|^2},$$

then it is easy to show that there is no radial velocity component on the circle boundary, hence the boundary condition (3.0.1) is satisfied. The azimuthal velocity of the vortex at $\mathbf{x}_1$ is given by

$$u_\theta = \frac{\Gamma}{2\pi} \cdot \frac{\|\mathbf{x}_1 - \mathbf{x}_c\|}{(a^2 - \|\mathbf{x}_1 - \mathbf{x}_c\|^2)}.$$

One could equivalently view the vortex at $\mathbf{x}_1$ as the image of the other and hence conclude that the above azimuthal velocity formula also corresponds to a point vortex of strength Γ located at $\mathbf{x}_1$ outside the circular cylinder. The Green's function associated with this configuration is given by

$$G_I(\mathbf{x}; \mathbf{x}_1) = -\frac{1}{2\pi} \left[\log \|\mathbf{x} - \mathbf{x}_1\| - \log \left(\frac{a}{\|\mathbf{x}_1 - \mathbf{x}_c\|} \cdot \frac{1}{\|\mathbf{x} - \mathbf{x}_1^{im}\|} \right) \right]$$

$$\equiv G(\mathbf{x}; \mathbf{x}_1) + G_H^{(1)}(\mathbf{x}). \qquad (3.2.3)$$

◊

We now outline a useful method (called the circle theorem) for constructing complex potentials for general flows in the presence of a circular boundary (Milne-Thomson (1968)).

Theorem 3.2.1 (Circle Theorem). *Consider a flowfield with complex potential $f(z)$, which has no singularities inside the circle $|z| = a$. Then the complex potential for that flow in the presence of the circular boundary $|z| = a$ is*

$$w(z) = f(z) + f^*(a^2/z)$$

Proof. The theorem is easy to prove if one notices that on the circle $|z|^2 = a^2$ we have $z^* = a^2/z$. Then, by the above, we have on the circle

$$w(z) = f(z) + f^*(z^*),$$

which is a real quantity[1]. Since $w(z) = \phi + i\psi$, we have $\psi = 0$ on the circle, which is therefore a solid boundary.   □

This theorem allows us to construct the flow for a point vortex outside or inside a circular domain, as in Example 3.6. The complex potential becomes, upon use of the circle theorem,

$$w(z) = \frac{\Gamma}{2\pi i} \log \left(\frac{z - z_0}{a^2/z - z_0^*} \right), \tag{3.2.4}$$

where z_0 is the point vortex location.

Remark. The complex potential (3.2.4) can be rewritten as

$$w(z) = \frac{\Gamma}{2\pi i} \log(z - z_0) - \frac{\Gamma}{2\pi i} \log \left(\frac{a^2}{z} - z_0^* \right).$$

If we add to this expression the constant term $\frac{\Gamma}{2\pi i} \log(-1/z_0^*)$, it becomes, after some algebra,

$$\tilde{w}(z) \equiv w(z) + \frac{\Gamma}{2\pi i} \log \left(\frac{-1}{z_0^*} \right)$$

$$= \frac{\Gamma}{2\pi i} \log(z - z_0) - \frac{\Gamma}{2\pi i} \log \left(z - \frac{a^2}{z_0^*} \right) + \frac{\Gamma}{2\pi i} \log z - \frac{\Gamma}{2\pi i} \log z_0^{*2},$$

the last term being an irrelevant constant. In this way, one can view the complex potential $\tilde{w}(z)$ as that due to:

1. A vortex of strength Γ located at $z = z_0$.

2. A vortex of strength $-\Gamma$ located at $z = a^2/z_0^*$.

3. A vortex of strength Γ at the origin.

---

[1]The notation $f^*(z)$ is equivalent to $f^*(z) \equiv (f(z^*))^* \equiv \phi(x, -y) - i\psi(x, -y)$. For example, if $f(z) = z$, then $f^*(z) = (z^*)^* = z = f(z)$, while for $f(z) = i \log(z - z_0)$, we have $f^*(z) = -i \log(z - z_0^*)$.

Dropping the constant term, we can rewrite the complex potential in an alternative form sometimes used in the literature,

$$\tilde{w}(z) = \frac{\Gamma}{2\pi i} \log\left(\frac{z - z_0}{z - a^2/z_0^*}\right) - \frac{\Gamma}{2\pi i} \log z.$$

See Milne-Thomson (1968) or Pozrikidis (1997) for more discussions, as well as the exercise at the end of the chapter. The recent papers of Kadtke and Novikov (1993) and Luithardt, Kadtke, and Pedrizzetti (1994) analyze the interactions of a circular cylinder with one and two vortices. In the case of a single vortex, Kadtke and Novikov (1993) show how chaotic motion is induced by oscillating the cylinder periodically, while for the case of two vortices outside a cylinder, chaotic motion occurs with no external perturbations.   ◇

3.3   Conformal Mapping Techniques

A much more general and powerful technique for constructing the Hamiltonian associated with point vortex flows in domains with boundaries, where the method of images might be complicated or inappropriate, is based on conformal mapping ideas. We review the main ideas here, as applied to the construction of the equations for vortex motion in general domains.

Let us start by considering a complex function $f(z)$ that is analytic in all or part of the complex z-plane. We view the function as a mapping from the $(z = x + iy)$-plane to the $(\zeta = \xi + i\eta)$-plane where

$$\zeta = f(z).$$

The inverse mapping f^{-1} is denoted

$$z = f^{-1}(\zeta) \equiv F(\zeta)$$

and we have assumed that branch cuts have been appropriately introduced in the z-plane and ζ-plane in the case of multivalued functions f or F, in order to render the mapping and its inverse unique.

How does a given complex potential $w(z)$ transform under the above mapping? By direct substitution, write

$$w(z) = w[F(\zeta)] \equiv W(\zeta) = \Phi(\xi, \eta) + i\Psi(\xi, \eta)$$

where Φ and Ψ are the real and imaginary parts of the transformed complex potential W. Since the composition of an analytic function with another analytic function is analytic (see, for example, Ahlfors (1979)), the functions Φ and Ψ both satisfy Laplace's equation

$$\frac{\partial^2 \Phi}{\partial \xi^2} + \frac{\partial^2 \Phi}{\partial \eta^2} = 0$$

$$\frac{\partial^2 \Psi}{\partial \xi^2} + \frac{\partial^2 \Psi}{\partial \eta^2} = 0$$

and are related to each other via the Cauchy–Riemann equations

$$\frac{\partial \Phi}{\partial \xi} = \frac{\partial \Psi}{\partial \eta}$$

$$\frac{\partial \Phi}{\partial \eta} = -\frac{\partial \Psi}{\partial \xi}.$$

Next, we ask how the flow due to a point vortex of strength Γ located at $z = z_0$ in the z-plane transforms under conformal mapping. We will denote the image of z_0 by ζ_0, i.e.,

$$f(z_0) = \zeta_0 \equiv \xi_0 + i\eta_0,$$

with the variable $\zeta = \xi + i\eta$ denoting the image of z in the transformed plane. Then

$$w(z) = -\frac{i\Gamma}{2\pi} \log(z - z_0)$$

and

$$
\begin{aligned}
W(\zeta) &= -\frac{i\Gamma}{2\pi} \log[F(\zeta) - F(\zeta_0)] \\
&\approx -\frac{i\Gamma}{2\pi} \log[F'(\zeta_0)(\zeta - \zeta_0)] \\
&= -\frac{i\Gamma}{2\pi} \log(\zeta - \zeta_0) - \frac{i\Gamma}{2\pi} \log[F'(\zeta_0)] \\
&\approx -\frac{i\Gamma}{2\pi} \log(\zeta - \zeta_0),
\end{aligned}
$$

where we have used a local Taylor series expansion about the point ζ_0 and kept the dominant term. This shows the well-known and important result that *the flow in the ζ plane corresponds to a point vortex of the same strength, located at the image point ζ_0,* assuming that z_0 is not a critical point of the conformal map, i.e., $F'(\zeta_0) \neq 0$.

Hence, to obtain the velocity at any point in the original z-plane one can transform the position of all the point vortices to the ζ-plane, calculate the fluid velocity at the transformed point, then map this velocity back to the original plane via the chain rule [2]

$$u(z) - iv(z) = \frac{dw}{dz} = \frac{dW}{d\zeta}\frac{d\zeta}{dz}.$$

If the point happens to be a vortex position, then the transformation rule becomes more complicated and leads to a useful formula known in the

---

[2] One can show that the boundary condition (3.0.1) is also satisfied in the ζ-plane under conformal transformations — for a discussion of this, see for example, Pozrikidis (1997).

literature as the **Routh rule**. To derive this formula, let $w_{z_0}(z)$ be the complex potential in the z-plane due to all causes, except for the vortex at site $z = z_0$, with $W_{\zeta_0}(\zeta)$ denoting the corresponding complex potential in the ζ-plane. To compute the velocity at the vortex site, we write

$$w_{z_0}(z) - \frac{i\Gamma}{2\pi}\log(z - z_0) = W_{\zeta_0}(\zeta) - \frac{i\Gamma}{2\pi}\log(\zeta - \zeta_0).$$

This can be rewritten as

$$w_{z_0}(z) = W_{\zeta_0}(\zeta) - \frac{i\Gamma}{2\pi}\log\left(\frac{\zeta - \zeta_0}{z - z_0}\right). \tag{3.3.1}$$

Next, use the fact that in the neighborhood of the vortex site, we have

$$\zeta = f(z) = f(z_0) + (z - z_0)f'(z_0) + \frac{1}{2}(z - z_0)^2 f''(z_0) + O(z - z_0)^3$$

giving

$$\frac{\zeta - \zeta_0}{z - z_0} = f'(z_0) + \frac{1}{2}(z - z_0)f''(z_0) + O(z - z_0)^2.$$

Using this expansion in (3.3.1) gives

$$w_{z_0}(z) = W_{\zeta_0}(\zeta) - \frac{i\Gamma}{2\pi}\log[f'(z_0) + \frac{1}{2}(z - z_0)f''(z_0) + O(z - z_0)^2].$$

From this we get the complex velocity formula

$$\frac{dw_{z_0}}{dz} = \frac{dW_{\zeta_0}}{d\zeta}\frac{d\zeta}{dz} - \frac{i\Gamma}{2\pi}\left[\frac{\frac{1}{2}f''(z_0) + O(z - z_0)}{f'(z_0) + O(z - z_0)}\right]$$

$$= \frac{dW_{\zeta_0}}{d\zeta}\frac{d\zeta}{dz} - \frac{i\Gamma}{4\pi}\frac{f''(z_0)}{f'(z_0)} + O(z - z_0).$$

Then, to evaluate the velocity at the vortex site we take the limit $z \to z_0$, $\zeta \to \zeta_0$,

$$\frac{dw_{z_0}}{dz}\Big|_{z=z_0} = \frac{dW_{\zeta_0}}{d\zeta}\Big|_{\zeta=\zeta_0} \cdot \frac{d\zeta}{dz}\Big|_{z=z_0} - \frac{i\Gamma}{4\pi} \cdot \frac{f''(z_0)}{f'(z_0)}. \tag{3.3.2}$$

Equation (3.3.2) is known as the Routh rule. From this one can write the relationship between the Hamiltonian in the transformed ζ-plane ($\hat{\mathcal{H}}_0(\xi, \eta)$) and the original Hamiltonian in the z-plane ($\mathcal{H}_0(x, y)$) (see exercises),

$$\mathcal{H}_0(x, y) = \hat{\mathcal{H}}_0(\xi, \eta) - \frac{\Gamma}{4\pi}\log\left|\frac{df}{dz}\right|_{z_0}. \tag{3.3.3}$$

We know that the Riemann mapping theorem says that any closed and simply connected region of the z-plane can be mapped conformally to the unit circle in the $(\zeta = \xi + i\eta)$-plane via an analytic function that we denote $\zeta(z)$. Then, upon using (3.3.3), the Hamiltonian $\mathcal{H}_{z_0}$ governing the motion of a vortex inside an arbitrary closed region can be obtained from that relative to the unit circle, $\hat{\mathcal{H}}_0$, namely,

$$\mathcal{H}_{z_0}(x_0, y_0) = \hat{\mathcal{H}}_0(\xi_0, \eta_0) - \frac{\Gamma}{4\pi} \log \left| \frac{d\zeta}{dz} \right|_{z_0}.$$

Using the known Hamiltonian for the unit circle which we write as

$$\hat{\mathcal{H}}_0(\xi_0, \eta_0) = -\frac{\Gamma}{4\pi} \log \left| \frac{1}{\zeta_0 \zeta_0^* - 1} \right|,$$

we have

$$\mathcal{H}_{z_0}(x_0, y_0) = -\frac{\Gamma}{4\pi} \log \left| \frac{1}{\zeta_0 \zeta_0^* - 1} \right| - \frac{\Gamma}{4\pi} \log \left| \frac{d\zeta}{dz} \right|_{z_0}$$

$$= -\frac{\Gamma}{4\pi} \log \left(\left| \frac{1}{\zeta_0 \zeta_0^* - 1} \right| \cdot \left| \frac{d\zeta}{dz} \right|_{z_0} \right). \tag{3.3.4}$$

This useful formula gives the Hamiltonian governing the motion of a single vortex in an arbitrary closed domain that is mapped to the unit circle via the mapping $\zeta(z)$. A corresponding expression for the Hamiltonian associated with the N-vortex problem is given in the exercises.

Example 3.7 (Single vortex in unit circle) For a vortex of strength Γ located at position $z_0 = x_0 + iy_0$ inside the unit circle centered at the origin, the trajectories correspond to level curves of the Hamiltonian

$$\mathcal{H}_{z_0} = -\frac{\Gamma}{4\pi} \log \left| \frac{1}{z_0 z_0^* - 1} \right|. \tag{3.3.5}$$

The vortex paths lie on concentric circles

$$z_0 = r_0 \exp(i\omega_0 t)$$

with rotation frequency given by

$$\omega_0 = \frac{\Gamma}{2\pi} \left(\frac{1}{1 - r_0^2} \right).$$

The motion of a fluid particle located at $z = x + iy$ inside the unit circle is governed by the Hamiltonian

$$\mathcal{H}_z = -\frac{\Gamma}{2\pi} \log \left| \frac{z - z_0}{z - 1/z_0^*} \right| - \frac{\Gamma}{2\pi} \log |z|. \tag{3.3.6}$$

It is useful to move in a rotating reference frame in which the vortex position is fixed. For this one can use polar coordinates to denote the particle position

$$z = r \exp(i\theta).$$

Then, employing the conjugate variables $Q = r^2/2$, $P = \theta - \omega_0$, one gets the Hamiltonian governing the particle motion

$$\mathcal{H}_0(Q, P) = -\frac{\Gamma}{4\pi} \log \left(\frac{2Q - 2r_0\sqrt{2Q} \cos P + r_0^2}{2Q - 2(1/r_0)\sqrt{2Q} \cos P + 1/r_0^2} \right) + Q\omega_0.$$

The level curves of this time-independent Hamiltonian giving the particle paths are shown in Figure 3.2.   ◊

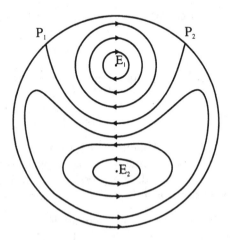

FIGURE 3.2. Particle paths in circular domain.

For simply connected domains there is a general theorem due to Flucher and Gustafsson (1997), which states that almost all orbits of a single point vortex are periodic and they move along level lines of the Robin function G_H. In particular, they prove the following.

Theorem 3.3.1 (Flucher and Gustafsson). *Consider a single point vortex of strength Γ located at position $z = x + iy$ inside a simply connected domain Ω:*

1. *The point vortex moves along level lines of the function $G_H(z)$ where*

$$\Delta G_H(z) = \frac{2}{\pi} \exp(4\pi G_H) \quad z \in \Omega.$$

2. *Near any smooth boundary point z the function $G_H(z)$ behaves like*

$$G_H(z - \epsilon\nu) = -\frac{1}{2\pi} \log(2\epsilon - h\epsilon^2 + o(\epsilon^2)) \quad as \ \epsilon \to 0,$$

where h denotes the curvature of the boundary at z with respect to the exterior normal.

3. *The vortex moves with speed*

$$\dot{z} = -\frac{\Gamma}{2}\mathbb{J}\nabla G_H(z)$$

where $\mathbb{J}$ is the symplectic matrix. Near the boundary its speed is

$$\dot{z} = -\frac{\Gamma}{4\pi\epsilon}\mathbb{J}\nu + O(1).$$

4. *Every critical point of $G_H(z)$ is a rest point for the vortex, and there is at least one inside Ω.*

They also prove the following theorem regarding the absence of vortex collisions in the interior of a planar domain. There are, a priori, two situations that one can consider: (i) vortex collisions with other vortices, and (ii) vortex collisions with boundaries. We state here one of their main theorems regarding collisions.

Theorem 3.3.2 (Flucher and Gustaffson). *The following types of collisions are excluded as long as no other collisions occur simultaneously:*

1. *Collisions of two vortices in the interior of the domain.*

2. *Collision of a single vortex with the boundary.*

3. *Collision of multiple vortices in the interior of the domain unless the expression $\sum_{i \neq j} \Gamma_i \Gamma_j \log |z_i - z_j|$ has a finite limit.*

More details can be found in their papers, as well as Chapter 15 of Flucher (1999).

3.4   Breaking Integrability

Vortex motion inside closed domains or in open domains with solid boundaries is often more complicated than the corresponding flow in the unbounded plane owing to the symmetry breaking effect of the boundary and the resulting loss of conserved quantities. For example, an isolated point vortex in the unbounded plane does not move, while fluid particles move in concentric circles centered at the vortex. Typically, a single vortex in a domain with boundaries undergoes some sort of motion, causing the particles to move and the fluid to mix. We will sketch this process in the context of a specific example, following the work of Zannetti and Franzese (1993, 1994) who have analyzed the symmetry-breaking effects of boundaries on vortex motion and fluid-mixing in perturbed circular domains.

   We can ask the question of what happens to the motion of a passive fluid particle in a circular domain under simple perturbations of the vortex trajectory. Zannetti and Franzese (1993) perturb the vortex trajectory in the following way

$$z_0 = (r_0 + \epsilon r_1(w_0 t)) \cdot \exp(iw_0 t) \equiv Z_0 + \epsilon Z_1$$

where r_1 is 2π-periodic. The perturbed particle Hamiltonian (3.3.6) can then be written

$$\begin{aligned}
\mathcal{H}_\epsilon &= -\frac{\Gamma}{2\pi} \log \left| \frac{z - z_0}{z - 1/z_0^*} \right| - \frac{\Gamma}{2\pi} \log |z| \\
&= -\frac{\Gamma}{2\pi} \log \left| \frac{z - Z_0}{z - 1/Z_0^*} \right| - \frac{\Gamma}{2\pi} \log |z| \\
&\quad - \frac{\Gamma}{2\pi} \left(\log \left| \frac{z - z_0}{z - 1/z_0^*} \right| - \log \left| \frac{z - Z_0}{z - 1/Z_0^*} \right| \right) \\
&= -\frac{\Gamma}{2\pi} \log \left| \frac{z - Z_0}{z - 1/Z_0^*} \right| - \frac{\Gamma}{2\pi} \log |z| + \mathcal{H}_1(q, p, t; \epsilon),
\end{aligned}$$

where $Z_0 \equiv r \exp(iw_0 t)$. From this one can derive the perturbed particle equations, where the perturbation terms are contained in $\mathcal{H}_1$. Zannetti and Franzese (1993) make the specific choice

$$r_1(w_0 t) = -\cos(2w_0 t)$$

and, at leading order, they derive the following perturbed system

$$\frac{dQ}{dt} = f_1 + \epsilon g_1 \tag{3.4.1}$$

$$\frac{dP}{dt} = f_2 + \epsilon g_2 \tag{3.4.2}$$

where

$$f_1 = -\frac{\Gamma}{2\pi} \sqrt{Q} \sin P \left(\frac{r_0}{D_1} - \frac{1}{r_0 D_2} \right)$$

$$g_1 = \frac{\Gamma}{2\pi} \cos(2w_0 t) \sqrt{Q} \sin P \left(\frac{Q - r_0^2}{D_1^2} + \frac{Q - 1/r_0^2}{r_0^2 D_2^2} \right)$$

$$f_2 = -\frac{\Gamma}{2\pi} \left(\frac{1 - \cos(P/r_0\sqrt{Q})}{D_2} - \frac{1 - r_0 \cos(P/\sqrt{Q})}{D_1} \right) - w_0$$

$$\begin{aligned}
g_2 &= \frac{\Gamma}{2\pi} \cos(2w_0 t) \left[\frac{\sqrt{Q} \cos P + r_0^2(\cos P/\sqrt{Q}) - 2r_0}{D_1^2} \right. \\
&\quad \left. + \frac{\sqrt{Q} \cos P + \cos(P/r_0^2\sqrt{Q}) - 2/r_0}{r_0^2 D_2^2} \right]
\end{aligned}$$

$$D_1 = Q + r_0^2 - 2r_0 \sqrt{Q} \cos P$$

$$D_2 = Q + \frac{1}{r_0^2} - \frac{2}{r_0} \sqrt{Q} \cos P.$$

The details of this are tedious and will be left as an exercise. The unperturbed particle orbits are shown in Figure 3.2. Since we are moving in a rotating reference frame at the vortex frequency, the vortex position is fixed, while the particle orbits are either periodic (such as those surrounding either E_1 or E_2), or have heteroclinic behavior such as the orbit connecting P_1 to P_2. We are now in a position to make use of the Melnikov formulas described in Chapter 1 in order to prove that the perturbation terms break the heteroclinic structure. For this we need to form the Melnikov integral

$$M(t_0) = \int_{-\infty}^{\infty} [f_1(\mathbf{q}^0(t))g_2(\mathbf{q}^0(t), t + t_0) - f_2(\mathbf{q}^0(t))g_1(\mathbf{q}^0(t), t + t_0)]dt,$$

where $\mathbf{q}^0(t) \equiv (Q, P)$ are the unperturbed heteroclinic coordinates. The Melnikov function $M(t_0)$ is computed numerically and shown in Zannetti and Franzese (1993) to have discrete zeroes periodically spaced. Therefore, the particle orbits are chaotic due to the perturbed vortex motion.

Remarks.

1. In their paper Zannetti and Franzese (1993) go one step further and use the conformal mapping transformation corresponding to a slightly perturbed circular boundary to derive the symmetry-breaking terms on the particle motion — this is in contrast to the approach outlined above in which the vortex motion is perturbed, keeping the boundary circular. Despite this difference, they show that for small enough perturbations, the two approaches are qualitatively similar and lead to the same mechanism of heteroclinic-breaking leading to chaotic particle motion. Perturbing the circle directly is slightly more complicated and makes the analysis associated with the Melnikov method more involved — see Zannetti and Franzese (1994) for more details adopting the second approach.

2. For other configurations involving boundaries that break symmetry and induce chaotic motion, see Luithardt, Kadtke, and Pedrizzetti (1994) for 2-vortex motion outside a circular cylinder, and Conlisk, Guezennec, and Elliott (1989) for 3-vortex motion above a flat wall. ◊

3.5 Bibliographic Notes

Interesting discussions on the boundary conditions imposed for inviscid and viscous flows can be found in Richardson (1973) or Pozrikidis (1997).

A general introduction to conformal mapping methods can be found in Ahlfors (1979), Dettman (1965), Ives (1976), Kober (1952), Nehari (1952), and Warschawski (1945). See also Jeans (1933), Kellogg (1954), or Marsden and Hoffman (1987). See Morse and Feshbach (1953), Courant and Hilbert (1953), or Kellogg (1954) for classical discussions of Green's functions, or for more general discussions, see Stakgold (1979) or Roach (1982). The best presentation of generalized functions can be found in Gelfand (1964). See also the beautifully concise book of Lighthill (1958). The motion of a single point vortex in a simply connected, bounded domain was treated in Routh (1881), Lagally (1921), Masotti (1931), and Paul (1934). A more general treatment for the case of multiple vortices in multiply connected domains is treated in Lin (1941a,b). A recent account is given in Saffman (1992) and Chapter 15 of Flucher (1999). The method of images is described in detail in Jeans (1933), Morse and Feshbach (1953), Jackson (1963), or Milne-Thompson (1968).

There are many examples of point vortex motion involving special domains sprinkled throughout the literature, some of which are described in Chapter 7 of Saffman (1992) and in Milne-Thompson (1968). Flows that have been analyzed include vortices approaching solid walls (Harvey and Perry (1971), Saffman (1979)), vortices in corners (Conlisk and Rockwell (1981)), channels (Singh (1954), Elcrat, Hu, and Miller (1997)), vortex pairs and rings approaching orifices (Karweit (1975), Miloh and Shlein (1977), Sheffield (1977)), vortices above flat walls (Conlisk, Guezennec, and Elliot (1989)), vortices in a semidisk (Okamoto and Kimura (1989)) and point vortex pairs behind a cylinder (called Föpple vortices) — see Milne-Thomson (1968) for a derivation of this flowfield. This last configuration is one of many examples of equilibrium vortex flows with boundaries. Other examples include a fixed point vortex over an airfoil analyzed originally in Saffman and Sheffield (1977) and used for control theoretic purposes in Cortelezzi and Leonard (1993), Cortelezzi, Leonard, and Doyle (1994) and Cortelezzi (1996); point vortices in channel geometries with obstacles analyzed in Elcrat, Hu, and Miller (1997), and in corner geometries as described in Elcrat and Miller (1995) and Miller (1996). We also mention work on distributed vorticity fields in bounded domains, such as that described in van Geffen, Meleshko, and van Heijst (1996). A fundamental work characterizing steady flows is that of Turkington (1983a,b). A general theory of vortex flows interacting with solid boundaries is currently being developed by Shashikanth, Burdick, Marsden, and Murray (2000).

3.6 Exercises

1. (a) Prove that the solution of the Dirichlet problem is solved via the generalized Poisson formula (3.1.3).

(b) Prove that the solution of the Neumann problem is solved via the formula (3.1.5).

2. Prove that $G_I(\mathbf{x}, \mathbf{x}_\alpha)$ satisfies the reciprocity property

$$G_I(\mathbf{x}, \mathbf{x}_\alpha) = G_I(\mathbf{x}_\alpha, \mathbf{x}).$$

3. Consider the Example 3.2 of an evenly spaced array of point vortices, each of strength Γ, separated by distance a. Prove that the fluid velocity is given by the formulas (3.2.1), and (3.2.2).

4. Consider the Example 3.4 of a vortex in the upper right quadrant with axes as solid boundaries:

   (a) Find the Hamiltonian for the vortex motion.

   (b) Compute the vortex trajectory.

   (c) Use the previous construction to find the trajectory of a pair of equal and opposite vortices (dipole) approaching a flat wall head-on.

5. Consider the Example 3.5 of a vortex in the upper right quadrant with stagnation point flow.

   (a) Derive the equations of motion for the vortex.

   (b) Derive the equations of motion for a passive particle.

   (c) Find the dimensional fixed point for the vortex and prove that it is elliptic, finding the (linearized) frequency in the neighborhood of the fixed point.

   (d) Find the curves in the (x, y)-plane for which *either* $\dot{x} = 0$ *or* $\dot{y} = 0$.

6. Derive the complex potential for a point vortex located at z_0 outside a circular boundary of radius a, which is centered at an arbitrary point z_c. In particular:

   (a) Generalize the circle theorem so that the circular boundary can be centered at an arbitrary point z_c.

   (b) Use part (a) to write the complex potential for a point vortex located at z_0 outside this circle.

7. Derive the Hamiltonian governing the motion of a particle in a unit circular domain, with one point vortex inside, as outlined in Example 3.7.

8. Prove that the Hamiltonian governing N vortices inside a circular domain of radius R is

$$
\mathcal{H} = \frac{1}{4\pi} \sum_{i=1}^{N} \Gamma_i^2 \log\left(1 - \mathbf{x}_i^2/R^2\right)
$$

$$
+ \frac{1}{8\pi} \sum_{i=1}^{N} \sum_{j \neq i}^{N} \Gamma_i \Gamma_j \log\left(1 + \frac{(1 - \mathbf{x}_i^2/R^2)(1 - \mathbf{x}_j^2/R^2)}{|\mathbf{x}_i/R - \mathbf{x}_j/R|^2}\right).
$$

9. From the Routh rule (3.3.2) derive the Hamiltonian relation (3.3.3).

10. (a) Prove that if $\zeta = f(z)$ is the analytic mapping from an arbitrary, simply connected domain in the z-plane to the unit circle in the ζ-plane, then the Hamiltonian governing the motion of N-point vortices of strength Γ_j can be written

$$
\mathcal{H} = -\frac{1}{4\pi} \sum_{j=1}^{N} \Gamma_j^2 \log \frac{|f'(z_j)|}{|1 - f_i f_j^*|} - \frac{1}{4\pi} \sum_{j=1}^{N} \sum_{k \neq j}^{N} \Gamma_j \Gamma_k \log \frac{|f_j - f_k|}{|1 - f_j f_k^*|}
$$

where $f_j \equiv f(z_j)$. The governing equations of motion are then

$$
\Gamma_j \dot{z}_j^* = 2i \frac{\partial \mathcal{H}}{\partial z_j}.
$$

(b) Prove that the trajectory of a single vortex is given by

$$
\frac{\Gamma^2}{4\pi} \log\left[\frac{(1 - ff^*)}{|f'(z)|}\right] = K_1,
$$

or equivalently

$$
\frac{1 - |f(z)|^2}{|f'(z)|} = K_2,
$$

where K_1 and K_2 are constants.

11. Derive the perturbed system (3.4.1), and (3.4.2) for a particle and perturbed vortex in a circular domain.

12. Derive the equations of motion for a system of three vortices above a flat wall, as analyzed in Conlisk, Guezennec, and Elliot (1989).

13. Show that the equations of motion for a leapfrogging pair, as treated in Example 2.4, are equivalent to the equations for two equal strength vortices above a flat wall.

14. Derive the equations of motion for two point vortices outside a circular cylinder, as treated in Luithardt, Kadtke, and Pedrizzetti (1994).

4
Vortex Motion on a Sphere

The problem of N-point vortices moving on the surface of a sphere is not as well understood as the corresponding planar problem, yet in some sense it is more general, since the planar problem can be obtained from the spherical problem in the limit as the radius of the sphere becomes large. This limit will be discussed in more detail later in the chapter. The spherical model is used in geophysical fluid dynamics when considering large-scale atmospheric and oceanographic flows with coherent structures that persist over long periods of time and move over such large distances that the spherical geometry of the earth's surface becomes important. In addition, when considering the streamline patterns generated by a given vorticity field, the spherical geometry is important to take into account if one wants to make statements that are relevant to atmospheric weather patterns. These issues are discussed later in the chapter. For a general introduction to geophysical fluid dynamics, see Pedlosky (1987) or Gill (1982).

In this chapter we formulate the equations of motion for N vortices moving on a spherical shell of radius R, as obtained in Bogomolov (1977, 1979); the initial reference to this model appears in Gromeka (1952). In Section 4.2 we examine the dynamics of the general 3-vortex problem in some detail, following Kidambi and Newton (1998, 1999, 2000a). All fixed and relative equilibria are described and categorized following Kidambi and Newton (1998), along with their stability properties as analyzed in Pekarsky and Marsden (1998). In Section 4.3 we introduce coordinates so that the spherical problem can be reduced to a phase plane analysis; we then locate all equilibria for the 3-vortex problem in the phase plane. In Section 4.4 we describe necessary and sufficient conditions for the collision of three vortices and study this process in detail, focusing on the differences between planar and spherical collisions. The equations of motion for vortices and particles in the stereographic plane are described in Section 4.5. This system is useful for understanding streamline patterns generated by N vortices, as well as treating problems with solid boundaries on the spherical shell. We then identify in Section 4.6 all possible integrable streamline topologies for the 3-vortex problem and give some general results for the N-vortex case. The streamline patterns arise as level sets of the particle Hamiltonian for a given vorticity distribution. Their connection with atmospheric weather patterns is discussed. The problem of treating solid boundaries on the spherical surface is addressed in Section 4.7. This model is relevant if one wants to apply the results to oceanographic situations where coastlines and shores play a role in the vorticity dynamics. We finish the chapter with a brief description of related work, including the use of symmetries and other literature involving the distribution of point charges on a conducting spherical shell.

4.1   General Formulation

The fact that the sphere is a compact surface places a topological constraint on the vorticity field. This can be expressed by Stokes's theorem, which implies that

$$\int_{\partial S} \omega \cdot dS = 0,$$

where ∂S denotes the spherical surface. To illustrate this, consider a simple closed loop on the surface of a sphere, which divides the surface into two components, each of finite area. When traversing the loop in a counterclockwise manner, let S_1 denote the area enclosed to the left of the boundary; let S_2 be the area enclosed to the right. Stokes's theorem says

$$\oint \mathbf{u} \cdot dl = \int_{\partial S} \omega \cdot dS_1 = -\int_{\partial S} \omega \cdot dS_2,$$

where the loop is traversed in the counterclockwise direction. This then yields

$$0 = \int_{\partial S} \omega \cdot dS_1 + \int_{\partial S} \omega \cdot dS_2 = \int_{\partial S} \omega \cdot dS.$$

Another manifestation of the compactness of the sphere will be described in Section 4.6, where we detail the consequences of the Poincaré index theorem.

We start by writing the planar and spherical equations in a unified way,

$$\dot{\mathbf{x}}_i = \sum_{j \neq i}^{N} \left(\frac{\Gamma_j}{2\pi} \right) \left(\frac{\hat{\mathbf{n}}_j \times (\mathbf{x}_i - \mathbf{x}_j)}{l_{ij}^2} \right),$$

where $i = 1, \ldots, N$. Here $\hat{\mathbf{n}}_j$ is the outward unit normal vector associated with the surface at the vortex location $\mathbf{x}_j$. For the sphere, we use

$$\hat{\mathbf{n}}_j = \mathbf{x}_j/R,$$

whereas for planar surfaces,

$$\hat{\mathbf{n}}_j = \hat{e}_z,$$

where $\hat{e}_z$ is the standard Cartesian unit vector perpendicular to both $\hat{e}_x$ and $\hat{e}_y$. The vector $\mathbf{x}_j$ points from the center of the sphere to the vortex location $\mathbf{x}_j = (x_j, y_j, z_j)$ on the spherical surface $\|x_j\| = R$. An equivalent way of writing the equations is

$$\dot{\mathbf{x}}_i = \frac{1}{4\pi R} \sum_{j \neq i}^{N} \Gamma_j \frac{(\mathbf{x}_j \times \mathbf{x}_i)}{(R^2 - \mathbf{x}_i \cdot \mathbf{x}_j)}, \tag{4.1.1}$$

where the denominator is the chord distance between vortex Γ_i and Γ_j,

$$l_{ij}^2 \equiv 2(R^2 - \mathbf{x}_i \cdot \mathbf{x}_j) = \|\mathbf{x}_i - \mathbf{x}_j\|^2.$$

In most cases the Cartesian formulation is simpler to use. However, there are some advantages to writing the equations in spherical coordinates (θ_i, ϕ_i), as was shown in (1.1.23), and (1.1.24)

$$\dot{\theta}_i = -\frac{1}{4\pi R^2} \sum_{j \neq i}^{N} \frac{\Gamma_j \sin\theta_i \sin(\phi_i - \phi_j)}{(1 - \cos\gamma_{ij})} \tag{4.1.2}$$

$$\sin(\theta_i)\dot{\phi}_i = \frac{1}{4\pi R^2} \sum_{j \neq i}^{N} \frac{\Gamma_j[\sin\theta_i \cos\theta_j - \cos\theta_i \sin\theta_j \cos(\phi_i - \phi_j)]}{(1 - \cos\gamma_{ij})}$$

$$\tag{4.1.3}$$

where

$$\cos\gamma_{ij} = \cos\theta_i \cos\theta_j + \sin\theta_i \sin\theta_j \cos(\phi_i - \phi_j).$$

Figure 4.1 shows the basic geometry associated with the 3-vortex configuration on a sphere.

The Hamiltonian for the system can be written

$$\mathcal{H} = -\frac{1}{4\pi R^2} \sum_{i<j} \Gamma_i \Gamma_j \log(l_{ij}^2).$$

With the coordinates

$$Q_i = \sqrt{|\Gamma_i|}\phi_i \qquad P_i = \sqrt{|\Gamma_i|}\cos\theta_i,$$

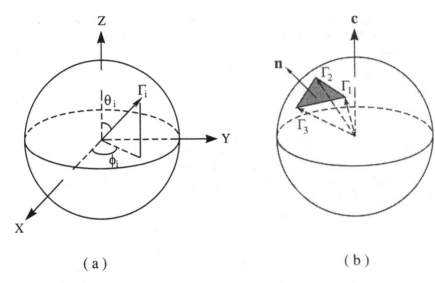

(a)   (b)

FIGURE 4.1. 3-vortex configuration on a sphere.

the system is put in canonical form[1]

$$\dot{Q}_i = \frac{\partial \mathcal{H}}{\partial P_i}, \quad \dot{P}_i = -\frac{\partial \mathcal{H}}{\partial Q_i}.$$

We now make use of the Poisson bracket defined by

$$\{f, g\} = \sum_{i=1}^{N} \frac{1}{\Gamma_i} \left(\frac{\partial f}{\partial \phi_i} \frac{\partial g}{\partial \cos \theta_i} - \frac{\partial f}{\partial \cos \theta_i} \frac{\partial g}{\partial \phi_i} \right),$$

which allows us to conveniently write the system

$$\frac{\partial \cos \theta_i}{\partial t} = \{\cos \theta_i, \mathcal{H}\}, \quad \frac{\partial \phi_i}{\partial t} = \{\phi_i, \mathcal{H}\}.$$

One can then define the following important vector-conserved quantity on the sphere.

Definition 4.1.1. *The **center of vorticity** vector for the system (1.1.23), and (1.1.24) is given by*

$$\mathbf{c} = \frac{\mathbf{M}}{\Gamma}, \tag{4.1.4}$$

---

[1]The reader should not be confused by the fact that in Kidambi and Newton (1998, 1999, 2000a, 2000b) the Hamiltonian was defined without the negative sign, and thus the canonical equations were written

$$\dot{Q}_i = -\frac{\partial \mathcal{H}}{\partial P_i}, \quad \dot{P}_i = \frac{\partial \mathcal{H}}{\partial Q_i}.$$

Either convention leads, of course, to the same equations (1.1.23), and (1.1.24).

where

$$\mathbf{M} \equiv \sum_{i=1}^{N} \Gamma_i \mathbf{x}_i$$

is the **moment of vorticity,** *and* $\Gamma \equiv \sum_{i=1}^{N} \Gamma_i$ *is the* **total vorticity.**
The three components of the center of vorticity vector are

$$Q = \frac{1}{R} \sum_{i=1}^{N} \Gamma_i x_i = \sum_{i=1}^{N} \Gamma_i \sin \theta_i \cos \phi_i$$

$$P = \frac{1}{R} \sum_{i=1}^{N} \Gamma_i y_i = \sum_{i=1}^{N} \Gamma_i \sin \theta_i \sin \phi_i$$

$$S = \frac{1}{R} \sum_{i=1}^{N} \Gamma_i z_i = \sum_{i=1}^{N} \Gamma_i \cos \theta_i.$$

Then we have the following fundamental theorem on integrability.

Theorem 4.1.1 (Kidambi and Newton (1998)). *The* 3-*vortex problem
on the sphere is completely integrable for all vortex strengths. If the center
of vorticity is zero, the* 4-*vortex problem is integrable as well.*

Proof. The proof follows closely that for the planar problem and is a direct
result of the fact that the center of vorticity vector is conserved (see exercise
3), giving three independent, involutive, conserved quantities $\mathcal{H}$, $P^2 + Q^2$,
and S. Hence,

$$\{\mathcal{H}, P^2 + Q^2\} = 0, \quad \{\mathcal{H}, S\} = 0, \quad \{P^2 + Q^2, S\} = 0,$$

and, in addition, we have

$$\{P, Q\} = S, \quad \{Q, S\} = P, \quad \{S, P\} = Q.$$

If we set $Q = P = S = 0$, then there are four involutive quantities making
the 4-vortex problem integrable as well. □

Remarks.

1. This theorem should be compared with Theorem 2.1.1 of Chapter 2.
   The special case where the vortices have equal strength is treated in
   Bogomolov (1979) following the planar formulation of Novikov (1975).
   For more details on the integrability of the general 3-vortex problem,
   see Kidambi and Newton (1998). The papers of Borisov and Pavlov
   (1998) and Borisov and Lebedev (1998) also contain information on
   3-vortex motion.

2. The condition for integrability of the 4-vortex problem in the plane is $\Gamma = 0$, as exploited in Eckhardt (1988). On the sphere, the analogous condition, $\mathbf{c} = 0$, is less restrictive in terms of the vortex strengths. Details of this special case for the sphere have not yet been worked out, but it is clear that, as in the plane, the result is sharp since Bagrets and Bagrets (1997) have proven nonintegrability for a particular 4-vortex configuration in which $\mathbf{c} \neq 0$.

3. It is clear from symmetry considerations alone that an isolated point vortex placed anywhere on the surface of a sphere, in the absence of all symmetry-breaking effects, will not move. This is because there is no preferred direction in which it can move. If the sphere rotates however, or if there are other complications that break the symmetry of the problem, such as boundaries, then in general a single point vortex will move (see, for example, Bogomolov (1985) or Klyatskin and Reznik (1989)).

4. One could scale the radius of the sphere so that $R = 1$ in the equations of motion; however, this obscures the limit $R \to \infty$. For purposes of comparisons with the planar problem, we prefer to leave the radius R as a parameter in the equations. $\Diamond$

Example 4.1   (2-vortex motion) The 2-vortex problem on the sphere is easily solvable. The invariant

$$\mathcal{H} = -\frac{\Gamma_1 \Gamma_2}{4\pi R^2} \log(l_{12}^2) = const$$

allows us to conclude that the distance between the two vortices remains fixed, so that all solutions form relative equilibria. The second invariant

$$\mathbf{c} = \frac{\Gamma_1 \mathbf{x}_1 + \Gamma_2 \mathbf{x}_2}{\Gamma_1 + \Gamma_2} = const$$

then allows us to separate the analysis into the cases in which $\mathbf{c} = 0$ or $\mathbf{c} \neq 0$.

(i) $\mathbf{c} = 0$.
In this degenerate case the vectors $\mathbf{x}_1$ and $\mathbf{x}_2$ are linear multiples of each other, so they point in opposite directions, which means the vortices lie on opposite sides of the sphere. Since the length of each vector is R, we must have $|\Gamma_1| = |\Gamma_2|$. It is easy to show that for any strength the two vortices must be in a fixed equilibrium. In addition, since each lies at an elliptic fixed point generated by the opposite vortex, the equilibrium is nonlinearly stable.

(ii) $\mathbf{c} \neq 0$.

For this case, there is no loss in orienting the vortices so that the $\mathbf{c}$ vector points to the North Pole. The dynamical formula (4.1.1) reduces to

$$\dot{\mathbf{x}}_i = \frac{\mathbf{M} \times \mathbf{x}_i}{2\pi R l_{12}^2}.$$

From this we have:

1. The two vortices must move on cones around the center of vorticity vector $\mathbf{c}$. If $\mathbf{M} \times \mathbf{x}_i = 0$, i.e., the vortices are directly opposite each other, the vortices are in a fixed, nonlinearly stable equilibrium. If $\mathbf{M} \times \mathbf{x}_i \neq 0$, the frequency of rotation is given by

$$\omega = \frac{\|\mathbf{M}\|}{2\pi R l_{12}^2}.$$

2. If $\Gamma_1 = \Gamma_2$, the vortices lie on the same latitude, but on opposite sides of the sphere.

3. If $\Gamma_1 + \Gamma_2 = 0$, the vortices are on the same longitude and symmetrically placed on equal latitudes on either side of the sphere.

For a specialized discussion of 2-vortex motion, particularly for the case $\Gamma = 0$, see DiBattista and Polvani (1998). We show in Figure 4.2 the general situation for the motion of two vortices.   ◇

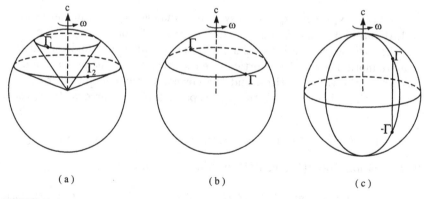

(a)   (b)   (c)

FIGURE 4.2. 2-vortex motion on the sphere: (a) In general, the two vortices move on fixed latitude curves whose planes are perpendicular to $\mathbf{c}$. (b) If the vortices have equal strength, they move on the same latitude on opposite sides of the sphere. (c) If they are equal but opposite strength, they move on the same longitude, symmetrically placed with respect to the equator.

4.2   Dynamics of Three Vortices

Following the work of Kidambi and Newton (1998), we describe the general motion of three vortices on a sphere. The equations for the relative dynamics can easily be derived from the original system and are given by

$$\frac{d}{dt}(l_{12}^2) = \frac{\Gamma_3 V}{\pi R}\left(\frac{1}{l_{23}^2} - \frac{1}{l_{31}^2}\right) \tag{4.2.1}$$

$$\frac{d}{dt}(l_{23}^2) = \frac{\Gamma_1 V}{\pi R}\left(\frac{1}{l_{31}^2} - \frac{1}{l_{12}^2}\right) \tag{4.2.2}$$

$$\frac{d}{dt}(l_{31}^2) = \frac{\Gamma_2 V}{\pi R}\left(\frac{1}{l_{12}^2} - \frac{1}{l_{23}^2}\right), \tag{4.2.3}$$

where V is the parallelpiped volume formed by the vectors $\mathbf{x}_1, \mathbf{x}_2, \mathbf{x}_3$, as obtained from the scalar triple product

$$V = \mathbf{x}_1 \cdot (\mathbf{x}_2 \times \mathbf{x}_3).$$

It is important to note that with these equations one can construct initial states with identical l_{ij}, but with different subsequent evolutions. This is because the sign of V can be positive or negative depending on whether the vectors form a right or left handed coordinate system. Note that if $V = 0$, the three vortices lie on a great circle. In fact, a similar system can be derived for the intervortical separations in the general N-vortex problem

$$\frac{d}{dt}\left(l_{ij}^2\right) = \frac{1}{\pi R}\sum_{k \neq i \neq j}^{N} \Gamma_k V_{ijk}\left(\frac{1}{l_{jk}^2} - \frac{1}{l_{ki}^2}\right),$$

where $V_{ijk} = \mathbf{x}_i \cdot (\mathbf{x}_j \times \mathbf{x}_k)$. This system is analogous to the planar system (2.1.5) and highlights the fundamental nature of 3-vortex interactions.

Other geometric quantities that help one visualize and understand the relative motion of the three vortices are the area $A(t)$ of the planar triangle formed by the three vortices, and the normal vector $\mathbf{n}$ pointing from the center of the sphere through the plane spanned by the three vortices, that is,

$$\mathbf{n} = (\mathbf{x}_1 - \mathbf{x}_2) \times (\mathbf{x}_2 - \mathbf{x}_3) = \mathbf{x}_1 \times \mathbf{x}_2 + \mathbf{x}_2 \times \mathbf{x}_3 + \mathbf{x}_3 \times \mathbf{x}_1.$$

The equations for $A(t)$ and $V(t)$, in terms of l_{ij} are

$$A = \pm\frac{1}{4}(2l_{12}^2 l_{23}^2 + 2l_{23}^2 l_{31}^2 + 2l_{31}^2 l_{12}^2 - l_{12}^4 l_{23}^4 - l_{31}^4)^{1/2} \tag{4.2.4}$$

$$V = \pm 2AR\left(1 - \frac{l_{12}^2 l_{23}^2 l_{31}^2}{16 A^2 R^2}\right)^{1/2}$$

$$= \pm 2AR\left(1 - \frac{\bar{a}^2}{R^2}\right)^{1/2} \tag{4.2.5}$$

where $\bar{a}$ is the radius of the circle in which the vortices are inscribed. In the limiting case where $0 < \bar{a}/R << 1$, it is easy to see that $V \sim \pm 2AR$, in which case our equations agree with the planar equations considered by Aref (1979) and Synge (1949). Hence, when $\bar{a}/R$ is small we expect the motion to correspond to planar motion. Conversely, the closer the vortices are to a great circle state at any point in their evolution ($\bar{a}/R \sim 1$), the more the dynamics should differ from planar dynamics. Another useful way of writing the volume is

$$V = \mathbf{c} \cdot \mathbf{n} = \mathbf{x}_i \cdot \mathbf{n},$$

which gives rise to the following constraint:

Lemma 4.2.1. *The vector* $(\mathbf{c} - \mathbf{x}_i)$ *must lie in the plane of the vortex triangle.*

Proof. The proof is an immediate consequence of the condition

$$(\mathbf{c} - \mathbf{x}_i) \cdot \mathbf{n} = 0.$$

$\square$

The system (4.2.1), (4.2.2), (4.2.3) has two fundamental invariants

$$C_1 = \sum_{i<j} \Gamma_i \Gamma_j l_{ij}^2 \tag{4.2.6}$$

$$C_2' = -\frac{1}{4\pi R^2} \sum_{i<j} \Gamma_i \Gamma_j \log(l_{ij}^2),$$

(arising from the conservation of momentum and energy), where the second quantity can more usefully be exponentiated and written

$$C_2 = \exp\left(-\frac{4\pi R^2 C_2'}{\Gamma_1 \Gamma_2 \Gamma_3}\right) = (l_{12}^2)^{1/\Gamma_3} \cdot (l_{23}^2)^{1/\Gamma_1} \cdot (l_{31}^2)^{1/\Gamma_2}. \tag{4.2.7}$$

We first state the following general theorem.

Theorem 4.2.1 (Discrete Symmetries). *The structure of the equations of motion (4.1.1) give rise to the following discrete symmetries:*

1. $\Gamma_j \to -\Gamma_j;\ \mathbf{x}_i \to -\mathbf{x}_i;\ \mathbf{x}_j \to -\mathbf{x}_j.$

2. $\Gamma_j \to -\Gamma_j;\ t \to -t.$

3. $\mathbf{x}_i \to -\mathbf{x}_i;\ \mathbf{x}_j \to -\mathbf{x}_j;\ t \to -t.$

4. Cyclic or anticyclic permutations of indices.

As in the planar case, by using these symmetries, one needs only to consider the two cases: $\Gamma_1, \Gamma_2, \Gamma_3 > 0$ and $\Gamma_1, \Gamma_2 > 0, \Gamma_3 < 0$. The invariance of the equations to cyclic or anticyclic permutations is easily verified. Cyclically permuting the indices corresponds to exchanging the positions of the vortices while maintaining the right hand rule. Anticyclically permuting them corresponds to exchanging the positions in such a way as to break the right hand rule. Although the permutation does not effect (4.1.1), it does have a significant effect on (4.2.1), (4.2.2), and (4.2.3), changing V to $-V$. This change will have important implications for the case of vortex collapse considered in Section 4.4.

4.2.1 Geometric Classification Scheme

As a first step toward classifying all possible motions on the sphere, we make use of Lemma 4.2.1. There are five qualitatively distinct families of motion based on the relative size of $\|c\|$ as compared to the sphere radius.

1. **Superradial states** $\|c\| > R$. For this case, shown in Figure 4.3, the tip of the center of vorticity vector, labeled C^*, lies outside the sphere. Since the vector c is conserved, the point C^* remains fixed in the plane P spanned by the three vortices. At any fixed time the vortices must lie on a curve that is a slice of the sphere by the plane P. From this figure we can conclude that

    (a) No fixed latitude states are possible.

    (b) The only great circle states are the family of states where c lies in the plane P, hence c and n are perpendicular.

    (c) No collapse is possible.

    (d) Because $C_1 < 0$, these states are only possible if $\Gamma_1, \Gamma_2 > 0$, $\Gamma_3 < 0$.

2. **Subradial states** $\|c\| < R$. For this case, shown in Figure 4.4, C^* lies inside the sphere. We can conclude that

    (a) Only one fixed latitude state is possible, where c is parallel to n, hence c is perpendicular to the plane P

    (b) The only great circle states are found in the family where c lies in the plane P, hence c and n are perpendicular

    (c) No collapse is possible.

3. **Radial states** $\|c\| = R$. For this case, shown in Figure 4.5, C^* lies on the sphere. We can conclude that

    (a) No fixed-latitudinal states are possible.

(b) The only great circle states are found in the family where **c** lies in P, hence **c** and **n** are perpendicular.

(c) Collapse is possible only at the point C^*, and only if $C_1 = 0$.

4. **Limiting superradial states** $\|\mathbf{c}\| = \infty$. For this case, shown in Figure 4.6, $C^* = \infty$. To make it hold we must have $\Gamma = 0$. We can conclude that

   (a) No fixed latitudinal states are possible. All states must lie on "vertical latitudes" as shown in the figure.

   (b) Great circle states are possible only if **M** lies in the plane P.

   (c) No collapse is possible.

5. **Degenerate states** $\|\mathbf{c}\| = 0$. For this case, shown in Figure 4.7, C^* lies at the center of the sphere. We can conclude that

   (a) No fixed latitude states are possible.

   (b) All states are great circle states, hence $\bar{a} = R$.

   (c) No collapse is possible.

   (d) $C_1 = \Gamma^2 R^2$.

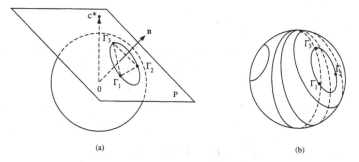

(a)   (b)

FIGURE 4.3. Geometric classification: superradial states $\|\mathbf{c}\| > R$: (a) Plane intersecting sphere at fixed angle. (b) Family of intersecting curves.

4.2.2   Equilibria

We can now classify all relative and fixed equilibrium states for three vortices on the sphere. We will refer to cases as **degenerate** where $\mathbf{c} = 0$, and as **nondegenerate** and where $\mathbf{c} \neq 0$.

Theorem 4.2.2 (Fixed Equilibria). *Fixed equilibria are given by the fixed points of system (4.1.1).*

1. *All fixed equilibrium states lie on great circles.*

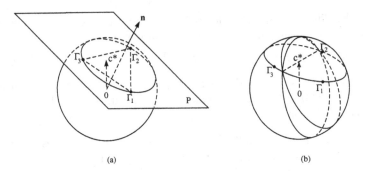

(a)   (b)

FIGURE 4.4. Geometric classification: Subradial states $\|\mathbf{c}\| < R$: (a) Plane intersecting sphere at fixed angle. (b) Family of intersecting curves.

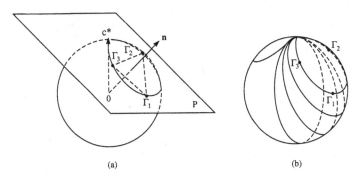

(a)   (b)

FIGURE 4.5. Geometric classification: Radial states $\|\mathbf{c}\| = R$. (a) Plane intersecting sphere at fixed angle. (b) Family of intersecting curves.

2. *A necessary and sufficient condition for fixed equilibria is that*

$$\sum_{i=1}^{3} \Gamma_i(\Gamma_j + \Gamma_k)\mathbf{x}_i = 0, \quad i \neq j \neq k.$$

3. *If $\Gamma_1 = \Gamma_2 = \Gamma_3$, the fixed equilibria form equilateral triangles and are degenerate great circle states.*

4. *For general vortex strengths the fixed equilibria are nondegenerate great circle states with positions and strengths satisfying the condition*

$$\Gamma_1 \tan \alpha_1 = \Gamma_2 \tan \alpha_2 = \Gamma_3 \tan \alpha_3.$$

The normal vector $\mathbf{n}$ is perpendicular to $\mathbf{c}$. If $\Gamma_1, \Gamma_2, \Gamma_3 > 0$, the vortex triangle is acute. If $\Gamma_1, \Gamma_2 > 0$, $\Gamma_3 < 0$, the vortex triangle is obtuse.

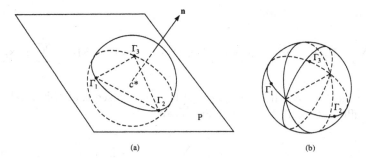

(a)   (b)

FIGURE 4.6. Geometric classification: Degenerate states $\|\mathbf{c}\| = 0$. (a) Plane intersecting sphere at fixed angle. (b) Family of intersecting curves.

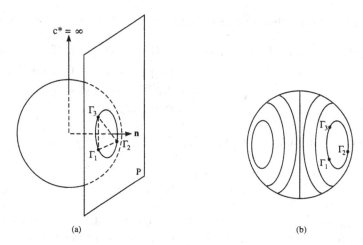

(a)   (b)

FIGURE 4.7. Geometric classification: Limiting radial states $\|\mathbf{c}\| = \infty$. (a) Plane intersecting sphere at fixed angle. (b) Family of intersecting curves.

Proof. To prove necessity of condition (2) we start by setting time derivatives in (4.1.1) to zero,

$$\sum_{j=1}^{3} \frac{\Gamma_j(\mathbf{x}_j \times \mathbf{x}_i)}{(R^2 - \mathbf{x}_i \cdot \mathbf{x}_j)} = 0,$$

and take the cross product with $\Gamma_i \mathbf{x}_i$,

$$\Gamma_i \mathbf{x}_i \times \sum_{j=1}^{3} \frac{\Gamma_j(\mathbf{x}_j \times \mathbf{x}_i)}{(R^2 - \mathbf{x}_i \cdot \mathbf{x}_j)} = 0.$$

Next, use the following manipulations

$$\mathbf{x}_i \times (\mathbf{x}_j \times \mathbf{x}_i) = (\mathbf{x}_i \cdot \mathbf{x}_i)\mathbf{x}_j - (\mathbf{x}_i \cdot \mathbf{x}_j)\mathbf{x}_i$$
$$= \|\mathbf{x}_i\|^2 \mathbf{x}_j - (\mathbf{x}_i \cdot \mathbf{x}_j)\mathbf{x}_i = R^2 \mathbf{x}_j - (\mathbf{x}_i \cdot \mathbf{x}_j)\mathbf{x}_i,$$

and sum over i to obtain

$$\sum_{i<j}^{3} \Gamma_i \Gamma_j (\mathbf{x}_i + \mathbf{x}_j) = 0 \Rightarrow \sum_{i=1}^{3} \Gamma_i \mathbf{x}_i (\Gamma - \Gamma_i) = 0 \Rightarrow \mathbf{M}\Gamma = \sum_{i=1}^{3} \Gamma_i^3 \mathbf{x}_i,$$

from which the condition in part (2) follows. To see that this condition is also sufficient assume that it holds. Then

$$\sum_{i=1}^{3} \Gamma_i \mathbf{x}_i (\Gamma - \Gamma_i) = 0 \Rightarrow \qquad (4.2.8)$$

$$\Gamma_1 (\Gamma_2 + \Gamma_3) \mathbf{x}_1 + \Gamma_2 (\Gamma_3 + \Gamma_1) \mathbf{x}_2 + \Gamma_3 (\Gamma_1 + \Gamma_2) \mathbf{x}_3 = 0. \qquad (4.2.9)$$

Taking the dot product of this equation with $\mathbf{x}_i$ for $i = 1, 2, 3$ gives

$$\begin{aligned}
\Gamma_2 (\Gamma_3 + \Gamma_1) l_{12}^2 + \Gamma_3 (\Gamma_1 + \Gamma_2) l_{31}^2 &= \Gamma_1 (\Gamma_2 + \Gamma_3) l_{12}^2 + \Gamma_3 (\Gamma_1 + \Gamma_2) l_{23}^2 \\
&= \Gamma_1 (\Gamma_2 + \Gamma_3) l_{31}^2 + \Gamma_2 (\Gamma_1 + \Gamma_3) l_{23}^2 \\
&= 2R^2 [\Gamma_1 (\Gamma_2 + \Gamma_3) + \Gamma_2 (\Gamma_1 + \Gamma_3) \\
&\quad + \Gamma_3 (\Gamma_1 + \Gamma_2)].
\end{aligned}$$

These equations allow us to write l_{12}^2 in terms of l_{23}^2 and l_{31}^2,

$$\begin{aligned}
(\Gamma_1 + \Gamma_2) l_{31}^2 &= (\Gamma_1 + \Gamma_3) l_{12}^2 \\
(\Gamma_2 + \Gamma_3) l_{12}^2 &= (\Gamma_1 + \Gamma_2) l_{23}^2. \qquad (4.2.10)
\end{aligned}$$

As a final step we substitute these equations into (4.1.1) and make use of (4.2.9) to obtain $\dot{\mathbf{x}}_i = 0$. It is clear that the vortices must lie on a great circle because $\mathbf{x}_1, \mathbf{x}_2, \mathbf{x}_3$ are linearly dependent, thus proving (1).

We next prove condition (3). For fixed equilibria we know from condition (2) that $\mathbf{M}\Gamma = \sum \Gamma_i^2 \mathbf{x}_i = 0$. Also, we have $\mathbf{M} = \sum \Gamma_i \mathbf{x}_i = 0$; therefore,

$$\begin{aligned}
\Gamma_1^2 \mathbf{x}_1 + \Gamma_2^2 \mathbf{x}_2 + \Gamma_3^2 \mathbf{x}_3 &= 0 \\
\Gamma_1 \mathbf{x}_1 + \Gamma_2 \mathbf{x}_2 + \Gamma_3 \mathbf{x}_3 &= 0.
\end{aligned}$$

Multiplying the second equation by Γ_1 and subtracting from the first gives

$$\Gamma_2 (\Gamma_2 - \Gamma_1) \mathbf{x}_2 = \Gamma_3 (\Gamma_1 - \Gamma_3) \mathbf{x}_3,$$

which means that either $\Gamma_1 = \Gamma_2 = \Gamma_3$, or that $\mathbf{x}_2$ is parallel to $\mathbf{x}_3$. Multiplying the second equation by Γ_2 and subtracting from the first gives

$$\Gamma_1 (\Gamma_1 - \Gamma_2) \mathbf{x}_1 = \Gamma_3 (\Gamma_2 - \Gamma_3) \mathbf{x}_3,$$

which means that either $\Gamma_1 = \Gamma_2 = \Gamma_3$ or that $\mathbf{x}_1$ and $\mathbf{x}_3$ are parallel. Since it is not possible for $\mathbf{x}_1, \mathbf{x}_2$, and $\mathbf{x}_3$ to all be parallel, we can conclude that

$\Gamma_1 = \Gamma_2 = \Gamma_3$ and that $x_1 + x_2 + x_3 = 0$. The second condition implies that the vortices must lie on an equilateral triangle. To see this write

$$\mathbf{x}_i = R(\cos\theta_i \hat{\mathbf{i}} + \sin\theta_i \hat{\mathbf{j}})$$

and let $\theta_1 = 0$ with no loss of generality. Using $x_1 + x_2 + x_3 = 0$ then gives

$$1 + \cos\theta_2 + \cos\theta_3 = 0$$
$$\sin\theta_2 + \sin\theta_3 = 0.$$

Solving these equations gives $\theta_2 = 2\pi/3$ and $\theta_3 = 4\pi/3$, proving that the triangle is equilateral.

To prove condition (4) we start with the relations

$$\Gamma_2 = \Gamma_1 \frac{l_{12}^2 + l_{23}^2 - l_{31}^2}{l_{12}^2 + l_{31}^2 - l_{23}^2}$$
$$\Gamma_3 = \Gamma_1 \frac{l_{23}^2 + l_{31}^2 - l_{12}^2}{l_{12}^2 + l_{31}^2 - l_{23}^2}.$$

We also have the following elementary relations between the sides and the angles of a triangle

$$l_{12}^2 = l_{23}^2 + l_{31}^2 - 2l_{23}l_{31}\cos\alpha_3$$
$$l_{23}^2 = l_{31}^2 + l_{12}^2 - 2l_{31}l_{12}\cos\alpha_1$$
$$l_{31}^2 = l_{12}^2 + l_{23}^2 - 2l_{12}l_{23}\cos\alpha_2.$$

Using these relations, the above equations can be written

$$\Gamma_2 l_{31}\cos\alpha_1 = \Gamma_1 l_{23}\cos\alpha_2 \qquad (4.2.11)$$
$$\Gamma_3 l_{12}\cos\alpha_1 = \Gamma_1 l_{23}\cos\alpha_3. \qquad (4.2.12)$$

We also have the sine formula for the sides and angles of a triangle,

$$\frac{l_{23}}{\sin\alpha_1} = \frac{l_{31}}{\sin\alpha_2} = \frac{l_{12}}{\sin\alpha_3}.$$

Using these in (4.2.11) and (4.2.12) gives the relation in condition (4). □

Remarks.

1. The necessity of the statements generalizes to N-vortex equilibria on the sphere, however, the proof of sufficiency is more complicated for $N > 3$. A general analysis of equilibrium states for $N > 3$ has not yet been carried out, either on the sphere or in the plane; however, see Lim, Montaldi, and Roberts (1999) for a start in this direction. Section 4.8 contains a brief discussion of related literature associated with the electrostatic problem on a conducting spherical shell.

2. In the plane, fixed equilibria imply that $h = 0$, where

$$h = \frac{1}{3}\left(\frac{1}{\Gamma_1} + \frac{1}{\Gamma_2} + \frac{1}{\Gamma_3}\right)$$

(see Theorem 2.2.1, condition (b)), whereas on the sphere it can be shown that fixed equilibria imply $h \neq 0$.

3. In the plane, only collinear configurations can be fixed (Theorem 2.2.1), whereas on the sphere, the vortices must be placed on a great circle. We explain this by (4.1.1), which implies that the three vectors x_1, x_2, x_3 are coplanar. Collinear configurations in the plane and great circle configurations on the sphere are both **geodesics** on their respective surfaces.

4. In the plane, fixed states are possible only for $\Gamma_1, \Gamma_2 > 0$, $\Gamma_3 < 0$, whereas on the sphere fixed states are possible for both $\Gamma_1, \Gamma_2, \Gamma_3 > 0$ and $\Gamma_1, \Gamma_2 > 0, \Gamma_3 < 0$. $\Diamond$

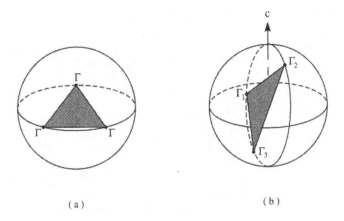

(a)   (b)

FIGURE 4.8. Fixed equilibrium states. (a) Equilateral triangle, degenerate great circle state with $\Gamma_1 = \Gamma_2 = \Gamma_3$; (b) Nondegenerate great circle state.

Next, we consider all relative equilibrium configurations, which are defined as fixed points of (4.2.1), (4.2.2), and (4.2.3). The starting point for our theorem on relative equilibria is the condition that $\mathbf{c} \cdot \mathbf{x}_i = const$ (this follows from the fact that $l_{ij} = const$), therefore each vortex moves on a cone around the center of vorticity vector, staying on a fixed latitude; the general situation is shown in Figure 4.9. Since each of the vortex triangle sides is constant, we can conclude that there is only one frequency of rotation around $\mathbf{c}$, which we label $\dot{\phi}_i \equiv \omega = const$ Then we have the following theorem on relative equilibrium states.

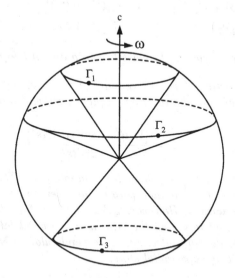

FIGURE 4.9. Vortices in relative equilibrium move on cones around the center of vorticity vector $\mathbf{c}$.

Theorem 4.2.3 (Relative Equilibria). *Relative equilibria are given by the fixed points of (4.2.1), (4.2.2), and (4.2.3).*

1. *For degenerate relative equilibria all states lie on great circles, with positions and strengths satisfying*

$$\Gamma_1 \csc 2\alpha_1 = \Gamma_2 \csc 2\alpha_2 = \Gamma_3 \csc 2\alpha_3.$$

The vortices rotate around the fixed vector

$$\mathbf{x} = -\frac{1}{2\pi R} \left(\frac{\Gamma_1 \mathbf{x}_1(0)}{l_{23}^2} + \frac{\Gamma_2 \mathbf{x}_2(0)}{l_{31}^2} + \frac{\Gamma_3 \mathbf{x}_3(0)}{l_{12}^2} \right),$$

with frequency

$$\omega = \frac{1}{2\pi R} \left[R^2 \left(\frac{\Gamma_1}{l_{23}^2} + \frac{\Gamma_2}{l_{31}^2} + \frac{\Gamma_3}{l_{12}^2} \right)^2 - \frac{\Gamma_1 \Gamma_2 l_{12}^4 + \Gamma_2 \Gamma_3 l_{23}^4 + \Gamma_1 \Gamma_3 l_{13}^4}{l_{12}^2 l_{23}^2 l_{31}^2} \right]^{1/2}.$$

2. *For nondegenerate relative equilibria the vortices rotate around the center of vorticity vector $\mathbf{c}$ with constant frequency*

$$\omega = \left(\frac{A}{2\pi RB} \right) \cdot \frac{\|\mathbf{M}\|}{l_{12}^2 l_{31}^2}, \qquad (4.2.13)$$

where

$$A = l_{12}^2 l_{31}^2 \left[\Gamma_2^2(4R^2 - l_{12}^2) + \Gamma_2\Gamma_3(4R^2 - l_{12}^2 - l_{31}^2) + \Gamma_3^2(4R^2 - l_{31}^2)\right]$$
$$+ 2R^2\Gamma_2\Gamma_3 \left[l_{12}^2(l_{12}^2 - l_{23}^2) + l_{31}^2(l_{31}^2 - l_{23}^2)\right]$$
$$B = (\Gamma_2 l_{12}^2 + \Gamma_3 l_{31}^2)\left[\Gamma_2(4R^2 - l_{12}^2) + \Gamma_3(4R^2 - l_{31}^2)\right] - 4R^2\Gamma_2\Gamma_3 l_{23}^2.$$

All such equilibria can be classified into one of two cases:

(a) *The vortices form an equilateral triangle, but do not lie on a great circle. The rotation frequency simplifies to*

$$\omega = \frac{\left[\Gamma^2 R^2 - 3h\Gamma_1\Gamma_2\Gamma_3 s^2\right]^{1/2}}{2\pi R s^2}, \qquad (4.2.14)$$

where s is the length of the triangle side. The normal vector, in general, is neither parallel nor perpendicular to the center of vorticity **c**. *However, if* $\Gamma_1 = \Gamma_2 = \Gamma_3$, *the normal vector is parallel to* **c** *and the case is a subradial latitudinal state. If* $\Gamma = 0$, *the normal vector is perpendicular to* **M** *and the case is a limiting superradial state.*

(b) *The vortices lie on a great circle, but form triangles of arbitrary shape. The sides of the triangle and the vortex strengths satisfy*

$$V(0) = R^2 \left[2\left(l_{12}^2 l_{23}^2 + l_{23}^2 l_{31}^2 + l_{31}^2 l_{12}^2\right) - l_{12}^4 - l_{23}^4 - l_{31}^4\right]$$
$$- l_{12}^2 l_{23}^2 l_{31}^2 = 0,$$
$$\dot{V}(0) = \frac{1}{8\pi}\left[2R \sum \frac{l_{ij}^2 - l_{jk}^2}{l_{ki}^2}(\Gamma_i + \Gamma_k) - \frac{1}{R}\sum l_{ij}^2(\Gamma_i - \Gamma_j)\right] = 0.$$

The normal vector **n** *is perpendicular to* **c**.

Proof. To prove the first part start with

$$\mathbf{c} = 0 \Rightarrow \Gamma_1\mathbf{x}_1 + \Gamma_2\mathbf{x}_2 + \Gamma_3\mathbf{x}_3 = 0. \qquad (4.2.15)$$

Taking the dot product of $\mathbf{x}_i$ with the above for $i = 1, 2, 3$ gives

$$\Gamma_2 l_{12}^2 + \Gamma_3 l_{31}^2 = \Gamma_3 l_{23}^2 + \Gamma_1 l_{12}^2 = \Gamma_1 l_{31}^2 + \Gamma_2 l_{23}^2 = 2\Gamma R^2. \qquad (4.2.16)$$

From these we can write Γ_2 and Γ_3 in terms of Γ_1 and the l_{ij} as

$$\Gamma_2 = \Gamma_1 \frac{l_{31}^2(l_{12}^2 + l_{23}^2 - l_{31}^2)}{l_{23}^2(l_{31}^2 + l_{12}^2 - l_{23}^2)}$$
$$\Gamma_3 = \Gamma_1 \frac{l_{12}^2(l_{23}^2 + l_{31}^2 - l_{12}^2)}{l_{23}^2(l_{31}^2 + l_{12}^2 - l_{23}^2)},$$

which, upon using the standard triangle relations, become

$$\Gamma_1 \csc 2\alpha_1 = \Gamma_2 \csc 2\alpha_2 = \Gamma_3 \csc 2\alpha_3.$$

To get expressions for $\mathbf{x}$ and ω, start with the equation

$$\dot{\mathbf{x}}_1 = \frac{1}{2\pi R} \left[\frac{\Gamma_2 \mathbf{x}_2 \times \mathbf{x}_1}{l_{12}^2} + \frac{\Gamma_3 \mathbf{x}_3 \times \mathbf{x}_1}{l_{31}^2} \right].$$

Using the relation (4.2.15) in the above and noting the fact that the vortices are in a relative equilibrium gives

$$\dot{\mathbf{x}}_1 = \frac{\Gamma_2 \mathbf{x}_2 \times \mathbf{x}_1}{2\pi R} \left(\frac{1}{l_{12}^2 - l_{31}^2} \right) = k_1 (\mathbf{x}_2 \times \mathbf{x}_1).$$

Likewise, one can show that

$$\dot{\mathbf{x}}_2 = k_2 (\mathbf{x}_1 \times \mathbf{x}_2).$$

Taken together, we have $k_2 \mathbf{x}_1 + k_1 \mathbf{x}_2 \equiv \mathbf{x} = const$, which gives

$$\dot{\mathbf{x}}_1 = \mathbf{x} \times \mathbf{x}_1,$$

from which the expressions for $\mathbf{x}$ and ω follow.

For the second part, the starting point is based on our previous observation that each vortex moves on a fixed latitude rotating around $\mathbf{c}$ with constant frequency. The fact that ω is constant follows from (4.2.13), whose right hand side is constant. Of course, if a vortex is on the axis of rotation, it stays fixed by symmetry and for it ω is undefined. Equation (4.2.14) follows on using the relations in (4.2.13):

$$\cos \theta_i = \frac{R\Gamma}{\|\mathbf{M}\|} - \frac{C_1 - \Gamma_j \Gamma_k l_{jk}^2}{2R \|\mathbf{M}\| \Gamma_i}$$

$$\cos(\phi_i - \phi_j) = \csc \theta_i \csc \theta_j \left[1 - \cos \theta_i \cos \theta_j - \frac{l_{ij}^2}{2R^2} \right].$$

Since $\omega_1 = \omega_2 = \omega_3$, any of the three equations for $\dot{\phi}$ could be used. We use the one for which $\theta \neq 0, \pi$ so that the vortex is not on the axis of rotation.

The two separate cases of part (2) are just the conditions giving the fixed points of (4.2.1), (4.2.2), and (4.2.3). We start with case (a) where $l_{12} = l_{23} = l_{31} \equiv s$ and $V \neq 0$. (4.2.13) can be specialized to yield (4.2.14). When the vortex strengths are equal, (4.2.14) further specializes to the result in Bogomolov (1979). Since $\mathbf{n} \cdot \mathbf{c} = V \neq 0$, in general it follows that the angle between the two vectors is not zero or π. Figure 4.10(a) shows the equilateral triangle state for general vortex strengths. If the vortex strengths are equal, we see that $\dot{\mathbf{n}} = 0$ which implies that $\mathbf{n} = const$. But we know that $\mathbf{c} = const$ and that $\mathbf{x}_i \cdot \mathbf{c} = const$. For both of these conditions to hold, $\mathbf{x}_i$ must describe a cone around both $\mathbf{c}$ and $\mathbf{n}$, which is possible only if they are parallel. This state in which the vortices just rotate on one fixed latitude is shown in 4.10(b). If $\Gamma = 0$, then $\mathbf{n} \cdot \mathbf{M} = \Gamma V = 0$, which implies that $\mathbf{n}$ and $\mathbf{M}$ are perpendicular. The frequency is obtained by setting $\Gamma = 0$ in (4.2.14):

$$\omega = \frac{\left[\frac{1}{2}(\Gamma_1^2 + \Gamma_2^2 + \Gamma_3^2)\right]^{1/2}}{2\pi R s}.$$

This case is shown in Figure 4.10(c). The planar limit $R \to \infty$, s fixed, gives $\omega = 0$, i.e., in the planar limit the configuration does not rotate. This is not surprising, since $\Gamma = 0$ leads to a rigid translation of the vortices in the plane. If we compute the linear velocity of each vortex as

$$v_i \equiv \omega R \sin \theta_i = \omega R \left[1 - \frac{2\Gamma_i^2 s^2}{(\Gamma_1^2 + \Gamma_2^2 + \Gamma_3^2)R^2}\right]^{1/2}$$

and take the limit $R \to \infty$, s fixed, we get the limiting value

$$v = \frac{\left[\frac{1}{2}(\Gamma_1^2 + \Gamma_2^2 + \Gamma_3^2)\right]^{1/2}}{2\pi s},$$

which is the velocity of a translating configuration in the plane, as shown in Rott (1989). All great circle configurations comprise the second case (b), so that $V = 0$. It is straightforward to prove that $V \equiv 0$ if $V(0) = 0$, $\dot{V}(0) = 0$. This can be seen by differentiating the volume equation repeatedly and using induction. Finally, $\mathbf{n}$ and $\mathbf{c}$ are perpendicular for this case because $\mathbf{x}_1, \mathbf{x}_2, \mathbf{x}_3$ are co-planar — Figure 4.10(d) depicts this case. □

We finish this subsection with a stability theorem of Pekarsky and Marsden (1998), who obtained the following criterion for nonlinear stability and instability of the equilibrium states.

Theorem 4.2.4 (Stability of Equilibria). *Consider the stability of the equilibrium 3-vortex states on the sphere.*

1. *The nondegenerate, non great circle, relative equilibria form equilateral triangles that are nonlinearly stable modulo SO(2) rotations around $\mathbf{c}$ if*

$$\sum_{i<j} \Gamma_i \Gamma_j > 0$$

   and unstable if

$$\sum_{i<j} \Gamma_i \Gamma_j < 0.$$

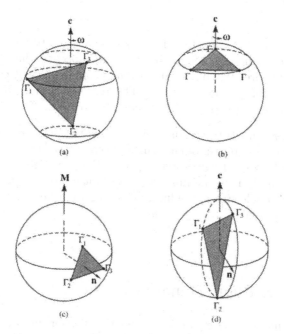

FIGURE 4.10. Nondegenerate relative equilibrium states: (a), (b), (c) are equilateral triangle non great circle states. (a) The general case for arbitrary $\Gamma_1, \Gamma_2, \Gamma_3$. (b) $\Gamma_1 = \Gamma_2 = \Gamma_3$: Vortices move on one fixed latitude. (c) $\Gamma = 0$: Vortices move on different latitudes. (d) Great circle state.

2. *The nondegenerate great circle relative equilibrium states are nonlinearly stable modulo SO(2) rotations around* **c** *if*

$$\Gamma_1^2 + \Gamma_2^2 > \sum_{i \neq j} \Gamma_i \Gamma_j$$

and unstable if

$$\Gamma_1^2 + \Gamma_2^2 < \sum_{i \neq j} \Gamma_i \Gamma_j.$$

3. *The degenerate great circle relative equilibria are nonlinearly stable modulo SO(3) rotations.*

Proof. The proof is based on computing the second variation of an augmented version of the Hamiltonian. See Pekarsky and Marsden (1998) for details. $\square$

Remarks.

1. Notice that the transitional case between stability and instability of the nondegenerate, non great circle states occurs when $\sum_{i<j} \Gamma_i \Gamma_j = 0$. This is precisely one of the necessary and sufficient conditions for 3-vortex collapse to occur (see Theorem 4.4.1) and probably corresponds to a degenerate Hamiltonian bifurcation.

2. The first part of the theorem generalizes the planar results of Synge (1949) to the sphere. In fact, since the condition is independent of the radius R, it must agree precisely with the planar result, since we are free to apply it in the limit $R \to \infty$. Since parts (2) and (3) pertain to great circle states, the limit $R \to \infty$ cannot be compared with the planar problem. $\Diamond$

4.3   Phase Plane Dynamics

We now formulate the equations of relative motion in the phase plane, allowing us more simply to identify all equilibria and make comparisons with the planar results described in Chapter 2. A two-dimensional phase plane system can be obtained by making use of the first invariant (4.2.6), as pointed out by Synge (1949) for the planar problem. First, we assume that $C_1 \neq 0$. To this end, we introduce the scaled length variables

$$b_1 = \frac{3l_{23}^2 \Gamma_2 \Gamma_3}{C_1}, \quad b_2 = \frac{3l_{31}^2 \Gamma_3 \Gamma_1}{C_1}, \quad b_3 = \frac{3l_{12}^2 \Gamma_1 \Gamma_2}{C_1}. \quad (4.3.1)$$

Then it is clear from the defining equation for C_1 that

$$b_1 + b_2 + b_3 = 3.$$

As in Aref (1979), we use trilinear coordinates to define an arbitrary point P in the plane, making the invariant (4.2.6) identically satisfied for any point P in the phase plane. The trilinear coordinates b_1, b_2, b_3 are related to rectangular coordinates x, y by

$$b_1 = y, \quad b_2 = \frac{1}{2}(3 - y - \sqrt{3}x), \quad b_3 = \frac{1}{2}(3 - y + \sqrt{3}x).$$

The second invariant (4.2.7) in the b_i variables becomes

$$f(b_1, b_2, b_3) = |b_1|^{1/\Gamma_1} |b_2|^{1/\Gamma_2} |b_3|^{1/\Gamma_3}.$$

Curves defined by $f(b_1, b_2, b_3) = const$ are called phase curves and this is precisely the same equation for the phase curves in the planar problem discussed by Aref (1979). This equation can be written in terms of Cartesian variables as

$$\mathcal{H}(x,y) = \frac{1}{\Gamma_1} \log |y| + \frac{1}{\Gamma_2} \log \frac{|3 - y - \sqrt{3}x|}{2} + \frac{1}{\Gamma_3} \log \frac{|3 - y - \sqrt{3}x|}{2}.$$

It is straightforward to verify that in terms of the Cartesian variables $x = (b_3 - b_2)/\sqrt{3}$, $y = b_1$, there is a Hamiltonian structure (see Marsden and Ratiu (1999)) for the equations of motion, which can be written

$$\dot{x} = \frac{6\sqrt{3}g^6 V(x,y)}{\pi RC_1^2} \cdot \frac{\partial \mathcal{H}}{\partial y} \tag{4.3.2}$$

$$\dot{y} = -\frac{6\sqrt{3}g^6 V(x,y)}{\pi RC_1^2} \cdot \frac{\partial \mathcal{H}}{\partial x}, \tag{4.3.3}$$

where $g = (\Gamma_1\Gamma_2\Gamma_3)^{1/3}$. It is interesting to note that the canonical Hamiltonian structure of the relative equations (4.2.1), (4.2.2), (4.2.3) remains hidden in their original form but becomes transparent when the Cartesian variables are used.

Though the phase plane is unbounded, the system obviously does not explore all of it because of the compact nature of the sphere. This is also clear from considering (4.3.1). Since the vortices lie on a sphere of radius R, the maximum value attained by l_{ij} is $2R$, which means the b_i lie between zero and $12R^2\Gamma_j\Gamma_k/C_1$. The region accessible to the system is the interior of a polygonal domain D, which can have anywhere from three to six sides, depending on the values of $\Gamma_1, \Gamma_2, \Gamma_3$, and C_1. Even within this domain, the physically relevant region is restricted by the condition $V^2 \geq 0$, which we call the physical region. We show this region in Figure 4.11 for fixed values of the parameters. The physical region boundary, defined by $V = 0$,

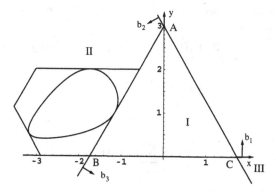

FIGURE 4.11. Five-sided domain for $\Gamma_1 = \Gamma_2 = \frac{1}{2}$, $\Gamma_3 = -\frac{1}{2}$, and $C_1 = -3R^2$. The physical region is also shown within the five-sided domain as the enclosed egg-shaped region.

is expressed in terms of the b_i as

$$3R^2 V_p - b_1 b_2 b_3 C_1 = 0, \tag{4.3.4}$$

where

$$V_p = 2(\Gamma_1 \Gamma_2 b_1 b_2 + \Gamma_2 \Gamma_3 b_2 b_3 + \Gamma_3 \Gamma_1 b_3 b_1) - (\Gamma_1 b_1)^2 - (\Gamma_2 b_2)^2 - (\Gamma_3 b_3)^2.$$

An important difference between the spherical and planar cases is that the boundary on the sphere is a function of the conserved quantity C_1, whereas in the plane it is independent of this quantity. This means that on the sphere, even for fixed values of the vortex strengths, we have a full range of boundary curves as we vary the parameter C_1. We refer the reader to the discussion in Kidambi and Newton (1998) for more details.

Since V can take on both positive or negative values during the course of the vortex motion, the phase plane can be thought of as two-sided, with the front side having all positive values and the back side all negative ones. The two sides are then joined together at the physical region boundary $V = 0$, representing a great circle. When the system, evolving along a phase curve, reaches the physical region boundary, its subsequent evolution will take place on the other side of the phase plane. An important relation for the volume evolution is given by

$$\dot{V} = \frac{3\sqrt{3}g^6}{\pi R C_1^2} \left(\frac{\partial V^2}{\partial x} \frac{\partial \mathcal{H}}{\partial y} - \frac{\partial V^2}{\partial y} \frac{\partial \mathcal{H}}{\partial x} \right). \tag{4.3.5}$$

This equation says that $\dot{V} = 0$ at all points where the curves $V^2 = const$ and the phase curve $\mathcal{H} = const$ are tangent.

We can now locate all equilibria as points in the phase plane.

Theorem 4.3.1 (Location of Equilibria). *All equilibrium points can be represented in the trilinear phase plane shown in Figure 4.12 as follows:*

1. *Fixed equilibria are represented in the trilinear plane by points P, whose coordinates are*

$$b_1 = \frac{3\Gamma_2\Gamma_3(\Gamma_2 + \Gamma_3)}{\Gamma_{ij}}, \quad b_2 = \frac{3\Gamma_3\Gamma_1(\Gamma_3 + \Gamma_1)}{\Gamma_{ij}},$$

$$b_3 = \frac{3\Gamma_1\Gamma_2(\Gamma_1 + \Gamma_2)}{\Gamma_{ij}}, \tag{4.3.6}$$

where

$$\Gamma_{ij} \equiv \sum_{i<j} \Gamma_i \Gamma_j (\Gamma_i + \Gamma_j),$$

and

$$C_1 = \frac{12\Gamma_1\Gamma_2\Gamma_3 h\Gamma_{ij}}{(\Gamma_1 + \Gamma_2)(\Gamma_2 + \Gamma_3)(\Gamma_3 + \Gamma_1)}.$$

2. (a) *Degenerate relative equilibria are located by the points U whose trilinear coordinates are given by*

$$b_1 = \frac{3(\Gamma_2 + \Gamma_3 - \Gamma_1)}{\Gamma},$$

$$b_2 = \frac{3(\Gamma_3 + \Gamma_1 - \Gamma_2)}{\Gamma},$$

$$b_3 = \frac{3(\Gamma_1 + \Gamma_2 - \Gamma_3)}{\Gamma},$$

with $C_1 = \Gamma^2 R^2$.

(b) *Nondegenerate relative equilibria are represented by the points Q or S where*

(i) *Q are points representing equilateral triangle configurations, with trilinear coordinates given by*

$$Q = \left(\frac{1}{\Gamma_1 h}, \frac{1}{\Gamma_2 h}, \frac{1}{\Gamma_3 h} \right).$$

(ii) *S are points at which the physical boundary $V = 0$ and the phase curves $\mathcal{H} = const$, are tangent.*

Proof. The trilinear coordinates of P are obtained using the definitions (4.3.1) in the expressions (4.2.10). Since for a fixed equilibrium $V = 0$, (4.3.4) with the b_i supplied by the expressions in (4.3.6) yields the expression for C_1. If $\Gamma_1 = \Gamma_2 = \Gamma_3$, then P represents a degenerate fixed equilibrium. The trilinear coordinates of U are obtained using the definitions (4.3.1) in (4.2.16). C_1 has the stated value because $\mathbf{c} = 0$ for a degenerate equilibrium. The condition for equilateral triangles yields the coordinates of Q. For further detail, see Kidambi and Newton (1998).   □

4.4   3-Vortex Collapse

Our analysis of vortex collapse is based on the equations of motion for the chord lengths (l_{12}, l_{23}, l_{31}) with the two invariants

$$C_1 = \Gamma_1 \Gamma_2 l_{12}^2 + \Gamma_2 \Gamma_3 l_{23}^2 + \Gamma_3 \Gamma_1 l_{31}^2 \equiv 0$$
$$C_2 = (l_{12}^2)^{1/\Gamma_3} \cdot (l_{23}^2)^{1/\Gamma_1} \cdot (l_{31}^2)^{1/\Gamma_2}.$$

The first invariant is zero due to the fact that the lengths vanish at collapse. This implies that the vortex strengths cannot all be of the same sign, so we adopt the convention $\Gamma_1 > 0$, $\Gamma_2 > 0$, $\Gamma_3 < 0$. The formula

$$\|\mathbf{c}\|^2 = R^2 - \frac{C_1}{\Gamma}$$

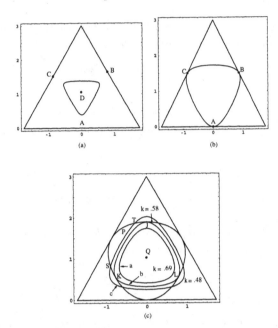

FIGURE 4.12. Trilinear phase plane for $\Gamma_1 = \Gamma_2 = \Gamma_3 = 1$, showing equilibrium locations and phase curves: (a) $C_1 = 9R^2$. (b) $C_1 = 6R^2$. (c) $C_1 = 3R^2$.

allows us to conclude that $\|\mathbf{c}\| = R$, assuming $\Gamma \neq 0$.

As in the planar problem, we start with the basic *ansatz* that the relative distances between the vortices remains constant throughout their motion

$$l_{12}^2 = \lambda_1 l_{31}^2$$
$$l_{23}^2 = \lambda_2 l_{31}^2,$$

where

$$\lambda_1 = \left(\frac{l_{12}(0)}{l_{31}(0)}\right)^2$$

$$\lambda_2 = \left(\frac{l_{23}(0)}{l_{31}(0)}\right)^2,$$

hence the collapse process is assumed to be self-similar. The second conserved quantity, C_2, then yields

$$\left(l_{31}^2(t)\right)^{\sum 1/\Gamma_i} = const$$

implying

$$\sum_{i=1}^{3} \frac{1}{\Gamma_i} = 0.$$

We can now formulate the following basic theorem stating necessary and sufficient conditions for 3-vortex, self-similar collapse on the sphere.

Theorem 4.4.1 (Spherical Collapse). *Necessary and sufficient conditions for the self-similar collapse of three vortices on the sphere are:*

1. $\|\mathbf{c}\| = R$.

2. $\sum_{i=1}^{3} 1/\Gamma_i = 0$.

3. *The vortices must not be in equilibrium.*

Proof. We outline here the main ideas behind the proof. The necessity of the first two conditions has been proven. Condition (3) is necessary because conditions (1) and (2) do not preclude the possibility that the vortices are in equilibrium. To understand why the conditions are sufficient, we need to describe the phase plane associated with the collapsing states. Since $C_1 = 0$, we redefine the b_i variables as

$$b_1 = \frac{l_{23}^2}{4\Gamma_1 R^2}, \qquad b_2 = \frac{l_{31}^2}{4\Gamma_2 R^2}, \qquad b_3 = \frac{l_{12}^2}{4\Gamma_3 R^2},$$

so that we have the identity

$$b_1 + b_2 + b_3 = \frac{C_1}{4R^2(\Gamma_1\Gamma_2\Gamma_3)} = 0.$$

Then one can eliminate the coordinate b_3 in favor of the other two. Based on the conserved quantity C_1, we now have

$$\left(\frac{b_1}{b_1 + b_2}\right)^{1/\Gamma_1} \left(\frac{b_2}{b_1 + b_2}\right)^{1/\Gamma_2} = const$$

which implies that $b_1/b_2 = const$ i.e., the phase curves are straight lines passing through the origin. The only physically accessible region is defined by the condition $V^2 \geq 0$ which one can write as

$$2(\Gamma_1\Gamma_2 b_1 b_2 + \Gamma_2\Gamma_3 b_2 b_3 + \Gamma_3\Gamma_1 b_3 b_1) - \Gamma_1^2 b_1^2 - \Gamma_2^2 b_2^2 - \Gamma_3^2 b_3^2$$
$$\geq 4\Gamma_1\Gamma_2\Gamma_3 b_1 b_2 b_3.$$

Figure 4.13 shows the (b_1, b_2)-phase plane with each collapsing trajectory lying on a ray going through the origin. For definiteness we show the case $\Gamma_1 = \Gamma_2 = 1, \Gamma_3 = -1/2$. Notice that for each collapsing state there is another state with identical l_{ij} that also collapses. For one of the states we have $V > 0$, while for the other, $V < 0$. In both these cases the sign of $\dot{V}$ is the same. For case (a) the trajectory evolves straight to the origin along the ray (on the front face $V > 0$), collapsing at time τ^-. State (c) evolves away from the origin (on the back face $V < 0$) until it hits the $V = 0$ curve, corresponding to a great circle configuration, then evolves to the origin on the front face $V > 0$, collapsing at the later time $\tau^+ > \tau^-$. The difference between this process and the corresponding one in the plane described in Aref (1979) is that for the planar problem, there is nothing to bound the coordinates from above, hence the accessible region is unbounded. As a result, the trajectory analogous to state (c) continues to travel away from the origin on the same ray, representing a self-similar expanding state. □

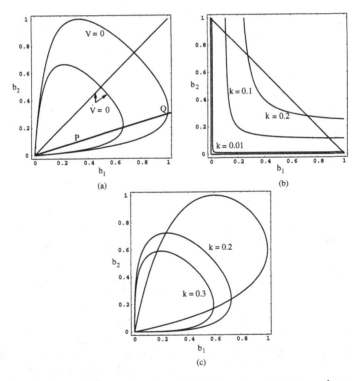

FIGURE 4.13. Phase diagram for $C_1 = 0$: (a) $\Gamma_1 = \Gamma_2 = 1$, $\Gamma_3 = -\frac{1}{2}$, all states evolve to the origin in finite time. (b) $\Gamma_1 = \Gamma_2 = -\Gamma_3$ ($h > 0$), phase curves are open and do not pass through the origin. (c) $\Gamma_1 = \Gamma_2 = 1$, $\Gamma_3 - -\frac{1}{4}$ ($h < 0$), phase curves are closed through the origin but are intercepted by $V = 0$ before reaching it.

More detailed information about the collapse process can be obtained from the equations governing the intervortical distances

$$\frac{d}{dt}(l_{31}^2) = \frac{\Gamma_1}{\pi R}\left(\frac{\lambda_1 - 1}{\lambda_1 \lambda_2}\right)\frac{V}{l_{31}^2}$$
$$= \frac{\Gamma_2}{\pi R}\left(\frac{\lambda_2 - \lambda_1}{\lambda_1 \lambda_2}\right)\frac{V}{l_{31}^2}$$
$$= \frac{\Gamma_3}{\pi R}\left(\frac{1 - \lambda_2}{\lambda_1 \lambda_2}\right)\frac{V}{l_{31}^2}.$$

This immediately yields a relationship that must be satisfied:

$$\Gamma_1(\lambda_1 - 1) = \Gamma_2(\lambda_2 - \lambda_1) = \Gamma_3(1 - \lambda_2).$$

We can then use the volume formula (4.2.5) to obtain the scalar equation

$$\frac{d}{dt}(l_{31}^2) = \pm\omega(1 - \rho l_{31}^2)^{1/2}, \tag{4.4.1}$$

where

$$\rho = \lambda_1 \lambda_2 / R^2 \gamma, \quad \omega = \frac{\Gamma_2}{2\pi}\left(\frac{\lambda_1 - \lambda_2}{\lambda_1 \lambda_2}\right)\sqrt{\gamma},$$
$$\gamma = 2(\lambda_1 + \lambda_2) - (\lambda_1 - \lambda_2)^2 - 1.$$

Solving (4.4.1), one obtains

$$l_{31}^2(t) = l_{31}^2(0) \pm \mu\omega t - \frac{\rho\omega^2}{4}t^2 \tag{4.4.2}$$

where

$$\mu = \sqrt{1 - \rho l_{31}^2(0)}.$$

Since $\rho\omega^2/4 > 0$, there must be two zeroes of $l_{31}^2(t)$, one positive, one negative. A more suggestive way to write (4.4.2) is

$$l_{ij}(t) = l_{ij}(0)(1 + \frac{t}{\tau^{\mp}})^{1/2}(1 - \frac{t}{\tau^{\pm}})^{1/2}, \tag{4.4.3}$$

where

$$\tau^{\pm} = \frac{4\pi R^2 \sqrt{\gamma}}{\Gamma_2|\lambda_1 - \lambda_2|}(1 \pm \mu) > 0$$

are the two collapse times associated with a given collapsing state.

Definition 4.4.1. *Related initial configurations that lead to the two collapse times* $t = \tau^+$ *and* $t = \tau^-$ *are called* **partner states.** *They are ordered so that* $0 < \tau^- < \tau^+ < \infty$.

It is interesting to compare the collapse formula (4.4.3) with the planar formula (2.2.15) derived in Chapter 2. Near the collapse time one would expect that the formulas should agree since $\bar{a}/R << 1$ when $t \to \tau^\pm$. To see this in more detail, we asymptotically expand the spherical formula (4.4.3) assuming we are near the collapse time, i.e., $t \approx \tau^\pm$. This yields

$$l_{ij}(t) \sim A\left(1 - \frac{t}{\tau^\pm}\right)^{1/2} + B\left(1 - \frac{t}{\tau^\pm}\right)^{3/2} + O\left((1 - \frac{t}{\tau^\pm})^{5/2}\right),$$

where

$$A = l_{ij}(0)\sqrt{1 + \frac{1 \pm \mu}{1 \mp \mu}}$$

$$B = \frac{A\Gamma_2|\lambda_1 - \lambda_2|\tau^\pm}{8\pi R^2 \sqrt{\gamma}}.$$

The power of the leading term agrees with that in the exact planar result shown in formula (2.2.15) of Chapter 2, while the coefficients result from the spherical geometry.

We can also calculate the trajectory of each vortex on the route toward collapse. A difference between the planar and spherical problems is that for the planar case there is only one frequency associated with the collapsing state, given by

$$\omega_p = \dot{\phi} = \frac{1}{4\pi}\sum_{i,j=1}^{3} \frac{(\Gamma_i + \Gamma_j)}{l_{ij}^2}.$$

On the sphere, however, since the orientation of the vortex triangle changes in time, generally, $\dot{\phi}_1 \neq \dot{\phi}_2 \neq \dot{\phi}_3$. The frequencies are given by the formulas

$$\omega_i \equiv \dot{\phi}_i = \frac{1}{2\pi}\left[\frac{\Gamma_j(\cos\theta_j - \cos\theta_i \sin\theta_j \cos(\phi_i - \phi_j))}{l_{ij}^2}\right.$$
$$\left. + \frac{\Gamma_k(\cos\theta_k - \cos\theta_i \sin\theta_k \cos(\phi_k - \phi_i))}{l_{ki}^2}\right], \tag{4.4.4}$$

where $i \neq j \neq k$. To write these frequencies directly as functions of time, we first need to calculate $\cos\theta_i$ in terms of the l_{ij}. It is straightforward to derive

$$\cos\theta_i = 1 + \frac{\Gamma_j\Gamma_k l_{jk}^2}{2R^2\Gamma_i\Gamma}.$$

To calculate $\cos(\phi_i - \phi_j)$ in terms of l_{ij} we start with

$$l_{ij}^2 = 2R^2(1 - \cos\theta_i \cos\theta_j - \sin\theta_i \sin\theta_j \cos(\phi_i - \phi_j)),$$

giving

$$\cos(\phi_i - \phi_j) = \csc\theta_i \csc\theta_j (1 - \cos\theta_i \cos\theta_j - \frac{l_{ij}^2}{2R^2}).$$

Using these formulas then gives the result

$$\omega_i = \frac{\omega_p - \frac{(\Gamma_j + \Gamma_k)}{8\pi R^2}}{1 + \left(\frac{\Gamma_j \Gamma_k l_{jk}^2}{4\Gamma_i R^2 \Gamma}\right)}, \tag{4.4.5}$$

where $i \neq j \neq k$. Note that in the above formula for ω_p, the expression for length l_{ij} should be that on the sphere, as given in (4.4.3). Finally, the equation for ω_i can be integrated to give expressions for the angles $\phi_i(t)$:

$$\phi_i(t) = D_i \log\left(\frac{\tau^\mp + t}{\tau^\pm - t}\right) + B_i \tan^{-1}(\gamma_i t + \delta_i).$$

Expressions for the coefficients $D_i, B_i, \gamma_i, \delta_i$ can be found in Kidambi and Newton (1998). Near collision we can use the asymptotic expansions for $l_{ij}(t)$ to get the following expressions for the frequencies near the collapse time:

$$\omega_1 \sim \frac{(1 \mp \mu)}{2}\omega_p$$
$$- \frac{1}{8\pi R^2}\left[\Gamma_2 + \Gamma_3 + R^2\left(\sum \frac{\Gamma_i + \Gamma_j}{l_{ij}^2(0)}\right)\left(\frac{\mu^2 - 1}{2} + 2\alpha_1 l_{23}^2(0)\right)\right]$$
$$+ O(1 - t/\tau^\pm)$$

$$\omega_2 \sim \frac{(1 \mp \mu)}{2}\omega_p$$
$$- \frac{1}{8\pi R^2}\left[\Gamma_1 + \Gamma_3 + R^2\left(\sum \frac{\Gamma_i + \Gamma_j}{l_{ij}^2(0)}\right)\left(\frac{\mu^2 - 1}{2} + 2\alpha_2 l_{31}^2(0)\right)\right]$$
$$+ O(1 - t/\tau^\pm)$$

$$\omega_3 \sim \frac{(1 \mp \mu)}{2}\omega_p$$
$$- \frac{1}{8\pi R^2}\left[\Gamma_1 + \Gamma_2 + R^2\left(\sum \frac{\Gamma_i + \Gamma_j}{l_{ij}^2(0)}\right)\left(\frac{\mu^2 - 1}{2} + 2\alpha_3 l_{12}^2(0)\right)\right]$$
$$+ O(1 - t/\tau^\pm).$$

Notice that the leading term of each frequency agrees (up to a multiplicative constant) with the planar frequency ω_p.

4.4.1   Geometry of Partner States

To further understand why there are two distinct collapse times associated with a set of initial conditions to equations (4.2.1), (4.2.2), (4.2.3) we make use of the constraint of Lemma 4.2.1. Figure 4.5 shows the relevant geometry, with the tip of the center of vorticity vector denoted C^* lying at the North Pole. The vortex plane, denoted P, always intersects this point, hence the vortices lie on circles formed by intersecting the plane with the sphere. Our first goal is to compute the angle between $\mathbf{c}$ and $\mathbf{n}$, which we denote $\alpha(t)$. For this we use the formula

$$\mathbf{c} \cdot \mathbf{n} = \|\mathbf{c}\|\|\mathbf{n}\| \cos \alpha = V.$$

Since $\|\mathbf{n}\| = 2|A| \geq 0$ and $\|\mathbf{c}\| = R$, we have

$$\cos \alpha = \frac{V}{2|A|R} = \pm\sqrt{1 - \frac{\bar{a}^2}{R^2}}. \qquad (4.4.6)$$

Differentiating this formula yields

$$\dot{\alpha} = \frac{1}{R^2 \sin 2\alpha} \cdot \frac{d}{dt}(\bar{a}^2),$$

giving rise to

$$\dot{\alpha} = \frac{\rho}{\sin 2\alpha} \cdot \frac{\Gamma_2 V}{\pi R} \left(\frac{1}{l_{12}^2} - \frac{1}{l_{23}^2} \right) \qquad (4.4.7)$$

where

$$\rho = \frac{\lambda_1 \lambda_2}{R^2[2(\lambda_1 + \lambda_2) - (\lambda_1 - \lambda_2)^2 - 1]} > 0.$$

From (4.4.6) we know that

- $V \geq 0 \Rightarrow 0 < \alpha \leq \frac{\pi}{2} \Rightarrow \sin 2\alpha \geq 0$,

- $V < 0 \Rightarrow \frac{\pi}{2} < \alpha < \pi \Rightarrow \sin 2\alpha < 0$,

while from (4.4.7) it is clear that

- $\dot{\alpha} > 0$ if $l_{12} < l_{23}$,

- $\dot{\alpha} < 0$ if $l_{12} > l_{23}$.

The collapsing configurations are shown in Figure 4.14. Suppose we have the collapsing state shown in Figure 4.14(a), which we call configuration I. It is set up so that $\Gamma_1, \Gamma_2 > 0$, $\Gamma_3 < 0$, $l_{12} > l_{23}$, and $V(0) > 0$. Then by (4.4.7), we have $\dot{\alpha} < 0$ and hence $\alpha \downarrow 0$ as $t \to \tau^-$. To obtain the partner state associated with I consider the same set-up, but with the signs of the

Γ's reversed. Because it is the partner state associated with I, we label this configuration I_p. We have $\dot{\alpha} > 0$ for this state, hence $\alpha \uparrow \pi$ as $t \to \tau^+$. The partner states are related to each other by the opposite directions in which the plane P must swing in order to become tangent to the sphere at the North Pole, thereby squeezing the three vortices to their ultimate collapse. Another way of obtaining the partner state related to I is by using the discrete symmetries inherent in the problem. Consider a configuration II obtained by reversing the signs of the x-and/or y-coordinates of configuration I. All the chord lengths l_{ij} and vortex strengths remain as in configuration I. Once again, $\dot{\alpha} < 0$, hence $\alpha \downarrow 0$ as $t \to \tau^+$. Then configuration II_p is obtained by reversing the signs of the Γ's, giving $\dot{\alpha} > 0$, $\alpha \uparrow \pi$ as $t \to \tau^-$. We show in Figure 4.15 the actual partner states on the spherical surface looking down from the North Pole.

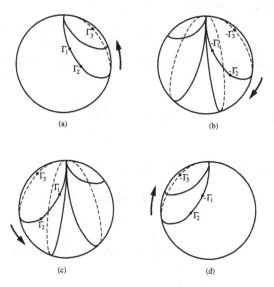

FIGURE 4.14. Collapsing configurations of (a) I. (b) I_p. (c) II. (d) II_p. Configurations I and II_p collapse at time τ^-, while I_p and II collapse at time τ^+. In all cases, the initial lengths are the same.

We now highlight the differences between the planar and spherical collapse process:

1. The spherical collapse has two distinct collapse times, whereas the planar collapse has only one. In the plane the analogue of the partner state is a self-similar expanding state (Aref (1979)) which cannot occur on the sphere because of the extra length scale R, which puts an upper bound on the maximum chord length.

2. The formulas for chord lengths are different for the sphere and the plane. However, the leading term near collapse in the spherical case

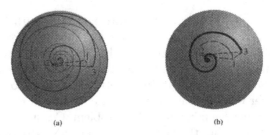

FIGURE 4.15. Actual trajectories of partner states on the spherical surface viewed from North Pole: (a) Counterclockwise orientation of vortices. (b) Clockwise oriented partner state.

agrees with the planar result, with higher order corrections due to the spherical geometry.

3. In the plane the vortices all rotate at the same angular velocity as they collapse, whereas on the sphere each has a distinct angular velocity. The formula for the angular velocity, $\dot{\phi}_i \equiv \omega_i$ associated with vortex Γ_i is

$$\omega_i = \frac{\omega_p - \frac{(\Gamma_j + \Gamma_k)}{8\pi R^2}}{1 + \left(\frac{\Gamma_j \Gamma_k l_{jk}^2}{4\Gamma_i R^2 \Gamma}\right)}$$

where ω_p is the planar angular velocity given by

$$\omega_p = \frac{1}{4\pi} \sum_{i,j=1}^{3} \frac{\Gamma_i + \Gamma_j}{l_{ij}^2}.$$

See Kidambi and Newton (1998, 1999) for further detail.

Remark. Very little is known about the collapse process for $N > 3$ and nothing is known about collapse processes that are not self-similar, although one can presume that they exist.   ◊

4.5 Stereographic Projection

It is useful to project the spherical equations (1.1.23), (1.1.24) onto the stereographic plane. This is accomplished by the coordinate transformation

$$r = \tan(\theta/2),$$

which results in a stereographic projection of a particle located at (θ, ϕ) onto the extended complex plane $\mathcal{C}$, which is tangent to the sphere at

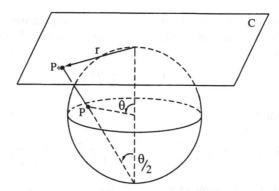

FIGURE 4.16. Stereographic projection of three vortices onto the plane tangent to the sphere at the North Pole.

the North Pole, as shown in Figure 4.16. This point of tangency is at the origin of the projected plane, while the South Pole ($\theta = \pi$) maps to the point at infinity. An important aspect of the stereographic projection is that it is conformal (Needham (1997)) and vector fields on the sphere are mapped in a one-to-one fashion to vector fields on the projected plane. The Hamiltonian for vortex motion on the projected plane is given by

$$\mathcal{H} = -\frac{1}{8\pi R^2} \sum_{i<j}^{N} \Gamma_i \Gamma_j \log \frac{r_i^2 + r_j^2 - 2r_i r_j \cos(\phi_i - \phi_j)}{(1 + r_i^2)(1 + r_j^2)},$$

in terms of which the equations of motion are

$$\Gamma_i \frac{d}{dt}(r_i^2) = (1 + r_i^2)^2 \frac{\partial \mathcal{H}}{\partial \phi_i}$$

$$\Gamma_i \frac{d\phi_i}{dt} = -(1 + r_i^2)^2 \frac{\partial \mathcal{H}}{\partial r_i^2},$$

where (r_i, ϕ_i) are the polar coordinates of the vortex Γ_i in the projected plane. The Hamiltonian for particle motion on the projected plane is

$$\mathcal{H}_p(r, \phi; r_i, \phi_i) = -\frac{1}{4\pi R^2} \sum_{i=1}^{N} \Gamma_i \log \left(\frac{r^2 + r_i^2 - 2rr_i \cos(\phi - \phi_i)}{(1 + r^2)(1 + r_i^2)} \right),$$

where (r, ϕ) are the polar coordinates of the particle in the projected plane. Note that both r and r_i are dimensionless. The equations of motion for a particle in the projected plane can then be written

$$(\dot{r})^2 = \frac{(1 + r^2)^2}{2} \frac{\partial \mathcal{H}_p}{\partial \phi}$$

$$\dot{\phi} = -\frac{(1 + r^2)^2}{2} \frac{\partial \mathcal{H}_p}{\partial r^2}.$$

It is also quite useful to write these equations in complex form so that they can more easily be compared with the planar equations. Most compactly, they are written

$$\dot{z}^* = i \frac{(1 + \|z\|^2)^2}{2} \frac{\partial \mathcal{H}}{\partial z},$$

with Hamiltonian

$$\mathcal{H}(z, z^*) = -\frac{1}{4\pi R^2} \sum_{j=1}^{N} \Gamma_j \log \frac{\|z - z_j\|^2}{(1 + \|z\|^2)(1 + \|z_j\|^2)}.$$

Expanded out, the equations read

$$\dot{z}^* = -i \frac{(1 + \|z\|^2)^2}{8\pi R^2} \left[\sum_{j=1}^{N} \frac{\Gamma_j}{(z - z_j)} - \frac{\Gamma z^*}{(1 + \|z\|^2)} \right]. \qquad (4.5.1)$$

Remarks.

1. When written out for the vortex motion, (4.5.1) becomes

$$\dot{z}_\alpha^* = -i \frac{(1 + \|z_\alpha\|^2)^2}{8\pi R^2}$$

$$\times \left[\sum_{\beta \neq \alpha}^{N} \frac{\Gamma_\beta}{(z_\alpha - z_\beta)} - \frac{\Gamma z_\alpha^*}{(1 + \|z_\alpha\|^2)} \right], \qquad (4.5.2)$$

which generalizes the one obtained by Hally (1980) and reduces to it when $\Gamma = 0$.

2. To understand how (4.5.1) behaves in the planar limit $R \to \infty$, it is useful to use the dimensional variable $w = 2Rz$, so that (4.5.1) transforms to

$$\dot{w}^* = -i \frac{(4R^2 + \|w\|^2)^2}{32\pi R^4} \left[\sum_{j=1}^{N} \frac{\Gamma_j}{(w - w_j)} - \frac{\Gamma w^*}{(4R^2 + \|w\|^2)} \right],$$

$$= -i \frac{(4 + \|w\|^2/R^2)^2}{32\pi} \left[\sum_{j=1}^{N} \frac{\Gamma_j}{(w - w_j)} - \frac{\Gamma w^*}{R^2(4 + \|w\|^2/R^2)} \right].$$

Then, in the limiting case where $\|w\|/R << 1$, to leading order we obtain the equations

$$\dot{w}^* = -\frac{i}{2\pi} \sum_{j=1}^{N} \frac{\Gamma_j}{(w - w_j)},$$

which exactly corresponds to the planar equations (1.1.22).

3. Although (4.5.1) gives the correct velocity at all finite points in the plane, it cannot be used to calculate the point at infinity, i.e., the image of the South Pole, since the stereographic projection is not continuous there. In general, to treat this special point, the Cartesian equations (4.1.1) are more useful. ◊

4.6 Integrable Streamline Topologies

In this section we briefly summarize work in Kidambi and Newton (2000a) in which the streamline patterns for integrable two and 3-vortex motion are categorized and the motion of stagnation points is studied. The analogous planar problem was recently treated by Aref and Bröns (1998), while a general program to topologically classify all integrable motion has been initiated by Fomenko and coworkers (1991). We start by writing the equations for the stagnation points $z \equiv z_s$ on the sphere, which are stationary points of (4.5.1),

$$\sum_{j=1}^{N} \frac{\Gamma_j}{(z_s - z_j)} = \frac{\Gamma z_s^*}{(1 + \|z_s\|^2)}.$$

In vector form the stagnation points $\mathbf{x} \equiv \mathbf{x}_s$ are obtained by solving the system

$$\sum_{j=1}^{N} \frac{\Gamma_j (\mathbf{x}_i \times \mathbf{x}_s)}{\|\mathbf{x}_s - \mathbf{x}_j\|^2} = 0.$$

The topological constraint that the flow lie on a spherical surface gives rise to the following general and useful *index* theorem.

Theorem 4.6.1 (Poincaré Index Theorem). *The index $I_f(S)$ of a two-dimensional surface S, relative to any C^1 vector field f on S with at most a finite number of critical points is equal to the Euler–Poincaré characteristic of S, denoted $\chi(S)$, i.e., $I_f(S) = \chi(S)$.*

Remarks.

1. For a full proof of the theorem, see Perko (1996). The book by Needham (1997) also gives a nice discussion of the theorem.

2. Critical points refer to points where the vector field vanishes or is singular. In our case these are the **stagnation points** and the point vortex locations, which generically are not stagnation points.

3. For a sphere $\chi(S) = 2$. The index of a center is $+1$, while that for a saddle is -1. This implies that if only centers and saddles are present in the flowfield, with c denoting the number of centers and s the number of saddles, then

$$c - s = 2.$$

4. Recall that each vortex is a center, which means there must be at least N centers, i.e., $c \geq N$. For the planar problem studied by Aref and Bröns (1998), $c = N$. $\Diamond$

To understand and classify the vector fields and streamline patterns we first need the following.

Definition 4.6.1. *A stationary point $\mathbf{x}_0$ of the system $\dot{\mathbf{x}} = f(\mathbf{x})$ is called* **nondegenerate** *if the Jacobian matrix $Df(\mathbf{x}_0)$ has no zero eigenvalues. Otherwise, it is called* **degenerate**.

Then we can state the main result concerning stagnation points (Kidambi and Newton (2000a)).

Theorem 4.6.2 (Stagnation Points). *In a flowfield of N-point vortices on a sphere,*

1. *There are at most $N^2 - 2N + 2$ stagnation points if $N > 2$.*

2. *The only possible nondegenerate stagnation points are centers (index $+1$) or saddles (index -1).*

3. *For $N = 1$, the only stagnation point is a center located at the antipodal point to the vortex.*

4. *For $N = 2$, all stagnation points must lie on a great circle passing through the two vortices.*

5. *If $\Gamma = 0$, there can be at most $N - 2$ stagnation points. If $N = 3$, the single stagnation point must lie on the vortex circle.*

We refer the reader to Kidambi and Newton (2000a) for details of the proof.

We now categorize the streamline patterns for the case of two and three vortices. Classifying the topologies according to a set of "building block" figures, which we call **primitives**, and using only continuous deformations (homotopies) and linear superposition of the primitives, all possible streamline topologies for the 3-vortex problem can be constructed.

4.6.1 Two Vortices

For this simple case, since the vortices necessarily lie on a great circle, we assume for convenience that they lie on the equator defined in the complex plane as the unit circle $\|z\| = 1$. Without loss, we can fix the position and strength of one of the vortices, so we take $\Gamma_1 = 1$, $\phi_1 = 0$, hence $z_1 \equiv \exp(\phi_1) = 1$. This leaves a two-parameter problem as we vary the vortex strength $\Gamma \in (-\infty, \infty)$, and angle $\phi \in (0, \pi]$ of the second vortex, i.e., we take $z_2 = \exp(i\phi)$. We treat the case where the vortices lie on opposite sides of the sphere separately.

If we let z_s denote the location of the stagnation points, we know that it must satisfy the algebraic equation

$$\frac{1}{z_s - 1} + \frac{\Gamma}{z_s - \exp(i\phi)} = \frac{(1 + \Gamma)z_s^*}{1 + z_s z_s^*}.$$

Since we also know that $\|z_s\|^2 = 1$, we can simplify this equation to obtain

$$F(z_s; \phi, \Gamma) \equiv (1 + \Gamma)z_s^2 + (1 - \Gamma)(1 - \exp(i\phi))z_s$$
$$- (1 + \Gamma)\exp(i\phi) = 0. \tag{4.6.1}$$

The solutions of this complex quadratic are summarized in the following.

Theorem 4.6.3 (2-Vortex Problem). *Consider the solutions of*

$$F(z_s; \phi, \Gamma) = 0$$

on the unit circle $\|z_s\| = 1$, as a function of the two parameters $\phi \in (0, \pi]$, $\Gamma \in [-\infty, \infty]$. In the (ϕ, Γ)-plane, define the following two curves shown in Figure 4.17:

$$\Gamma \equiv \Gamma_+ = \frac{\sin(\phi/2) - 1}{\sin(\phi/2) + 1}, \quad \Gamma \equiv \Gamma_- = \frac{1}{\Gamma_+}.$$

Then for any fixed value $\phi \in (0, \pi]$ we have:

1. *For $\Gamma > \Gamma_+$ or $\Gamma < \Gamma_-$, there are two stagnation points $z_s = z_\pm$ given by*

$$z_+ = \exp[i(\frac{\phi}{2} + \alpha)], \quad z_- = \exp[i(\frac{\phi}{2} + \pi - \alpha)], \tag{4.6.2}$$

   where

$$\sin\alpha = \frac{1 - \Gamma}{1 + \Gamma}\sin(\frac{\phi}{2}), \quad -\frac{\pi}{2} \leq \alpha \leq \frac{\pi}{2}.$$

The stagnation point at z_+ is a saddle, while the one at z_- is a center.

2. *At the endpoints* $\Gamma = \Gamma_{\pm}$, *the roots coalesce and there is only one stagnation point, which is a cusp located at*

$$z_s = \exp[i(\frac{\phi}{2} \pm \frac{\pi}{2})].$$

3. *For* $\Gamma \in (\Gamma_-, \Gamma_+)$, *there are no stagnation points in the flow.*

Proof. The proof is straighforward and is based on an examination of the solutions to the quadratic equation (4.6.1). See Kidambi and Newton (2000a) for more details. A picture of the cusp topology is shown in Figure 4.18. □

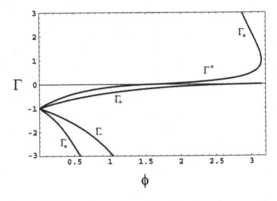

FIGURE 4.17. Special bifurcation curves for the 2-vortex problem. Shown are the curves $\Gamma_{\pm}$ and Γ^*, Γ_* in the (Γ, ϕ)-plane.

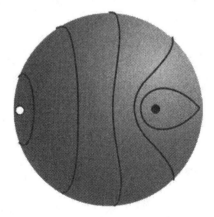

FIGURE 4.18. Streamline topology showing a cusp for the 2-vortex problem.

The case where the vortices lie on opposite sides of the sphere is summarized as:

Lemma 4.6.1. *Suppose the two vortices are at antipodes on the sphere, i.e., let $\Gamma_1 = 1$, $z_1 = 1$, $z_2 = -1$. Then*

1. *For $\Gamma > 0$, there is a continuum of stagnation points located on the vertical latitude of the sphere at a distance $2R/(1 + \Gamma)$ from z_1.*

2. *For $\Gamma = 0$, there is one stagnation point at $z_3 = -1$. This is produced by the vortex located at $z_1 = 1$.*

3. *For $\Gamma < 0$, there are no stagnation points.*

The proof is left as an exercise. We show in Figures 4.19 and 4.20 two special cases of interest. Figure 4.19 depicts the streamline topologies in a corotating frame of reference for the case of two equal strength vortices, both on the sphere and in the stereographic plane. Figure 4.20, by contrast, depicts the dipole case in which the vortices have equal and opposite strength.

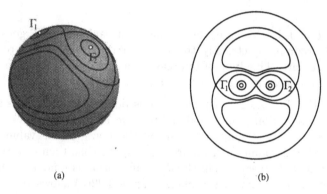

(a)   (b)

FIGURE 4.19. Streamline topologies for 2-vortex configuration, $\Gamma_1 = \Gamma_2$: (a) Corotating frame. (b) Stereographic projection.

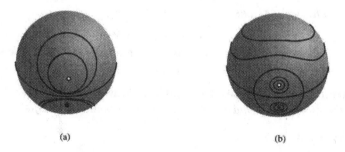

(a)   (b)

FIGURE 4.20. Streamline topologies for 2-vortex dipole, $\Gamma_1 + \Gamma_2 = 0$: (a) Fixed frame. (b) Corotating frame.

It is interesting to study the bifurcations that occur from one pattern to another as the parameters vary. We emphasize that these are not dynamical bifurcations for the 2-vortex problem since all solutions form relative equilibria. Bifurcations in streamline topologies can occur as the parameters ϕ and Γ are varied. The changes fall into two general classes: (1) homotopies due to the spherical topology; (2) changes in the number of stagnation points in the flow.

1. To understand the first type of bifurcation, consider Figure 4.21. Shown in this figure is a continuous deformation of one of the primitives (lemniscate as shown in Figure 4.21(a)) on the sphere to its homotopic equivalent, the limacon, as shown in Figure 4.21(d). As the sphere is pushed through the left loop of the lemniscate, it deforms to a limacon. For the planar problem considered in Aref and Bröns (1998), these two figures cannot be homotopic equivalents since it is not possible to continuously deform one to the other.

2. Changes in the number of stagnation points in the flow occur in two cases:

   (i) $\Gamma = 0$: In this case, the saddle point vanishes. As Γ goes through the origin parametrically, the lemniscate-limacon switches to limacon-lemniscate. This type of bifurcation occurs in the planar case as well and is shown in Figure 4.22, marked as point A.

   (ii) $\Gamma = \Gamma_\pm$: At $\Gamma = \Gamma_+$ (marked as point C in Figure 4.22(a)), the two stagnation points coalesce to form a cusp. For $\Gamma_- < \Gamma < \Gamma_+$, there are no stagnation points in the flow. At the value $\Gamma = \Gamma_-$ (marked as point D in Figure 4.22(a)) the cusp then gives birth to a saddle and a center. This kind of bifurcation is unique to the sphere and cannot occur in the plane. Figure 4.22(b) shows the topologies associated with each of the distinct intervals.

4.6.2   Three Vortices

We next consider the more complicated case of stagnation point structure and streamline patterns for three vortices of general strength. We start by classifying the number of stagnation points in the flow.

Theorem 4.6.4 (3-Vortex Stagnation Points). *For the case $N = 3$ let M be the number of nondegenerate stagnation points in the flow.*

1. $M = 1, 3$, *or* 5.

2. *If $M = 1$, it must be a saddle. If $M = 3$, there must be 2 saddles and 1 center. If $M = 5$, there must be 3 saddles and 2 centers.*

Proof. We know from Theorem 4.6.2 that $M \leq 5$. If we let s denote the number of saddles and c the number of centers that are stagnation points,

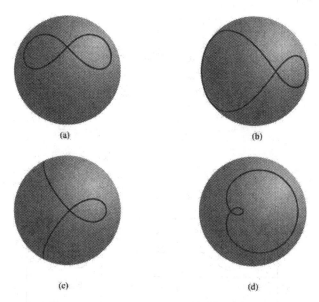

FIGURE 4.21. Continuous deformation of a lemniscate to a limacon. In the four steps (a)-(d), the sphere is pushed through what begins as the left loop.

then

$$s + c = M$$
$$-s + c + N = 2.$$

From this we have

$$s = \frac{M+1}{2}, \quad c = \frac{M-1}{2},$$

implying that $M = 1, 3, 5$. The second statement follows from this, together with the Poincaré Index Theorem. $\qquad \square$

We can now classify all possible primitive topologies for the 3-vortex case in the spirit of Fomenko and coworkers (1991), who initiated a general program to classify all possible topologies associated with integrable motion. They have achieved this in the limited case of integrable systems with two degrees of freedom. Our classification into 12 primitive topologies is shown in Figure 4.24. The left column lists the number of saddle points occuring in the figure. The numbers along the bottom of each figure indicate the number of homoclinic-heteroclinic-triheteroclinic loops in each figure. It is important to understand that each primitive can be continuously deformed on the surface of the sphere to a visually distinct but topologically equivalent figure — a *homotopic equivalent* of the original. Hence, each of the 12 figures represents a homotopy equivalence class (Schwarz (1994)). We

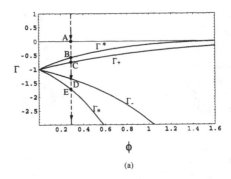

(a)

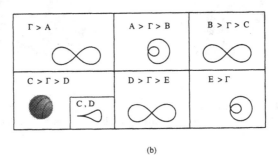

(b)

FIGURE 4.22. Changes in the number of stagnation points: (a) Bifurcation curves $\Gamma_\pm, \Gamma^*, \Gamma_*$ in the (Γ, ϕ)-plane. As the parameter Γ decreases, the topology changes at the points marked A, B, C, D, E. (b) Primitive topologies in each region — degenerate cusps occur at points marked C and D.

show in Figure 4.25 the primitives along with their topologically equivalent figures, of which there are 23. For example, to understand how a figure is continuously deformed to another, see Figure 4.21, which shows the deformation of a lemniscate to a limacon. In the four steps (a) through (d), the sphere is pushed through what is at first the left loop.

We now show the streamline patterns that occur when the three vortices form a relative equilibrium. The number of stagnation points in the flowfield of three vortices in equilibrium could be 1, 3, or 5, so that a variety of topologies are possible. We depict several of the more interesting cases. Figures 4.26 and 4.27 show two different topologies when the vortices are in a fixed equilibrium, hence lie on a great circle with strengths satisfying the appropriate relations as summarized in Theorem 4.2.2. Both configurations are isoceles triangles but with different angles. In the first case the angle is 40 degrees and there are three stagnation points (two saddles and one center); in the second the equal angle is 50 degrees and there are five stagnation points (three saddles and two centers). Both configurations are stable according to the stability criterion of Theorem 4.2.4. Figure 4.28

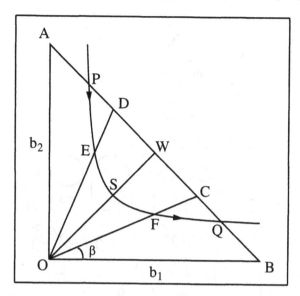

FIGURE 4.23. Symmetric phase plane diagram for special periodic solution.

shows four different topologies when the vortices are in a relative equilibrium state. Figure 4.28(a) and 4.28(b) are great circle states. The two configurations are identical isosceles equilateral triangles but with different vortex strengths ($\Gamma_1 = \Gamma_2 = 1$ in both cases, $\Gamma_3 = 1$ in Figure 4.28(a), $\Gamma_3 = -4$ in Figure 4.28(b)), leading to different topologies. Again, both are stable configurations. Figures 4.28(c) and (d) pertain to the other type of relative equilibria, the equilateral triangle configuration. Figure 4.28(c) shows a symmetric situation where $\Gamma_1 = \Gamma_2 = \Gamma_3$ and the vortices are on a fixed latitude of 45 degrees. There are three saddles and two centers and the topology is made up of the primitives (0,3,1) and (0,0,0). In Figure 4.28(d) $\Gamma_1 = 1, \Gamma_2 = 2, \Gamma_3 = 3$. Since c is aligned with the z-axis, the vortices are not on a fixed latitude. The topology of this case is simpler and consists of a pair of nested lemniscates (primitives (2,0,0) and (0,0,0)). Both equilateral triangles are stable. Figure 4.29 shows the degenerate case in which the total vorticity vanishes and the vortices are place in an equilateral triangle configuration. This configuration is the 3-vortex analogue of the corresponding 2-vortex dipole case shown in Figure 4.20.

We end this section by illustrating two examples of streamline patterns for dynamical states that are not in equilibrium. Shown in Figure 4.30 are the streamline patterns for a collapsing state at three successive times. As the vortices spiral towards the collapse point at the tip of the center of vorticity vector, the streamline pattern retains its topology throughout. An example of a nonequilibrium state that exhibits one bifurcation in its streamline topology is described next.

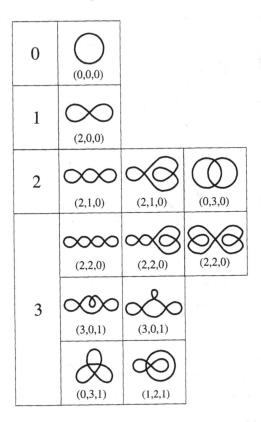

FIGURE 4.24. Primitive chart for the 3-vortex problem.

Example 4.2 (Periodic motion) A special periodic state can be derived exactly. The vortex strengths and separations satisfy

$$\Gamma_1 = \Gamma_2 = -\Gamma_3 \equiv \Gamma$$
$$\sum \Gamma_i \Gamma_j l_{ij}^2 = 0.$$

These conditions imply

$$l_{12}^2 = l_{23}^2 + l_{31}^2,$$

showing that the vortices form a right triangle with hypotenuse l_{12}. We can write the other two sides as

$$l_{23} = l_{12} \cos \alpha$$
$$l_{31} = l_{12} \sin \alpha.$$

Since $C_1 = 0$, we can define phase plane variables b_i by

$$b_1 = \frac{l_{23}^2}{4\Gamma_1 R^2}, \qquad b_2 = \frac{l_{31}^2}{4\Gamma_2 R^2}, \qquad b_3 = \frac{l_{12}^2}{4\Gamma_3 R^2},$$

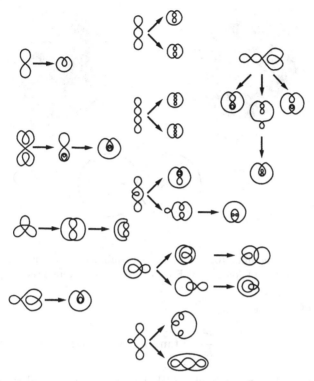

FIGURE 4.25. Homotopic equivalents of each of the primitives.

which then yields the identity

$$b_1 + b_2 + b_3 = \frac{C_1}{4R^2(\Gamma_1\Gamma_2\Gamma_3)} = 0.$$

We first can examine solutions in the (b_1, b_2)-plane, where the motion is restricted to lie in the first quadrant, since $b_1 > 0$, $b_2 > 0$. The phase plane is shown in Figure 4.23, where the physical region is shown by the triangle OAB with $A = (0, 1/\Gamma)$ and $B = (1/\Gamma, 0)$. The lines OA and OB are inaccessible to the motion since they correspond to 2-vortex collapse, which cannot occur. However, AB is accessible and all states on this line are great circle states with $l_{12} = 2R$. From the fact that $l_{23} = 0$ at A and then increases monotonically to $2R$ at B, it is clear that α decreases monotonically from $\pi/2$ at A to zero at B. Phase curves are given by $1/b_1 + 1/b_2 = const$, a typical one is marked PSQ in Figure 4.23. The curves are hyperbolas that are symmetric with respect to the line $b_1 = b_2$. On the front side of the plane, where $V > 0$, the vector field has the direction shown. On the backside it will be in the opposite direction.

To deduce further properties of the solution, note that we have $\alpha = const$ on any ray that goes through the origin, such as the one marked OC, since

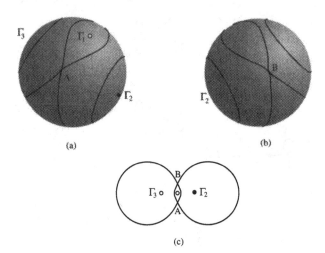

FIGURE 4.26. Streamline topology for fixed equilibria — vortices are on a fixed longitude. (a) Front of sphere. (b) Back of sphere. (c) Stereographic projection.

on this line we have

$$\frac{b_2}{b_1} = \frac{l_{31}^2}{l_{23}^2} = \tan^2 \alpha = \lambda = const$$

In particular, the angle bisector OW has $\alpha = \pi/4$, i.e., all states on this line are isosceles right triangles. Moreover, $\dot{V} = 0$ on OW which makes W a nondegenerate great circle relative equilibrium, as described in Theorem 4.2.3. From the phase plane analysis, we can deduce the following.

Lemma 4.6.2. *The hypotenuse $l_{12}(t)$ evolves periodically, with time period T, whereas the other triangle sides l_{23} and l_{31} have periods $2T$. The vortex triangle has the same shape and size at exactly four different times during one period.*

We can now compute the relative as well as the absolute motion of the vortices. Constancy of the Hamiltonian implies that

$$\frac{l_{23}^2 l_{31}^2}{l_{12}^2} = const$$

This allows us to write a simple expression for the volume,

$$V = \pm \frac{1}{2} l_{23} l_{31} \sqrt{4R^2 - l_{12}^2},$$

which when used in the relative equations of motion yields

$$\frac{d}{dt}(l_{12}^2) = \pm \frac{\Gamma \csc 2\alpha_0}{2\pi R^2} \mathrm{sgn}(l_{31}^2 - l_{23}^2)\sqrt{(4R^2 - l_{12}^2)(l_{12}^2 - 4R^2 \sin^2 2\alpha_0)},$$

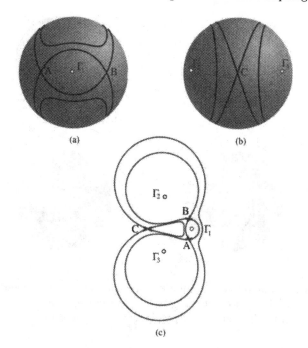

FIGURE 4.27. Fixed equilibrium on the equator. (a) Front of sphere. (b) Back of sphere. (c) Stereographic projection.

where α_0 is the initial value of α and the initial state is chosen to lie on the line AB in the phase plane. For convenience and without loss, we choose $l_{12}(0) = 2R$. From these equations, we can integrate for the side lengths

$$l_{12}(t) = 2R[\cos^2(\frac{ut}{2}) + \sin^2 2\alpha_0 \sin^2(\frac{ut}{2})]^{1/2},$$

$$l_{23}(t) = \left[\frac{l_{12}}{2}(l_{12} \pm \sqrt{l_{12}^2 - 4R^2 \sin^2 2\alpha_0})\right]^{1/2},$$

$$l_{31}(t) = \left[\frac{l_{12}}{2}(l_{12} \mp \sqrt{l_{12}^2 - 4R^2 \sin^2 2\alpha_0})\right]^{1/2},$$

where

$$u = \Gamma \frac{\csc 2\alpha_0}{2\pi R^2},$$
$$\alpha_0 = \alpha(0).$$

The appropriate sign is chosen depending on whether $l_{23} > l_{31}$ or $l_{23} < l_{31}$. It is interesting to note that even though the side lengths l_{12}, l_{23}, l_{31} can never equal zero, the vortices can be made to approach within an ϵ radius ball of each other for arbitrarily small ϵ. This is because l_{12}, which is the

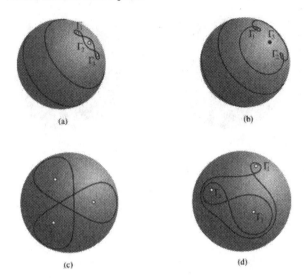

FIGURE 4.28. Four relative equilibria exhibiting distinct topologies. (a) Great circle state. (b) Great circle state. (c) Non great circle state, fixed latitude equilateral triangle. (d) Non great circle equilateral triangle state.

longest side, has a minimum value of $2R \sin 2\alpha_0$, which is attained at times $t = (2n+1)\pi/u, n = 0, 1, 2, \ldots$. Hence, by choosing

$$\alpha_0 = \frac{1}{2}\sin^{-1}\left(\frac{\epsilon}{2R}\right),$$

for example, the closest distance of approach of the vortices can be made to be $O(\epsilon)$ an event termed " ϵ-collapse" in Marchioro and Pulvirenti (1994).

The absolute motion can be obtained from

$$\cos\theta_i = 1 + \frac{\Gamma_j \Gamma_k l_{jk}^2}{2R^2 \Gamma_i \Gamma}$$

$$\phi_1 = \phi_3 + \cos^{-1}\frac{l_{23}}{l_{12}}\sqrt{\frac{4R^2 - l_{12}^2}{4R^2 - l_{23}^2}}$$

$$\phi_2 = \phi_3 - \cos^{-1}\frac{l_{31}}{l_{12}}\sqrt{\frac{4R^2 - l_{12}^2}{4R^2 - l_{31}^2}}$$

$$\phi_3 = \tan^{-1}\sin 2\alpha_0 \tan\left(\frac{ut}{2}\right).$$

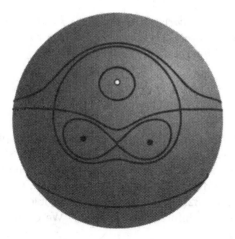

FIGURE 4.29. Equilateral degenerate 3-vortex configuration with vanishing total vorticity, shown in a corotating frame.

The formulas for the Cartesian coordinates associated with the vortex positions are

$$x_1 = \frac{l_{23}}{l_{12}^2} \left[l_{23} \sqrt{4R^2 - l_{12}^2} \cos(\frac{ut}{2}) - 2R l_{31} \sin 2\alpha_0 \sin(\frac{ut}{2}) \right]$$

$$y_1 = \frac{l_{23}}{l_{12}^2} \left[l_{23} \sqrt{4R^2 - l_{12}^2} \sin(\frac{ut}{2}) \sin 2\alpha_0 + 2R l_{31} \cos(\frac{ut}{2}) \right]$$

$$z_1 = \frac{1}{2R} [2R^2 - l_{23}^2]$$

$$x_2 = \frac{l_{31}}{l_{12}^2} \left[l_{31} \sqrt{4R^2 - l_{12}^2} \cos(\frac{ut}{2}) + 2R l_{23} \sin 2\alpha_0 \sin(\frac{ut}{2}) \right]$$

$$y_2 = \frac{l_{31}}{l_{12}^2} \left[l_{31} \sqrt{4R^2 - l_{12}^2} \sin(\frac{ut}{2}) \sin 2\alpha_0 - 2R l_{23} \cos(\frac{ut}{2}) \right]$$

$$z_2 = \frac{1}{2R} \left(2R^2 - l_{31}^2 \right)$$

$$x_3 = \sqrt{4R^2 - l_{12}^2} \cos(\frac{ut}{2})$$

$$y_3 = \sqrt{4R^2 - l_{12}^2} \sin 2\alpha_0 \sin(\frac{ut}{2})$$

$$z_3 = \frac{1}{2R} \left(2R^2 - l_{12}^2 \right).$$

Based on these formulas, we can draw the following conclusions regarding the absolute motion of the vortices.

Lemma 4.6.3. *The absolute vortex motion is periodic, with period $T = 4\pi/u$. The hypotenuse l_{12} has period $T/2$, whereas l_{23} and l_{31} have periods T. If $\alpha_0 = \pi/4$, we have the nondegenerate great circle relative equilibrium*

*state mentioned earlier, corresponding to the point W in the phase plane.
Γ_3 stays fixed on the **c**-axis, whereas the other two rotate around the axis
on the equator, with frequency ω, where*

$$\omega = \frac{\Gamma}{4\pi R^2}.$$

Figure 4.31 shows the periodic trajectories on the sphere for the case
where $\Gamma_1 = \Gamma_2 = -\Gamma_3 = 1$, $C_1 = 0$. The corresponding streamline patterns
are shown in Figure 4.32. For this solution there is one bifurcation in the
streamline topology during the course of one period, as seen by comparing
Figure 4.32(a) and 4.32(c). In more general problems one would expect
many more dynamical bifurcations to occur. We comment on this in the
chapter summary.   ◊

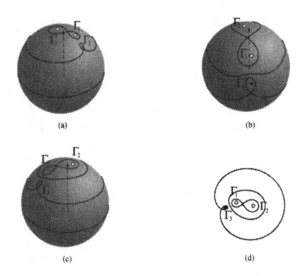

FIGURE 4.30. Collapsing state shown at three successive times with the final
state stereographically projected. Note that the streamline topology does not
change.

4.6.3 Rotating Frames

When the vortices are in relative equilibrium, they rotate rigidly around
their center of vorticity vector with a fixed angular velocity ω. In this
subsection we study the streamline patterns in a frame that is corotating
at the same angular velocity, so that the vortices are stationary and the
flow is steady.

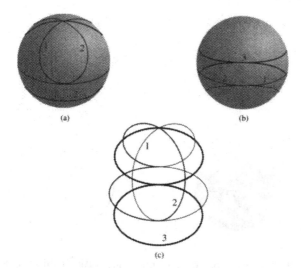

FIGURE 4.31. Vortex trajectories associated with periodic motion, where vortex configuration remains in a right triangle.

If we designate the coordinates in this rotating frame (Θ, Φ), it is easy to see that they are related to the old coordinates by

$$\Theta = \theta$$
$$\Phi = \phi - \omega t.$$

Implicit in this formulation is the assumption that the vortices rotate around the z-axis, or equivalently, around the center of vorticity vector **c**. The equations of motion then take the form

$$\dot{\Theta} = \frac{1}{\sin \Theta} \frac{\partial \mathcal{H}}{\partial \Phi}$$
$$\dot{\Phi} = -\frac{1}{\sin \Theta} \frac{\partial \mathcal{H}}{\partial \Theta},$$

where

$$\mathcal{H} = -\frac{1}{4\pi R^2} \left[\sum \Gamma_i \log l_i^2 + \mu \cos^2\left(\frac{\Theta}{2}\right) \right]$$
$$\mu = 8\pi R^2 \omega,$$

with l_i being the distance between a fluid particle at (θ, ϕ) and the vortex Γ_i. In the complex plane the equations of motion are governed by the Hamiltonian

$$\mathcal{H} = -\frac{1}{4\pi R^2} \left[\sum \Gamma_i \log \left(\frac{\|z - z_i\|^2}{(1 + \|z\|^2)(1 + \|z_i\|^2)} \right) + \frac{\mu}{1 + \|z\|^2} \right].$$

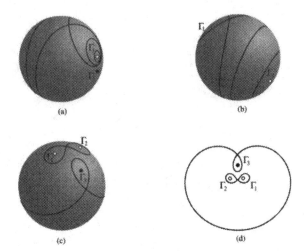

FIGURE 4.32. Streamline patterns for periodic solution exhibiting a single topological bifurcation through one period T. (a) Front of sphere. (b) Back of sphere. (c) Time $t = T/4$. (d) Stereographic projection of (c).

The stagnation points are given by solutions of $\dot{z}^* = 0$, or equivalently, $\partial \mathcal{H}/\partial z = 0$. Using the above Hamiltonian, we see that the stagnation points are solutions to

$$\sum \frac{\Gamma_i}{z - z_i} - \frac{\Gamma z^*}{1 + \|z\|^2} = \frac{\mu z^*}{(1 + \|z\|^2)^2}.$$

It is useful to write the governing equations in vector form as well. Since the stagnation points in the rotating frame are precisely those points at which the fluid particles rotate with angular velocity ω, these points, denoted $\mathbf{x}_s$, are solutions to

$$\dot{\mathbf{x}}_s = \omega \times \mathbf{x}_s = \frac{1}{2\pi R} \sum_{i=1}^{N} \frac{\Gamma_i \mathbf{x}_i \times \mathbf{x}_s}{l_i^2}.$$

This can be written

$$\left[\sum_{i=1}^{N} \frac{\Gamma_i \mathbf{x}_i}{l_i^2} - 2\pi R \omega \right] \times \mathbf{x}_s = 0,$$

which implies

$$\sum_{i=1}^{N} \frac{\Gamma_i \mathbf{x}_i}{l_i^2} - 2\pi R \omega = k \mathbf{x}_s,$$

where k is a scalar.

We demonstrate some of the results in Kidambi and Newton (2000a) for the case $N = 3$. Figure 4.33 shows a case with seven stagnation points in the corotating frame. The topology is a combination of the primitives 1, 2, and 5, and the same configuration in the inertial frame was shown in Figure 4.28. In this figure, there are five stagnation points on the great circle and two off it. Figure 4.34 depicts a non great circle relative equilibrium in the corotating frame.

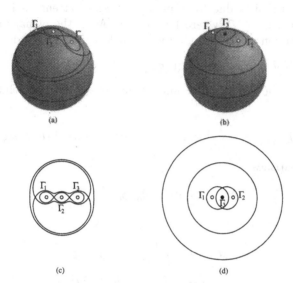

FIGURE 4.33. A 7-stagnation point topology in the corotating frame.

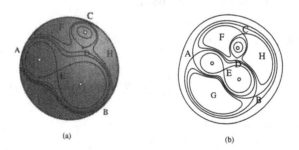

FIGURE 4.34. Non great circle relative equilibrium in corotating frame.

4.7 Boundaries

We now formulate the problem of vortex motion on a sphere with a solid boundary on its surface. This problem is analogous to the planar problem considered in Chapter 3 and is relevant for modeling closed natural basins such as lakes and oceans, as well as for modeling coastlines and shores in large scale oceanographic flows involving vortex motion. We assume that the domain $\mathcal{D}$ on the spherical surface is simply connected and that the condition of no fluid penetration on the boundary $\partial \mathcal{D}$, as specified in (3.0.1), must be satisfied. The discussion in this section is a summary of the work contained in Kidambi and Newton (2000b).

Consider the flow due to N-point vortices of strengths Γ_α located at positions $(\theta_\alpha, \phi_\alpha) \in \mathcal{D}$, where $1 \leq \alpha \leq N$. As in the planar problem, the Green's function of the first kind for a single source can be decomposed as

$$G(\theta, \phi; \theta_\alpha, \phi_\alpha) = G_I(\theta, \phi; \theta_\alpha, \phi_\alpha) + G_{II}(\theta, \phi; \theta_\alpha, \phi_\alpha),$$

where G_I is the Green's function for the sphere with no solid boundary, i.e.,

$$G_I(\theta, \phi; \theta_\alpha, \phi_\alpha) = -\frac{1}{4\pi} \log \left[1 - \cos \theta \cos \theta_\alpha - \sin \theta \sin \theta_\alpha \cos(\phi - \phi_\alpha) \right],$$

where G_I satisfies

$$\Delta G_I = -\delta(\theta, \phi; \theta_\alpha, \phi_\alpha) + \frac{1}{4\pi R^2}.$$

Then G_{II} must satisfy

$$\Delta G_{II} = -\frac{1}{4\pi R^2}, \quad (\theta, \phi) \in \mathcal{D}$$

$$G_{II} = -G_I, \quad (\theta, \phi) \in \partial \mathcal{D}.$$

The Green's function for N sources located at $(\theta_\alpha, \phi_\alpha) \in \mathcal{D}$ is obtained in the usual way, by linear superposition:

$$G^{(N)}(\theta, \phi; \theta_1, \phi_1, \ldots, \theta_N, \phi_N) = \sum_{\alpha=1}^{N} G(\theta, \phi; \theta_\alpha, \phi_\alpha)$$

$$= \sum_{\alpha=1}^{N} G_I(\theta, \phi; \theta_\alpha, \phi_\alpha) + \sum_{\alpha=1}^{N} G_{II}(\theta, \phi; \theta_\alpha, \phi_\alpha).$$

The streamfunction is then

$$\psi = \sum_{\alpha=1}^{N} \Gamma_\alpha G(\theta, \phi; \theta_\alpha, \phi_\alpha),$$

and the equations governing the motion of a passive particle located at $(\theta, \phi) \in \mathcal{D}$ are

$$\dot{\theta} = \frac{1}{R^2 \sin \theta} \frac{\partial \psi}{\partial \phi}$$

$$\dot{\phi} = -\frac{1}{R^2 \sin \theta} \frac{\partial \psi}{\partial \theta}.$$

In standard canonical form the equations can be written

$$\dot{Q} = \frac{\partial \mathcal{H}_p}{\partial P}$$

$$\dot{P} = -\frac{\partial \mathcal{H}_p}{\partial Q},$$

where $P \equiv \cos\theta$, $Q \equiv \phi$, and $\mathcal{H}_p \equiv (1/R^2)\psi$. The motion of the αth vortex is given by the same equations with the vortex Hamiltonian $\mathcal{H}_v$ in place of $\mathcal{H}_p$ and with canonical variables $P_\alpha \equiv \cos\theta_\alpha$, $Q_\alpha \equiv \phi_\alpha$.

4.7.1 Images

We now formulate the method of images on the sphere in such a way that known results for image systems in the planar vortex problem can be used. Keller (1953) gives a nice discussion of the scope of the classical image method. The main idea is to project the spherical problem up to the stereographic plane $\mathcal{C}$, then use a version of the planar image method in the stereographic plane. The final step is to then project the system and boundary back down to the sphere. In order to carry out this process, we first consider the equations in the stereographic plane, as described in Section 4.5 . To establish notation, we write the streamfunction equation in $\mathcal{C}$ as

$$\Delta\hat{\psi} = -\sum_{\alpha=1}^{N}\Gamma_\alpha\delta(x,y;x_\alpha,y_\alpha), \quad (x,y)\in\mathcal{D}'$$

$$\hat{\psi} = 0, \qquad\qquad\qquad (x,y)\in\partial\mathcal{D}',$$

where $\mathcal{D}'$ is the stereographic projection of $\mathcal{D}$, with boundary $\partial\mathcal{D}'$. Recall that the basic particle Hamiltonian $\mathcal{H}_p$ and vortex Hamiltonian $\mathcal{H}_v$ in the stereographic plane are

$$\mathcal{H}_p = -\frac{1}{4\pi R^2}\sum_{\alpha=1}^{N}\Gamma_\alpha\log\frac{|z-z_\alpha|^2}{(1+|z|^2)(1+|z_\alpha|^2)}$$

$$\mathcal{H}_v = -\frac{1}{4\pi R^2}\sum_{\alpha<\beta}\Gamma_\alpha\Gamma_\beta\log\frac{|z_\alpha-z_\beta|^2}{(1+|z_\alpha|^2)(1+|z_\beta|^2)}.$$

In complex form, the particle and vortex equations are

$$\dot{z}^* = -i\frac{(1+r^2)^2}{8\pi R^2}\left[\sum_{\alpha=1}^{N}\frac{\Gamma_\alpha}{z-z_\alpha} - \frac{\Gamma z^*}{1+|z|^2}\right]$$

$$\dot{z}_\alpha^* = -i\frac{(1+r_\alpha^2)^2}{8\pi R^2}\left[\sum_{\beta\neq\alpha}^{N}\frac{\Gamma_\beta}{z_\alpha-z_\beta} - \frac{(\Gamma-\Gamma_\alpha)z_\alpha^*}{1+|z_\alpha|^2}\right],$$

where

$$\Gamma \equiv \sum_{\alpha=1}^{N} \Gamma_\alpha.$$

For comparison, recall that in the physical plane the corresponding Hamiltonians are

$$\hat{\mathcal{H}}_p = -\frac{1}{2\pi} \sum_{\alpha=1}^{N} \Gamma_\alpha \log |z - z_\alpha|$$

$$\hat{\mathcal{H}}_v = -\frac{1}{2\pi} \sum_{\alpha<\beta} \Gamma_\alpha \Gamma_\beta \log |z_\alpha - z_\beta|,$$

with corresponding equations of motion

$$\dot{z}^* = -\frac{i}{2\pi} \sum_{\alpha=1}^{N} \frac{\Gamma_\alpha}{(z - z_\alpha)}$$

$$\dot{z}_\alpha^* = -\frac{i}{2\pi} \sum_{\beta\neq\alpha}^{N} \frac{\Gamma_\beta}{(z_\alpha - z_\beta)}.$$

In order to implement the image method on the sphere we first need to establish two important results.

Theorem 4.7.1 (Image System). *Consider a system of N vortices on the sphere in a simply connected, closed domain $\mathcal{D}$, which stereographically projects to the domain $\mathcal{D}'$. The corresponding boundaries are denoted $\partial\mathcal{D}$ and $\partial\mathcal{D}'$, respectively.*

1. *Assume the domain is such that $G(\theta, \phi)$ can be constructed by the method of images. Then the total vorticity, including both the vortices and their images, vanishes, i.e.,*

$$\Gamma \equiv \sum_{\alpha=1}^{M} \Gamma_\alpha = 0,$$

   where Γ_α, $1 \leq \alpha \leq N$, are the original vortices, with the remaining $(M - N)$ being their images.

2. *The system of vortices in the stereographic plane and the system in the physical plane have the same streamlines.*

Proof. To prove the first part, we need only consider the case of a single vortex of strength γ. By the Riemann mapping theorem, there exists a mapping $\xi = f(z)$ such that any domain $\mathcal{D}'$ can be analytically mapped to the upper half-plane. The complex potential transforms as $w(\xi) = w(f(z))$, which implies that $\psi(\xi) = \psi(f(z))$. The image system in the upper half-plane consists of just one vortex with strength $-\gamma$ located at position ξ_v^*, where ξ_v places the vortex in the upper half-plane. Then

$$\psi(\xi) = -\frac{\gamma}{2\pi} \log \left| \frac{\xi - \xi_v}{\xi - \xi_v^*} \right|.$$

Expressing $\xi = f(z)$, we have

$$\hat{\psi}(z) = -\frac{\gamma}{2\pi} \log \left| \frac{f(z) - f(z_v)}{f(z) - f^*(z_v)} \right|.$$

Thus, for a single vortex of strength γ, its image is of equal and opposite strength. The fact that the total vorticity vanishes for the general case of N vortices follows easily from linear superposition.

For the second part of the theorem, consider the streamlines associated with the system. To show that they are identical to the streamlines in the stereographic plane, consider first the particle equations in the physical plane

$$\dot{x} = -\frac{1}{2\pi} \sum_{\alpha=1}^{M} \frac{\Gamma_\alpha (y - y_\alpha)}{(x - x_\alpha)^2 + (y - y_\alpha)^2} \equiv \frac{A}{2\pi}$$

$$\dot{y} = \frac{1}{2\pi} \sum_{\alpha=1}^{M} \frac{\Gamma_\alpha (x - x_\alpha)}{(x - x_\alpha)^2 + (y - y_\alpha)^2} \equiv \frac{B}{2\pi}.$$

The streamlines are given by

$$\frac{dy}{dx} = \frac{B}{A}.$$

In the stereographic plane, since $\Gamma = 0$, the particle velocity is given by

$$\dot{z}^* = -i \frac{(1 + |z|^2)^2}{8\pi R^2} \sum_{\alpha=1}^{M} \frac{\Gamma_\alpha}{z - z_\alpha},$$

which in component form is

$$\dot{x} = -\frac{(1 + x^2 + y^2)^2}{8\pi R^2} \sum_{\alpha=1}^{M} \frac{\Gamma_\alpha (y - y_\alpha)}{(x - x_\alpha)^2 + (y - y_\alpha)^2}$$

$$\dot{y} = \frac{(1 + x^2 + y^2)^2}{8\pi R^2} \sum_{\alpha=1}^{M} \frac{\Gamma_\alpha (x - x_\alpha)}{(x - x_\alpha)^2 + (y - y_\alpha)^2}.$$

From this, one obtains the same streamlines by forming $dy/dx = B/A$. Thus, if $\Gamma = 0$, the streamlines in the planar and stereographic equations are identical for identical vortex positions. $\square$

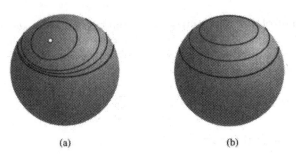

(a)   (b)

FIGURE 4.35. Vortex motion in a circular cap: (a) Instantaneous streamlines for $\theta_v = 30^0, \phi_v = 270^0$. (b) Vortex paths lie on constant latitudes.

Since the streamlines in the stereographic plane and the physical plane are identical, the images in the physical plane that are used for a given boundary can also be used in the stereographic plane, since both guarantee that the streamline is constant on the boundary. Hence, given a domain $\mathcal{D}$ on the sphere along with one vortex of strength γ located at position $(\theta_v, \phi_v) \in \mathcal{D}$, the particle and vortex Hamiltonians are constructed as follows:

1. We first map $\mathcal{D}$ stereographically to $\mathcal{D}'$, hence the coordinate map is given by $(\theta_v, \phi_v) \mapsto (r_v, \phi_v)$.

2. We then consider the identical problem for planar vortex motion in $\mathcal{D}'$ which we assume is solvable by the image method. The streamlines are

$$\left| \frac{f(z) - f(z_v)}{f(z) - f^*(z_v)} \right| = const \tag{4.7.1}$$

Here, $z = r \exp(i\phi)$ and $\xi = f(z)$ maps $\mathcal{D}'$ to the upper half-plane.

3. By the second part of the theorem, (4.7.1) describes the streamlines in the stereographic plane as well. The particle Hamiltonian is given by

$$\mathcal{H}_p = -\frac{\gamma}{4\pi R^2} \log \left| \frac{f(z) - f(z_v)}{f(z) - f^*(z_v)} \right|^2.$$

4. To obtain the Hamiltonian for the vortex motion, we locate the positive image vortices at ρ_j and the negative ones at η_j. Thus, we have

$$\mathcal{H}_v = -\frac{\gamma^2}{4\pi R^2} \log \left(\left| \frac{df}{dz} \right|^2_{z=z_v} \left| \frac{1}{f(z) - f^*(z_v)} \right|^2 \right)$$
$$- \frac{\gamma^2}{4\pi R^2} \log \left[\frac{(1 + |z_v|^2)(1 + |\eta_0|^2) \cdots (1 + |\eta_m|^2)}{(1 + |\rho_1|^2)(1 + |\rho_2|^2) \cdots (1 + |\rho_m|^2)} \right].$$

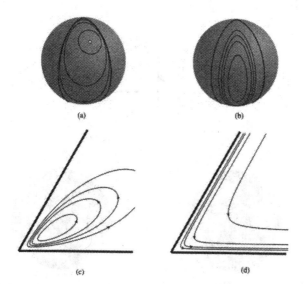

FIGURE 4.36. Vortex motion in a spherical sector: (a) Instantaneous stream-lines for $\theta_v = 45^0, \phi_v = 45^0$. (b) Vortex paths on the sphere. (c) Stereographic projection of the vortex paths. (d) Vortex paths in a planar domain.

We present several examples where the Green's function can be worked out explicitly. In the following examples a vortex of strength γ is located at $(\theta_v, \phi_v) \in \mathcal{D}$.

Example 4.3 (Spherical cap) Take $\mathcal{D}$ to be the spherical cap bounded by the latitude $\theta = \theta_0$. The stereographic projection of $\mathcal{D}$, denoted $\mathcal{D}'$, is the disk of radius $r_0 = \tan(\theta_0/2)$. The image system consists of just one vortex of strength $-\gamma$ at the inverse point $r_i = r_0^2/r_v$, $\phi_i = \phi_v$. The Hamiltonian is

$$
\begin{aligned}
\mathcal{H}_p &= -\frac{\gamma}{4\pi R^2} \log \left| \frac{z - z_v}{z - z_i} \right|^2 \\
&= -\frac{\gamma}{4\pi R^2} \log \left[\frac{(r^2 - 2rr_v \cos(\phi - \phi_v) + r_v^2)r_v^2}{r^2 r_v^2 - 2rr_v r_0^2 \cos(\phi - \phi_v) + r_0^4} \right] = const
\end{aligned}
$$

which give the streamlines. $\mathcal{H}_v$ is easily calculated as

$$
\mathcal{H}_v = -\frac{\gamma^2}{4\pi R^2} \log \left[\frac{(1 + |z_v|^2)(1 + |z_i|^2)}{|z_v - z_i|^2} \right],
$$

giving the vortex paths

$$
\frac{(1 + r^2)(r^2 + r_0^4)}{(r_0^2 - r^2)^2} = const
$$

Representative streamline patterns and vortex paths are shown in Figure 4.35(a),(b) for the case $\theta_0 = 60^0$, $\theta_v = 30^0$, $\phi_v = 270^0$. $\Diamond$

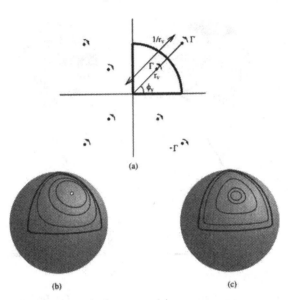

(a)

(b)   (c)

FIGURE 4.37. Vortex motion in a half-sector: (a) Image system. (b) Instantaneous streamlines for $\theta_v = 45^0, \phi_v = 60^0$. (c) Vortex paths on the sphere.

Example 4.4   (Longitudinal wedge) In this example, we take $\mathcal{D}$ to be the sector bounded by the longitudes 0 and π/m, $m \in N$. Then $\mathcal{D}'$ is a wedge of angle π/m. Since $f(z) = z^m$ maps $\mathcal{D}'$ to the upper half-plane, we have for the particle Hamiltonian

$$\mathcal{H}_p = -\frac{\gamma}{4\pi R^2} \log \left| \frac{z^m - z_v^m}{z^m - z_v^{*m}} \right|^2.$$

It is not difficult to see that there are $(2m - 1)$ image vortices, with the positive vortices located at

$$\rho_\alpha = z_v \exp\left(\frac{2i\pi\alpha}{m}\right), \quad 1 \le \alpha \le m - 1.$$

The negative ones are located at

$$\eta_\beta = z_v^* \exp\left(\frac{2i\pi\beta}{m}\right), \quad 1 \le \beta \le m.$$

The streamlines are expressed by

$$\frac{r^{2m} - 2r^m r_v^m \cos m(\phi - \phi_v) + r_v^{2m}}{r^{2m} - 2r^m r_v^m \cos m(\phi + \phi_v) + r_v^{2m}} = const$$

Figure 4.36(a) shows several streamlines for the case $m = 3$ and $\theta_v = \phi_v = 45^0$. The vortex paths are obtained from

$$\frac{(1 + r^2)^2}{r^2 \sin^2 m\phi} = const$$

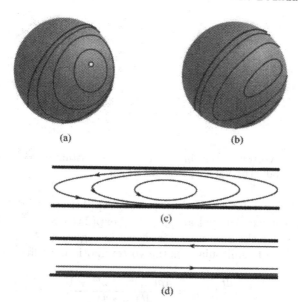

(a)   (b)

(c)

(d)

FIGURE 4.38. Vortex motion in a channel: (a) Instantaneous streamlines for $\theta_v = 45^0, \phi_v = 90^0$. (b) Vortex paths on the sphere. (c) Stereographic projection of the vortex paths. (d) Vortex paths in an identical planar domain.

Vortex paths for different initial positions (θ_v, ϕ_v) are shown in Figure 4.36(b), with their stereographic projections in Figure 4.36(c). For comparison, the vortex paths for the corresponding planar problem are shown in Figure 4.36(d) as open curves where the vortex goes to infinity.

$\Diamond$

Example 4.5   (Half longitudinal wedge) $\mathcal{D}$ is the half-sector bounded by the longitudes zero and $\pi/m, m \in N$. $\mathcal{D}'$ is the sector of a circle. The function $f(z) = ((1 + z^m)/(1 - z^m))^2$ maps the sector to the upper half-plane, so

$$\mathcal{H}_p = -\frac{\gamma}{4\pi R^2} \log \left[\left| \frac{((1 + z^m)/(1 - z^m))^2 - ((1 + z_v^m)/(1 - z_v^m))^2}{((1 + z^m)/(1 - z^m))^2 - ((1 + z_v^{*m})/(1 - z_v^{*m}))^2} \right|^2 \right].$$

The streamlines for the flow are

$$\frac{[r^{2m} - 2r^m r_v^m \cos m(\phi - \phi_v) + r_v^{2m}]}{[r^{2m} - 2r^m r_v^m \cos m(\phi + \phi_v) + r_v^{2m}]} \times$$
$$\frac{[r^{2m} r_v^{2m} - 2r^m r_v^m \cos m(\phi + \phi_v) + 1]}{[r^{2m} r_v^{2m} - 2r^m r_v^m \cos m(\phi - \phi_v) + 1]} = const$$

There are exactly $(4m-1)$ image vortices located at the following locations: $(m - 1)$ positive vortices located at the points $\rho_\alpha = z_v \exp(2\pi i\alpha/m)$, $1 \leq \alpha \leq m - 1$; m positive vortices located at $\rho_\beta = 1/z_v \exp(2\pi i\beta/m)$, $1 \leq \beta \leq$

(a)   (b)

FIGURE 4.39. Vortex motion in a rectangle: (a) Instantaneous streamlines for $\theta_v = 45^0, \phi_v = 30^0$. (b) Vortex paths on the sphere.

m; m negative vortices located at $\eta_\alpha = z_v^* \exp(2\pi i\alpha/m)$, $1 \leq \alpha \leq m$; and m negative vortices located at $\eta_\beta = 1/z_v^* \exp(2\pi i\beta/m)$, $1 \leq \beta \leq m$. The image system for the case $m = 2$ is shown in Figure 4.37(a).

To obtain the Hamiltonian for the vortex motion, we first compute

$$\left|\frac{df}{dz}\right|_{z=z_v} = \frac{4m|z_v|^{m-1}|z_v^m + 1|}{|(1 - z_v^m)^3|}.$$

This gives

$$\mathcal{H}_v = -\frac{\gamma^2}{4\pi R^2} \log\left[\frac{16m^2|z_v|^{2m-2}|z_v^m + 1|^2}{|(1 - z_v^m)^3|^2}\right.$$
$$\left.\times \frac{(1 + |z_v|^2)^2}{|((1 + z_v^m)/(1 - z_v^m))^2 - ((1 + z_v^{*m})/(1 - z_v^{*m}))^2|^2}\right].$$

The vortex paths are given by

$$\frac{[(1 + r^{2m})^2 - 4r^{2m}\cos^2 m\phi](1 + r^2)^2}{r^2(r^{2m} - 1)^2 \sin^2 m\phi} = const$$

The instantaneous streamlines are shown in Figure 4.37(b), while the vortex paths are shown in Figure 4.37(c).   ◊

Example 4.6   (Channel) $\mathcal{D}$ is the channel configuration bounded by the longitudes zero and π, and the curve $\tan(\theta/2)\sin\phi = c$, where c is a constant. $\mathcal{D}'$ is an infinite channel whose boundaries are $y = 0$ and $y = c$. The function $f(z) = \exp(\pi z/c)$ maps $\mathcal{D}'$ to the upper half-plane, so

$$\mathcal{H}_p = -\frac{\gamma}{4\pi R^2} \log\left|\frac{\exp(\pi z/c) - \exp(\pi z_v/c)}{\exp(\pi z/c) - \exp(\pi z_v^*/c)}\right|^2.$$

After some algebraic simplification, the streamlines are given by the curves

$$\frac{\sinh^2(\pi/2c)(x - x_v) + \sin^2(\pi/2c)(y - y_v)}{\sinh^2(\pi/2c)(x - x_v) + \sin^2(\pi/2c)(y + y_v)} = const$$

Several streamlines for the case $c = 1$, $x_v = 0$, $y_v = .4142$ are shown in Figure 4.38(a). The number of image vortices for this case is infinite. The vortex Hamiltonian and their paths are given by the formulas

$$\mathcal{H}_v = -\frac{\gamma^2}{4\pi R^2} \log \frac{\pi^2}{c^2} \cdot \frac{(1+x^2+y^2)^2}{\sin^2(\pi y_v/c)}$$

$$\frac{(1+x^2+y^2)^2}{\sin^2(\pi y/c)} = const$$

Figure 4.38(b) shows the vortex paths for this example, while Figure 4.38(c) and Figure 4.38(d) depict the stereographically projected vortex paths for the spherical problem and the corresponding planar problem for comparison.   ◇

Example 4.7   (Rectangle) $\mathcal{D}$ is the rectangle bounded by the longitudes zero and ϕ_1 and latitudes θ_1 and θ_2. $\mathcal{D}'$ is the annular sector bounded by the circular arcs of radii $r_1 = \tan(\theta_1/2)$ and $r_2 = \tan(\theta_2/2)$, and by the radial lines $\phi = 0$ and $\phi = \phi_1$. $\mathcal{D}'$ maps to the rectangle

$$\mathcal{D}'' = \{(x, y)| \log r_1 \le x \le \log r_2, \quad 0 \le y \le \phi_1\}$$

under the map $u = \log z$, where u is the plane of the rectangle $\mathcal{D}''$.

The rectangle is mapped to the lower half-plane by the map

$$w = \mathcal{P}(u - \log r_1)$$

where $\mathcal{P}(u) = \mathcal{P}(u; g2, g3)$ is the Weierstrass $\mathcal{P}$ function, and $g2, g3$ are the Weierstrass invariants, which are related to the half-periods w and w'. In the present case, $w = \log(r_2/r_1)$ and $w' = i\phi_1$. Thus, the annular sector $\mathcal{D}'$ is mapped to the lower half-plane by the function $f(z) = \mathcal{P} \log(z/r_1)$.

The image system for a vortex in a rectangle consists of a doubly infinite lattice whose corners are occupied by image vortices (see Kunin et al (1994)). Thus, even for a vortex in an annular sector, we will have a doubly infinite lattice of image vortices. The particle Hamiltonian for the annular sector is

$$\mathcal{H}_p = -\frac{\gamma}{2\pi R^2} \log \left| \frac{\mathcal{P} \log(z/r_1) - \mathcal{P} \log(z_v/r_1)}{\mathcal{P} \log(z_v/r_1) - \mathcal{P}(\log(z_v^*/r_1)} \right|.$$

The vortex Hamiltonian is

$$\mathcal{H}_v = -\frac{\gamma^2}{2\pi R^2} \log \left| \frac{(1+r_v^2)\mathcal{P} \log(z_v/r_1)}{\mathcal{P} \log(z_v/r_1) - \mathcal{P} \log(z_v^*/r_1)} \right|.$$

Several streamlines for the case $\phi_1 = \pi/4$, $\theta_1 = \pi/6$, $\theta_2 = \pi/3$, $\theta_v = \pi/4$, $\phi_v = \pi/6$ are shown in the Figure 4.39(a), while the vortex paths are shown in Figure 4.39(b).   ◇

Remark. Some of these flow configurations *with* boundaries can be used to construct more complex periodic and equilibrium solutions on the sphere *without* boundaries in the following way. Consider the configurations shown in Figures 4.36 and 4.37 for sector openings that are integer divisors of 2π. When the 1-vortex motion in each of these basic cells is periodic, then periodically extending the cells to tile the sphere leads to a periodic solution of an N-vortex problem with no boundaries. For example, the basic octant cell shown in Figure 4.37, when extended to cover the surface of the sphere, yields a continuous family of periodic 8-vortex solutions. In the limiting case when the vortex in each cell is placed at the elliptic point in the cell, the solution corresponds to an equilibrium obtained by inscribing a cube inside the sphere and placing a vortex at each of the eight vertices. The fact that each vortex sits at an elliptic point implies that the equilibrium is nonlinearly stable.   ◊

4.8 Bibliographic Notes

We finish this chapter by mentioning ongoing work on related problems, which seem promising for understanding the spherical vortex problem. There is a large literature concerning the distribution of points on the surface of a sphere, as reviewed recently in Saff and Kuijlaars (1997). This classical problem has been studied in several contexts, including the search for stable chemical molecules (Buckminsterfullerenes), the construction of computational algorithms for approximating integrals, and the attempt to place electric charges on a conducting spherical shell (also known as the dual problem for stable molecules) so that they remain in equilibrium with respect to coulomb interactions. Related to the electrostatic problem is the classical Tammes's problem, which calls for maximizing the smallest distance among N points on the sphere subject to an energy constraint. These problems, both for small numbers of points as well as the asymptotic problem $N \to \infty$, have been extensively studied and it is likely that methods developed in this context can be used to identify new relative vortex equilibria. Several groups are currently using symmetry ideas to study various aspects of vortex motion on the sphere. In addition to the work of Pekarsky and Marsden (1998) regarding the nonlinear stability of the relative 3-vortex equilibria, there is recent work of Lim (1998a) regarding relative equilibria on the sphere under more general interaction laws, as well as work of Lim, Montaldi, and Roberts (1999) regarding the use of symmetries in classifying and identifying relative N-vortex equilibria. See also the related work of Montaldi and Roberts (1998) on relative equilibria of molecular motion. While these methods rely on symmetries, there are other methods currently being developed that can be used to identify asymmetric states, such as the spiral states shown in Saff and Kuijlaars

(1997) and the recent method of Aref and Vainchtein (1998) for finding asymmetric equilibrium states in the plane.

Work on chaotic vortex motion and particle transport on the sphere is only in its infancy, with Bagrets and Bagrets (1997) being the first to generalize some of the most straightforward planar results to the sphere. The problem of particle dispersion in vortex-dominated flows is of great interest to atmospheric scientists and oceanographers, and many interesting and physically important problems still remain to be worked out. The classification of the integrable streamline topologies in Kidambi and Newton (2000a) would be a starting point for a Melnikov analysis of motion of particles in these flows under various perturbations. Regarding the classification of streamline topologies for the 3-vortex problem, one might reasonably ask the following questions:

- How representative are these patterns for more general flows?

- How do the streamline topologies evolve dynamically?

- What role do the bifurcations of patterns and the evolution of instantaneous streamline structures play in the mixing and transport of Lagrangian particles and the stretching of passive interfaces?

In general terms, these questions remain to be answered, although we do have indications that the patterns identified for the integrable cases do show up prominantly in much more complex flows. For example, Figure 4.40 shows a picture of spherical atmospheric streamline patterns from weather data, taken from Gill (1982) (see his Figure 2.3 for the original "unprocessed" patterns). This flow can be "nonlinearly decomposed" into 10 connected components, each containing a distinct streamline topology. Because there are 22 centers (shaded gray regions) and 20 saddles (marked as x's) in this figure, by our estimate from Theorem 4.6.2, we can demonstrate that it requires a minimum of 7 vortices to produce such a topological pattern. [2] However, despite the fact that 7 vortices are required to produce the flow, the vast majority [3] of topological structures one finds are identified in our 3-vortex primitive chart (Figure 4.24), and the corresponding homotopy chart shown in Figure 4.25. While it is somewhat surprising at first glance that the streamline patterns generated from a turbulent atmosphere would be so simple, an important fact associated with Figure 4.40 is that it is obtained as an *ensemble average* collected over several months.

---

[2] $(N^2 - 2N + 2)$ is an upper bound on the number of stagnation points in a flowfield generated by N vortices, hence $(N^2 - 2N + 2) + N$ gives an upper bound for the total number of centers and saddles. In Figure 4.40 there are a total of 42 centers plus saddles, which requires at least $N = 7$.

[3] The ones that are not can be produced with integrable 4-vortex structures. For more information, see the downloadable lecture "Decomposing atmospheric weather patterns using vortex dynamics," http:// online.itp.ucsb.edu/online/hydrot00/newton/

One might then begin to suspect that an equilibrium statistical mechanics theory would be more appropriate. We take up this topic in Chapter 6.

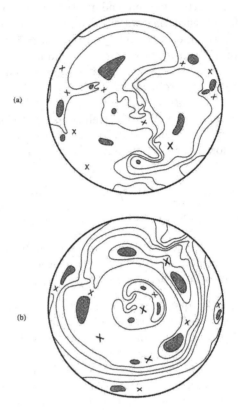

FIGURE 4.40. Global atmospheric streamline patterns: (a) Northern hemisphere (11 centers, 10 saddles). (b) Southern hemisphere (11 centers, 10 saddles).

4.9   Exercises

1. Use Stokes's theorem on a compact two-dimensional surface ∂S to prove that

$$\oint \mathbf{u} \cdot dl = \int_{\partial S} \omega \cdot dS = 0.$$

2. Derive the spherical equations of motion (1.1.23), (1.1.24) from the Cartesian equations (4.1.1).

3. (a) Prove that $\dot{\mathbf{c}} = 0$.

(b) Prove the relations

$$\{\mathcal{H}, P^2 + Q^2\} = 0, \qquad \{\mathcal{H}, S\} = 0, \qquad \{P^2 + Q^2, S\} = 0.$$

4. For the case of 2-vortex motion with $\mathbf{c} \neq 0$,

   (a) Prove that the two vortices must move on cones around the center of vorticity vector $\mathbf{c}$.

   (b) Derive the frequency of rotation formula

   $$\omega = \frac{\|\mathbf{M}\|}{2\pi R l_{12}^2}.$$

5. Prove that if two single-point vortices of any strength lie on opposite sides of the sphere, the configuration is both linearly and nonlinearly stable.

6. (a) Derive the relative equations of motion (4.2.1), (4.2.2), (4.2.3) starting from the absolute equations (4.1.1).

   (b) Prove that the relative equations have the invariants (4.2.6), (4.2.7).

7. (a) Derive the planar Cartesian equations (4.3.2), (4.3.3).

   (b) Derive the volume evolution equation (4.3.5).

8. Derive the area and volume formulas (4.2.4), (4.2.5).

9. Prove the discrete symmetry Theorem 4.2.1.

10. (a) Prove that for the spherical 3-vortex problem, fixed equilibria can occur only if

    $$h \equiv \frac{1}{3}\left(\frac{1}{\Gamma_1} + \frac{1}{\Gamma_2} + \frac{1}{\Gamma_3}\right) \neq 0.$$

    (b) Construct a 3-vortex fixed equilibrium state where $\Gamma_1, \Gamma_2 > 0$, $\Gamma_3 < 0$, and one where $\Gamma_1, \Gamma_2, \Gamma_3 > 0$.

11. For the case of 3-vortex collapse on a sphere:

    (a) Derive the formula (4.4.5) for the frequencies ω_i.

    (b) Derive the asymptotic expansions for these frequencies near the blow-up time.

12. Prove Theorem 4.6.3 and Lemma 4.6.1 regarding streamline patterns for 2-vortex configurations.

13. (a) Derive equation (4.5.1) for the motion of a particle in the stereo-graphic plane.

    (b) Derive equation (4.5.2) for the motion of a vortex in the stereo-graphic plane.

    (c) Explicitly construct a 3-vortex collapsing state on the sphere. It might be useful to look at the planar constructions of Kimura (1988) and Aref (1979).

14. Prove that on a sphere with N vortices, there are at most $(N^2 - 2N + 2)$ stagnation points.

15. Go through the details of Example 4.2 deriving the special periodic solution for three vortices, where the vortices remain in a right triangle configuration. In particular, prove Lemmas (4.6.2) and (4.6.3).

16. (a) Carry out the details of Example 4.3, deriving the equations of motion for a point vortex inside a spherical cap.

    (b) Carry out the details of Example 4.4, deriving the equations of motion for a point vortex in a longitudinal wedge.

    (c) Carry out the details of Example 4.5, deriving the equations of motion for a point vortex in a half-longitudinal wedge.

    (d) Carry out the details of Example 4.6, deriving the equations of motion for a point vortex inside a channel.

    (e) Carry out the details of Example 4.7, deriving the equations of motion for a point vortex inside a spherical rectangle.

5
Geometric Phases

In this chapter we introduce the Hannay angle, adiabatic Hannay angle, or geometric phase [1] for classical systems and describe how it arises in vortex dynamics problems. The chapter opens with three simple examples exhibiting a classical Hannay angle. We first outline the use of multiscale perturbation methods to compute phases in slowly varying Hamiltonian systems. These kinds of calculations go back at least to Volosov (1963), who developed averaging techniques in this and other contexts. In Section 5.3 we give a formal definition of the Hannay angle and adiabatic Hannay angle following the work of Golin (1988), Golin and Marmi (1990), and Golin, Knauf, and Marmi (1990). Our main example in this section is the restricted 3-vortex problem described in Newton (1994), where multiscale perturbation methods are used to derive a geometric phase formula. Section 5.4 details the Hannay angle calculation for the general 3-vortex problem following the work of Shashikanth and Newton (1998). We also give an interpretation based on the limiting adiabatic geometry of the problem, which leads to a formula for the geometric phase reminiscent of the bead-on-hoop formula. The final Section 5.5 describes two applications of these ideas. We first describe work developed in Shashikanth and Newton (1999) in which the growth rate of a passive fluid interface in certain vortex con-

---

[1] The term "Berry phase" is used commonly in the context of adiabatic quantum problems, while the term "Hannay angle" is used more conventionally for adiabatic classical problems. In general, the usage of the term "geometric phase" covers both contexts and is also used in the nonadiabatic setting when a geometric interpretation can be given. We also use the term "Hannay–Berry" phase to cover both contexts.

figurations can be related to the geometric phase. The second application, called "pattern evocation" by Marsden and Scheurle (1995), arises when one is interested in finding special rotating frames from which to view orbits of physical systems that are periodic, quasiperiodic, or chaotic. The frequency of the rotating frame can be related to geometric phase ideas, as explained in Marsden and Scheurle (1995), Marsden, Scheurle, and Wendlandt (1996). The original concept was developed in the context of the planar 4-vortex problem, as discussed in Kunin, Hussain, Zhou, and Kovich (1990), and Kunin, Hussain, Zhou, and Prishepionok (1992).

5.1 Geometric Phases in Various Contexts

Through a series of examples we now describe various settings in which a geometric phase has been derived.

Example 5.1   (Foucault pendulum) Consider the equations of motion for a Foucault pendulum in a rectangular coordinate system (x, y, z) fixed to a rotating Earth, rotating with angular frequency ω. The pendulum mass m is at the origin of the Cartesian system in its rest position, the z-axis points outward from Earth along the pendulum axis, while the x-axis points south and the y-axis points to the east. The Lagrangian for the system is given by

$$\mathcal{L}(x, y, \dot{x}, \dot{y}) = \frac{m}{2}(\dot{x}^2 + \dot{y}^2) - \frac{m\Omega^2}{2}(x^2 + y^2) + m\omega \cos\theta(x\dot{y} - y\dot{x}). \quad (5.1.1)$$

We have linearized the system for small oscillations, l is the pendulum length, g is the gravitational acceleration, $\Omega = (g/l)^{1/2}$ is the natural frequency of oscillation of the pendulum; we use the fact that $\omega << \Omega$ to drop the centrifugal force term, which is $O(\omega^2)$. The term $\omega \cos\theta$ is the component of angular frequency of the Earth at colatitude θ, where the pendulum is positioned (see Meirovitch (1970) for a nice general discussion of the Foucault pendulum, and Khein and Nelson (1993) for more details specifically related to this example).

The system is put into Hamiltonian form by defining the momentum variables in the usual way as

$$p_x = \frac{\partial \mathcal{L}}{\partial \dot{x}} = m(\dot{x} - \omega(\cos\theta)y)$$

$$p_y = \frac{\partial \mathcal{L}}{\partial \dot{y}} = m(\dot{y} - \omega(\cos\theta)x)$$

with Hamiltonian

$$\mathcal{H}(x, y; p_x, p_y) = \frac{1}{2\pi}(p_x + m\omega(\cos\theta)y)^2 + \frac{1}{2m}(p_y - m\omega(\cos\theta)x)^2$$
$$+ \frac{m\Omega^2}{2}(x^2 + y^2). \quad (5.1.2)$$

Written in polar coordinates

$$x = r \cos \phi, \qquad y = r \sin \phi,$$

the Lagrangian then becomes

$$\mathcal{L}(r, \phi, \dot{r}, \dot{\phi}) = \frac{m}{2}(\dot{r}^2 + r^2 \dot{\phi}^2) - (\frac{m\Omega^2}{2})r^2 + m\omega(\cos \theta)r^2 \dot{\phi}, \quad (5.1.3)$$

while the Hamiltonian is given by

$$\mathcal{H}(r, \phi, p_r, p_\phi) = \frac{p_r^2}{2m} + \frac{p_\phi^2}{2mr^2} - \omega(\cos \theta)p_\phi + \frac{m}{2}(\Omega^2 + \omega^2 \cos^2 \theta)r^2$$
$$\equiv E, \qquad\qquad (5.1.4)$$

with

$$p_r = \frac{\partial \mathcal{L}}{\partial \dot{r}} = m\dot{r}$$

$$p_\phi = \frac{\partial \mathcal{L}}{\partial \dot{\phi}} = mr^2(\dot{\phi} + \omega \cos \theta).$$

The action variables for the system are defined by

$$I_\phi = \frac{1}{2\pi} \oint p_\phi d\phi \;\; = \;\; p_\phi$$

$$I_r = \frac{1}{2\pi} \oint p_r dr.$$

Solving (5.1.4) for p_r gives

$$p_r = \pm \left(\frac{1}{r}\right) [-p_\phi^2 + 2m(E + \omega(\cos \theta)p_\phi)r^2 - m^2(\Omega^2 + \omega^2 \cos^2 \theta)r^4]^{1/2}.$$

As usual, $+$ is used as r increases from its minimum to its maximum over part of the closed cycle, while $-$ is used for the remainder of the cycle. Defining the new variables

$$R \equiv r^2$$

$$\alpha \equiv \frac{p_\phi^2}{m^2(\Omega^2 + \omega^2 \cos^2 \theta)}$$

$$\beta \equiv \frac{(E + (\omega \cos \theta)p_\phi)}{m(\Omega^2 \omega^2 \cos^2 \theta)}$$

gives the action expression

$$I_r = \frac{m}{4\pi}(\Omega^2 + \omega^2 \cos^2 \theta)^{1/2} \oint (-\alpha + 2\beta R - R^2)^{1/2} \frac{dR}{R}$$
$$= \frac{m}{2}(\Omega^2 + \omega^2 \cos^2 \theta)^{1/2}(\beta - \sqrt{\alpha})$$
$$= \frac{1}{2}\left[\frac{E + \omega(\cos \theta)p_\phi}{(\Omega^2 + \omega^2 \cos^2 \theta)^{1/2}} - |p_\phi|\right].$$

Then the Hamiltonian becomes

$$\mathcal{H}(I_r, I_\phi) = E = (2I_r + |I_\phi|)(\Omega^2 + \omega^2 \cos^2 \theta)^{1/2} - I_\phi \omega \cos \theta,$$

with equations of motion

$$\dot{I}_\phi = 0$$
$$\dot{I}_r = 0$$
$$\dot{\phi} = \frac{\partial \mathcal{H}}{\partial I_\phi} = \pm(\Omega^2 + \omega^2 \cos^2 \theta)^{1/2} - \omega \cos \theta$$
$$\dot{\psi} = \frac{\partial \mathcal{H}}{\partial I_r} = 2(\Omega^2 + \omega^2 \cos^2 \theta)^{1/2},$$

whose solution can be written

$$I_\phi = I_\phi(0)$$
$$I_r = I_r(0)$$
$$\phi = [\pm(\Omega^2 + \omega^2 \cos^2 \theta)^{1/2} - \omega \cos \theta]t + \phi_0$$
$$\psi = 2(\Omega^2 + \omega^2 \cos^2 \theta)^{1/2}t + \psi_0.$$

After Earth has rotated through one period, $T = 2\pi/\omega$, the angle $\phi(t)$ is given by

$$\phi(\frac{2\pi}{\omega}) = [\pm(\Omega^2 + \omega^2 \cos^2 \theta)^{1/2} - \omega \cos \theta] \cdot \frac{2\pi}{\omega} + \phi_0$$
$$= \pm(\frac{2\pi}{\omega})[\Omega^2 + \omega^2 \cos^2 \theta]^{1/2} + 2\pi(\frac{\phi_0}{2\pi} - \cos \theta)$$
$$\equiv \phi_d + \phi_g.$$

The first term, ϕ_d, is called the "dynamic phase"; the second term, ϕ_g, called the "geometric phase," or Hannay angle, is independent of the duration of the cycle, and is the angle swept out by the pendulum axis during one Earth revolution. Notice that even in the limit in which $\omega \to 0$, the geometric phase term remains. It is instructive to asymptotically expand this exact formula in powers of the dimensionless parameter $\omega/\Omega \equiv \epsilon \ll 1$. Hence, the singular limit $\epsilon \to 0$ is an adiabatic one in which Earth's rotation frequency becomes small relative to the pendulum frequency. The expansion becomes

$$\phi(\frac{2\pi}{\omega}) \sim \pm\left(\frac{2\pi}{\epsilon}\right) + 2\pi\left(\frac{\phi_0}{2\pi} - \cos \theta\right)$$
$$\pm \pi\epsilon \cos^2 \theta \mp \frac{\pi}{4}\epsilon^3 \cos^4 \theta + O(\epsilon^5), \qquad (5.1.5)$$

which, after rearrangement of terms, gives

$$\phi(\frac{2\pi}{\omega}) \sim \left[\pm\left(\frac{2\pi}{\epsilon}\right) \pm \epsilon\pi \cos^2 \theta + O(\epsilon^3)\right] + 2\pi\left[\frac{\phi_0}{2\pi} - \cos \theta\right].$$

The ϵ-dependent terms are the asymptotic expansion of the dynamic phase ϕ_d, while the $O(1)$ term is the geometric phase ϕ_g. Without having known the exact result, we could have derived the geometric phase by using a direct asymptotic expansion procedure on the equations of motion, taking care to capture the $O(1)$ effect. The singular limit $\epsilon \to 0$ and its associated asymptotics will be discussed more thoroughly in the context of the 3-vortex problem. $\Diamond$

Example 5.2   (Generalized harmonic oscillator) In this example we use

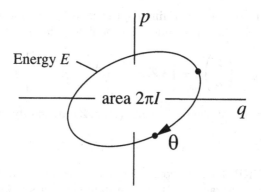

FIGURE 5.1. Fixed elliptic energy contour.

an asymptotic expansion directly on the equations of motion to derive the geometric phase. Consider the generalized harmonic oscillator (see Berry (1985) and Hannay (1985)) whose Hamiltonian is defined as

$$\mathcal{H}(q, p) = \frac{1}{2}[X(t)q^2 + 2Y(t)qp + Z(t)p^2],$$

where $(X(t), Y(t), Z(t))$ are time-dependent periodic coefficients with period T. When the coefficients are fixed in time, the system oscillates periodically around an elliptic contour defined by $\mathcal{H} = E$ in the (q, p)-phase plane, as shown in Figure 5.1, as long as the condition

$$XZ > Y^2 \tag{5.1.6}$$

holds, which we assume throughout. The angle variable θ shown in Figure 5.1 is defined so that the origin coincides with the maximum value of q. The area $\mathcal{A}$ enclosed by a contour $\mathcal{H} = E$ is given by the formula

$$\mathcal{A} = 2\pi I = \frac{2\pi E}{(XZ - Y^2)^{1/2}}$$

with I being the action variable. Hence, the Hamiltonian $\mathcal{H}$ and frequency ω can be written

$$\mathcal{H} = E \qquad = I(XZ - Y^2)^{1/2}$$

$$\omega = \frac{\partial \mathcal{H}}{\partial I} \qquad = (XZ - Y^2)^{1/2}.$$

The direct transformation $(q, p) \mapsto (I, \theta)$ can be written

$$q = \left(\frac{2ZI}{\omega}\right)^{1/2} \cos\theta, \qquad p = -\left(\frac{2ZI}{\omega}\right)^{1/2} \left(\frac{Y}{Z}\cos\theta + \frac{\omega}{Z}\sin\theta\right).$$

In terms of the position coordinate q the equation of motion corresponding to the system is a damped, linear harmonic oscillator

$$\ddot{q} - \left(\frac{\dot{Z}}{Z}\right)\dot{q} + \left[XZ - Y^2 + \frac{(\dot{Z}Y - \dot{Y}Z)}{Z}\right]q = 0.$$

Assuming then that the coefficients (X, Y, Z) are slowly varying, i.e.,

$$X = X(\tau), \qquad Y = Y(\tau), \qquad Z = Z(\tau); \qquad \tau = \epsilon t, \qquad 0 < \epsilon \ll 1,$$

we can use WKB theory to construct an approximate solution, as described in Bender and Orszag (1978). The system simplifies if we use the change of variable

$$q(t) = [Z(\tau)]^{1/2}Q(\tau),$$

putting it in standard WKB form,

$$Q'' + \epsilon^{-2}\left[XZ - Y^2 + \epsilon\frac{(Z'Y - Y'Z)}{Z} + \epsilon^2(\frac{1}{2}(\frac{Z'}{Z})' - \frac{1}{4}(\frac{Z'}{Z})^2)\right]Q$$
$$= 0, \tag{5.1.7}$$

where $' = d/d\tau$. Then, for ϵ sufficiently small, the potential term in brackets never vanishes due to the restriction (5.1.6), hence there is no need for turning point corrections to the basic theory. At leading order, the solution is given by Berry (1985):

$$Q(\tau) \sim [1 + O(\epsilon)] \cos(\theta(\tau))(XZ - Y^2)^{-1/2}, \tag{5.1.8}$$

where

$$\theta(\tau) = \theta(0) + \frac{1}{\epsilon}\int_0^\tau d\tau'(XZ - Y^2)^{1/2} + \frac{1}{2}\int_0^\tau d\tau' \frac{(Z'Y - Y'Z)}{Z(XZ - Y^2)^{1/2}}. \tag{5.1.9}$$

The first term in this formula is the initial phase, while the second is the instantaneous frequency integrated over a period, leading to the dynamic phase , θ_d. The third term, when integrated over one period, T, gives rise to the geometric phase

$$\theta_g = \frac{1}{2} \int_0^T \frac{(\dot{Z}Y - \dot{Y}Z)}{Z(XZ - Y^2)^{1/2}} dt = \frac{1}{2} \oint \frac{Y\,dZ - Z\,dY}{Z(XZ - Y^2)^{1/2}}. \quad (5.1.10)$$

The final formula shows that θ_g is independent of the time period T and is a purely geometric quantity since it is written as a contour integral over the closed path in parameter space. This example is worked out by Anderson (1992) using multiscale methods, and also by Bhattacharjee and Sen (1988).

$\Diamond$

Example 5.3 (Geometric phases in celestial mechanics) This third example shows how the geometric phase arises in restricted three body problems, as described in Berry and Morgan (1996). Shown in Figure 5.2 (reprinted with permission from Berry and Morgan (1996)) is an Earth (E), Sun (S), Jupiter (J) configuration. Several approximations are made in order to arive at a simple formula for the geometric phase associated with the Earth's orbit. First, it is assumed that Jupiter is far from the Sun, compared with the distance from the Earth to the Sun. Second, both the Earth and Jupiter are assumed to orbit in a circular way around the Sun in a plane. Hence, the orbital period of Jupiter, T_J, is large compared with the Earth's orbital period, T_E. We then ask what the Earth's angular position is after one period, T_J. This angle, as usual, decomposes into a dynamic part and a geometric part. The dynamic angle is simply the angular position that the Earth would have at time $t = T_J$ in the absence of Jupiter's effect. The geometric part takes into account the perturbative influence of Jupiter acting over the long time period T_J, and Berry and Morgan (1996) derive the simple (approximate) formula for the geometric phase

$$\theta_g = -3\pi \left(\frac{R_E}{R_J}\right)^4 \left(\frac{M_J}{M_S}\right)^2,$$

where R_E and R_J are the respective radii (i.e., distance from the Sun) associated with the Earth and Jupiter, and M_J and M_S are the respective masses of Jupiter and the Sun. As estimated in Berry and Morgan (1996), this corresponds to an annual displacement of the Earth's orbit by an amount $d_E \approx -150$m. A table of additional three-body problems (Mars-Jupiter-Sun; geosynchronous satellite-Moon-Earth) is given in Berry and Morgan (1996), as is a brief discussion of the influence of other physical effects that are comparable or larger in size.

$\Diamond$

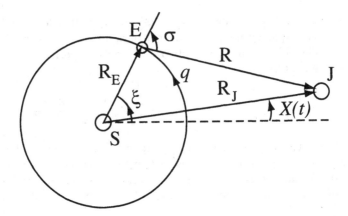

FIGURE 5.2. Three-body configuration exhibiting geometric phase.

Remarks.

1. The first example of the Foucault pendulum is ideal in that the action-angle formalism can be worked out exactly to construct the geometric phase, which serves as one of the prototypical examples in the literature (see, for example, Marsden (1992), Marsden and Ratiu (1999), as well as Berry (1988)).

2. The second example is somewhat unusual since the full equations are linear, hence WKB theory could be used to compute the geometric phase term. While the early developments of the geometric phase ideas focused on the Schrödinger equation (Simon (1983), Berry (1984)), the classical developments have focused more heavily on nonlinear examples.

3. In many cases of interest, the exact action-angle formalism is difficult to work out for the full problem, which may be slowly varying and slightly perturbed. In these problems, multiscale perturbation techniques have proven to be more useful, not only for construction of the full approximate solution (see Kevorkian (1987)), but also for calculating the geometric phase (see Anderson (1992), Newton (1994), Shashikanth and Newton (1998, 1999, 2000)). This is the approach we will develop in the following sections.   ◊

5.2   Phase Calculations For Slowly Varying Systems

As preparation for the calculations to follow, we first will outline the construction of an approximate solution for a Hamiltonian system in action-angle form using the multiscale technique as described in Kevorkian and

Cole (1996). The method is useful, as it provides a direct way to compute geometric phases for certain adiabatic problems.

Example 5.4   (Multiscale formalism) Consider the slowly varying ($\epsilon <<$ 1) one-degree-of-freedom system

$$\frac{dI}{dt} = \epsilon A(I,\tau)\cos\theta$$

$$\frac{d\theta}{dt} = \omega(I,\tau) + \epsilon B(I,\tau)\sin\theta,$$

where

$$I(0,\epsilon) = I^{(0)}$$

$$\theta(0,\epsilon) = \theta^{(0)}.$$

We can identify a slow time scale $\tau = \epsilon t$, and a fast time scale t^+, which we define by

$$\frac{dt^+}{dt} \equiv \Omega(\tau),$$

where Ω is as yet undetermined. Let

$$I(t,\epsilon) = I_0(t^+,\tau) + \epsilon I_1(t^+,\tau) + \epsilon^2 I_2(t^+,\tau) + O(\epsilon^3)$$
$$\theta(t,\epsilon) = \theta_0(t^+,\tau) + \epsilon\theta_1(t^+,\tau) + \epsilon^2\theta_2(t^+,\tau) + O(\epsilon^3).$$

Then

$$\frac{dI}{dt} = \Omega\frac{\partial I_0}{\partial t^+} + \epsilon\left(\frac{\partial I_0}{\partial\tau} + \Omega\frac{\partial I_1}{\partial t^+}\right) + \epsilon^2\left(\frac{\partial I_1}{\partial\tau} + \Omega\frac{\partial I_2}{\partial t^+}\right) + O(\epsilon^3)$$

$$\frac{d\theta}{dt} = \Omega\frac{\partial\theta_0}{\partial t^+} + \epsilon\left(\frac{\partial\theta_0}{\partial\tau} + \Omega\frac{\partial\theta_1}{\partial t^+}\right) + \epsilon^2\left(\frac{\partial\theta_1}{\partial\tau} + \Omega\frac{\partial\theta_2}{\partial t^+}\right) + O(\epsilon^3).$$

This gives the sequence of equations

$$O(1): \quad \Omega\frac{\partial I_0}{\partial t^+} = 0$$

$$\Omega\frac{\partial\theta_0}{\partial t^+} = \omega(I_0,\tau)$$

$$O(\epsilon): \quad \Omega\frac{\partial I_1}{\partial t^+} = -\frac{\partial I_0}{\partial\tau} + A(I_0,\tau)\cos\theta_0$$

$$\Omega\frac{\partial\theta_1}{\partial t^+} = -\frac{\partial\theta_0}{\partial\tau} + B(I_0,\tau)\sin\theta_0 + \frac{\partial\omega}{\partial I}(I_0,\tau)I_1.$$

Solving the leading order equations gives

$$I_0 = I_0(\tau)$$

$$\theta_0 = \left[\frac{\omega(I_0(\tau),\tau)}{\Omega(\tau)}\right]t^+ + \phi_0(\tau).$$

At this order, the unknowns are Ω, I_0, and ϕ_0 and our goal will be to derive a simple formula for the leading order phase correction term $\phi_0(\tau)$. It is this term, when integrated over a period, that leads to the geometric phase. We go to next order and integrate I_1:

$$I_1(t^+, \tau) = -\frac{t^+}{\Omega(\tau)} \frac{dI_0}{d\tau} + \frac{A(I_0, \tau)}{\omega(I_0, \tau)} \sin\theta_0 + J_1(\tau).$$

To eliminate secular effects at this order (see Kevorkian and Cole (1996)), we require

$$\frac{dI_0}{d\tau} = 0 \quad \Rightarrow \quad I_0 = const \equiv I^{(0)},$$

which then gives

$$J_1(0) = -\frac{A(I^{(0)}, 0)}{\omega(I^{(0)}, 0)} \sin I^{(0)}.$$

The phase equation becomes

$$\Omega \frac{\partial\theta_1}{\partial t^+} = \left[-\frac{(\partial\omega/\partial\tau)(I_0, \tau)}{\Omega} + \frac{(d\Omega/d\tau)\omega(I_0, \tau)}{\Omega^2} \right] t^+ - \frac{d\phi_0}{d\tau} \quad (5.2.1)$$
$$+ \frac{\partial\omega}{\partial I}(I_0, \tau)J_1(\tau) + \left[B(I_0, \tau) + \frac{(\partial\omega/\partial I)(I_0, \tau)A(I_0, \tau)}{\omega(I_0, \tau)} \right] \sin\theta_0.$$

Eliminating the first secular term gives the condition

$$\frac{\partial\omega/\partial\tau}{\Omega} = \frac{(d\Omega/d\tau)\omega}{\Omega^2},$$

which, without loss of generality, leads to

$$\Omega(\tau) = \omega(I_0, \tau).$$

The leading order phase correction then becomes

$$\frac{d\phi_0}{d\tau} = \frac{\partial\omega}{\partial I}(I_0, \tau)J_1(\tau),$$

which we can integrate

$$\phi_0(\tau) = \int_0^\tau \frac{\partial\omega}{\partial I}(I_0, s)J_1(s)ds + \phi_0(0), \quad (5.2.2)$$

where $\phi_0(0) \equiv \theta^{(0)}$. From this formula one can see that in order to determine this phase correction term, it is necessary to go to higher order in ϵ, since J_1 is not yet known. Integrating (5.2.2) then gives the formula for θ_1,

$$\theta_1(t^+, \tau) = -\frac{1}{\omega(I_0, \tau)} \left[B(I_0, \tau) + \frac{\partial\omega/\partial I(I_0, \tau)A(I_0, \tau)}{\omega(I_0, \tau)} \right] \cos\theta_0 + \phi_1(\tau),$$

where

$$\phi_1(0) = \frac{1}{\omega(I^{(0)}, 0)} \left[B(I^0, 0) + \frac{\partial\omega/\partial I(I^0, 0)}{\omega(I^0, 0)} A(I^0, 0) \right] \cos\theta^{(0)}.$$

To complete the $O(\epsilon)$ solutions we need formulas for the slowly varying functions $J_1(\tau)$ and $\phi_1(\tau)$. For this we write the $O(\epsilon^2)$ system

$$\Omega\frac{\partial I_2}{\partial t^+} = -\frac{\partial I_1}{\partial \tau} + \frac{\partial A}{\partial I} I_1 \cos\theta_0 - A\theta_1 \sin\theta_0$$

$$\Omega\frac{\partial \theta_2}{\partial t^+} = -\frac{\partial \theta_1}{\partial \tau} + \frac{\partial B}{\partial I} I_1 \sin\theta_0 + B\theta_1 \cos\theta_0 + \frac{\partial\omega}{\partial I} I_2 + \frac{1}{2}\frac{\partial^2\omega}{\partial I^2}(I_1)^2.$$

Substituting the known expressions into the right side of the first equation gives

$$\Omega\frac{\partial I_2}{\partial t^+} = -\frac{dJ_1}{d\tau} - \left[\left(\frac{A}{\omega}\right)_\tau + A\phi_1 \right] \sin\theta_0 + J_1 \left(\frac{\partial A}{\partial I} - A\frac{\partial\omega/\partial I}{\omega} \right) \cos\theta_0$$

$$+ \frac{1}{2\omega}\left(A\frac{\partial A}{\partial I} + AB + A^2\frac{\partial\omega/\partial I}{\omega} \right) \sin 2\theta_0.$$

This requires us to set

$$\frac{dJ_1}{d\tau} = 0 \Rightarrow J_1 = const = -\frac{A(I^{(0)}, 0)}{\omega(I^{(0)}, 0)} \sin\theta^{(0)}.$$

Using this in (5.2.2) then gives the main result:

$$\phi_0(\tau) = -\frac{A(I^{(0)}, 0)}{\omega(I^{(0)}, 0)} \sin\theta^{(0)} \int_0^\tau \frac{\partial\omega}{\partial I}(I_0, s)ds + \theta^{(0)}.$$

This expression for the leading order phase correction is essential in computing formulas for the Hannay–Berry phase in the following sections. Higher order solutions are carried out in Kevorkian and Cole (1996) Section 4.5.2, but are not needed for our purposes. Phase correction calculations of this type in many different contexts have been systematically developed recently by Bourland and Haberman (1988, 1989, 1994), Bourland, Haberman, and Kath (1991), Haberman (1988, 1991, 1992), and Haberman and Bourland (1988). See also the very comprehensive survey article by Volosov (1963) with discussions of phase calculations using averaging methods.   ◊

The computations carried out in this example are similar to those needed to evaluate the geometric phase contribution for the adiabatic vortex configurations described in the next section. Since the basic structure of the vortex equations we will encounter all take the same general form when prepared appropriately, it is convenient to lay the groundwork for some of the future calculations by treating one last example.

Example 5.5 (Vortex motion — general form) The equations we will treat for vortex motion all take the general form

$$\frac{dr}{dt} = \epsilon^2 f(r, \theta, D(\epsilon^2 t), \phi(\epsilon^2 t); \epsilon) \tag{5.2.3}$$

$$\frac{d\theta}{dt} = \frac{\Omega}{r^2} + \epsilon^2 g(r, \theta, D(\epsilon^2 t), \phi(\epsilon^2 t); \epsilon), \tag{5.2.4}$$

where $D(\epsilon^2 t)$ and $\phi(\epsilon^2 t)$ are slowly varying, D is periodic with period T, and ϕ is an angle variable that completes one revolution in period T. Hence,

$$D(T) - D(0) = 0$$
$$\phi(T) - \phi(0) = 2\pi.$$

We take as initial conditions $r(0) = r_i$ and $\theta(0) = \theta_i$, and assume that the functions f and g are infinitely smooth and can be expanded in a Taylor series in any of their arguments. The fast time scale t^+ and slow time scale τ are defined by

$$\frac{dt^+}{dt} = \omega(\tau), \qquad \tau = \epsilon^2 t,$$

where $\omega(\tau)$ is as yet undetermined. We then expand each function in powers of ϵ:

$$r(t, \epsilon) = r_0(t^+, \tau) + \epsilon r_1(t^+, \tau) + \epsilon^2 r_2(t^+, \tau) + O(\epsilon^3)$$
$$\theta(t, \epsilon) = \theta_0(t^+, \tau) + \epsilon \theta_1(t^+, \tau) + \epsilon^2 \theta_2(t^+, \tau) + O(\epsilon^3),$$

which leads to

$$\frac{dr}{dt} = \frac{\partial r}{\partial t^+}\frac{\partial t^+}{\partial t} + \frac{\partial r}{\partial \tau}\frac{\partial \tau}{\partial t}$$

$$= \omega\frac{\partial r_0}{\partial t^+} + \epsilon\omega\frac{\partial r_1}{\partial t^+} + \sum_{n=0}^{\infty} \epsilon^{n+2}\left(\frac{\partial r_n}{\partial \tau} + \omega\frac{\partial r_{n+2}}{\partial t^+}\right)$$

$$\frac{d\theta}{dt} = \omega\frac{\partial \theta_0}{\partial t^+} + \epsilon\omega\frac{\partial \theta_1}{\partial t^+} + \sum_{n=0}^{\infty} \epsilon^{n+2}\left(\frac{\partial \theta_n}{\partial \tau} + \omega\frac{\partial \theta_{n+2}}{\partial t^+}\right).$$

Substituting these expansions into (5.2.3), (5.2.4) and equating to zero coefficients of like powers of ϵ gives the sequence of equations (through

$O(\epsilon^2))$

$$O(1): \quad \omega\frac{\partial r_0}{\partial t^+} = 0$$

$$\omega\frac{\partial\theta_0}{\partial t^+} = \frac{\Omega}{r_0^2}$$

$$O(\epsilon): \quad \omega\frac{\partial r_1}{\partial t^+} = 0$$

$$\omega\frac{\partial\theta_1}{\partial t^+} = -\frac{2\Omega r_1}{r_0^3}$$

$$O(\epsilon^2): \quad \omega\frac{\partial r_2}{\partial t^+} = -\frac{\partial r_0}{\partial\tau} + f_0$$

$$\omega\frac{\partial\theta_2}{\partial t^+} = -\frac{\partial\theta_0}{\partial\tau} + \frac{\Omega}{r_0^2}\left(3\frac{r_1^2}{r_0^2} - 2\frac{r_2}{r_0}\right) + g_0,$$

where

$$f_0 \equiv f(r_0, \theta_0, D(\tau), \phi(\tau); 0)$$
$$g_0 \equiv g(r_0, \theta_0, D(\tau), \theta(\tau); 0).$$

The leading order equation gives

$$r_0 \equiv \tilde{r}_0(\tau), \qquad\qquad \tilde{r}_0(0) = r_i$$
$$\theta_0 = \left(\frac{\Omega}{\omega\tilde{r}_0^2}\right)t^+ + \tilde{\theta}_0(\tau); \quad \tilde{\theta}_0(0) = \theta_i.$$

The tilde notation indicates that the function depends only on the slow time scale τ.

At next order, we have

$$r_1 = \tilde{r}_1(\tau), \qquad\qquad \tilde{r}_1(0) = 0$$
$$\theta_1 = -\left(\frac{2\Omega}{\omega\tilde{r}_0^3}\tilde{r}_1\right)t^+ + \tilde{\theta}_1(\tau), \quad \tilde{\theta}_1(0) = 0.$$

To eliminate the secular term in the θ_1 solution we require that $\tilde{r}_1 = 0$. At $O(\epsilon^2)$ we then have

$$r_2 = -\left(\frac{1}{\omega}\frac{d\tilde{r}_0}{d\tau}\right)t^+ + \frac{1}{\omega}[F_0(t^+) - F_0(0)] + \tilde{r}_2(\tau)$$
$$= -\left(\frac{1}{\omega}\frac{d\tilde{r}_0}{d\tau}\right)t^+ + \frac{1}{\omega}F_0(t^+) + [\tilde{r}_2(\tau) - \frac{1}{\omega}F_0(0)].$$

Our notation here is that F_0 is the antiderivative of f_0, i.e., $dF_0/dt^+ \equiv f_0$. The first term requires us to choose

$$\frac{d\tilde{r}_0}{d\tau} = 0 \quad \Rightarrow \quad \tilde{r}_0 = r_i,$$

hence

$$r_2 = \frac{1}{\omega} F_0(t^+) + \left[\tilde{r}_2(\tau) - \frac{1}{\omega} F_0(0) \right], \qquad \tilde{r}_2(0) = 0.$$

The phase equation leads to

$$\frac{\partial \theta_2}{\partial t^+} = -\frac{1}{\omega} \frac{\partial}{\partial \tau} \left[\left(\frac{\Omega}{\omega r_i^2} \right) t^+ + \tilde{\theta}_0(\tau) \right]$$

$$- 2 \frac{\Omega}{\omega r_i^3} \left[\frac{1}{\omega} F_0(t^+) + (\tilde{r}_2(\tau) - \frac{1}{\omega} F_0(0)) \right] + g_0$$

$$= -\frac{1}{\omega} \left[-\frac{d\omega}{d\tau} \left(\frac{\Omega}{\omega^2 r_i^2} \right) t^+ + \frac{d\tilde{\theta}_0}{d\tau} \right]$$

$$- \frac{2\Omega}{\omega r_i^3} \left(\tilde{r}_2(\tau) - \frac{1}{\omega} F_0(0) \right) - \frac{2\Omega}{\omega^2 r_i^3} F_0(t^+) + g_0.$$

The secularity conditions that must be enforced at this order are

$$\frac{d\omega}{d\tau} = 0 \tag{5.2.5}$$

$$\frac{d\tilde{\theta}_0}{d\tau} = -2 \frac{\Omega}{r_i^3} (\tilde{r}_2(\tau) - \frac{1}{\Omega} F_0(0)). \tag{5.2.6}$$

From (5.2.5) we can conclude that $\omega = $ const, so without loss we can choose

$$\omega = \Omega.$$

The second condition (5.2.6) is the phase correction equation, which will ultimately lead to a formula for the Hannay–Berry phase.

To solve (5.2.6) for $\tilde{\theta}_0$ we must have a formula for $\tilde{r}_2$. It is clear from the structure of the expanded equations that the solvability condition for $\tilde{r}_2$ will come at $O(\epsilon^4)$. Hence, in the problems to be considered, to compute the Hannay–Berry phase, we will need to carry the expansions through this order. In all of the examples we treat, the governing equation for $\tilde{r}_2$ is

$$\frac{d\tilde{r}_2}{d\tau} = 0 \Rightarrow \tilde{r}_2(\tau) = \tilde{r}_2(0) = 0.$$

The solution for the slow phase is

$$\boxed{\tilde{\theta}_0(\tau) = \frac{2}{r_i^3} F_0(0)\tau + \theta_i.}$$

Solving the last equation for θ_2 gives

$$\theta_2 = -\frac{2}{\Omega r_i^3} F_0^+(t^+) + G_0(t^+) + \left[\tilde{\theta}_2(\tau) + \frac{2}{\Omega r_i^3} F_0^+(0) - G_0(0) \right],$$

where

$$\tilde{\theta}_2(0) = 0.$$

The notation here is that F_0^+ and G_0 are the antiderivatives of F_0 and g_0, respectively, i.e., $dF_0^+/dt^+ = F_0$, $dG_0/dt^+ = g_0$.   ◊

5.3 Definition of the Adiabatic Hannay Angle

We now follow the work of Golin (1988), Golin and Marmi (1990), and Golin, Knauf, and Marmi (1990) to give the formal definition of the adiabatic Hannay angle in the context of perturbed, slowly varying, single frequency Hamiltonian systems of the form

$$\dot{I}_\epsilon = \epsilon f(\theta_\epsilon, I_\epsilon, X(\epsilon t)) \tag{5.3.1}$$

$$\dot{\theta}_\epsilon = \omega(I_\epsilon, X(\epsilon t)) + \epsilon g(\theta_\epsilon, I_\epsilon, X(\epsilon t)), \tag{5.3.2}$$

where $(I_\epsilon, \theta_\epsilon)$ are the action-angle variables, f and g are 2π-periodic functions of θ_ϵ and $\epsilon \geq 0$ is a small parameter. Here $X(\epsilon t)$ is a slowly varying external parameter, which is periodic with period $t = T = 1/\epsilon$, hence $X(0) = X(1)$. We would like to consider the time evolution of the angle variable θ_ϵ when the exterior parameters undergo a closed adiabatic loop by comparing its exact evolution with its evolution due to the instantaneous frequency of the "frozen" problem. More specifically, at any time t we can define an angle change

$$\Delta\theta(t) = \theta_\epsilon(t) - \theta_\epsilon(0) - \int_0^t \omega(I_\epsilon(s), X(\epsilon s))ds.$$

Then we can evaluate this quantity at time $t = T = 1/\epsilon$:

$$\Delta\theta(T) \equiv \theta_\epsilon(T) - \theta_\epsilon(0) - \int_0^T \omega(I_\epsilon(s), X(\epsilon s))ds$$

$$= \theta_\epsilon(1/\epsilon) - \theta_\epsilon(0) - \int_0^{1/\epsilon} \omega(I_\epsilon(s), X(\epsilon s))ds$$

$$= \epsilon \int_0^{1/\epsilon} g(I_\epsilon(s), \theta_\epsilon(s), X(\epsilon s))ds.$$

We can now formally define the adiabatic Hannay angle.

Definition 5.3.1. *The **Hannay angle** for the system (5.3.1), (5.3.2) is defined as*

$$\Delta\theta_0 \equiv \lim_{\epsilon \to 0} \Delta\theta(1/\epsilon) = \lim_{\epsilon \to 0} \left[\theta_\epsilon(1/\epsilon) - \theta_\epsilon(0) - \int_0^{1/\epsilon} \omega(I_\epsilon(s), X(\epsilon s))ds \right]$$

*provided that this limit exists. The **adiabatic Hannay angle** is defined for $0 < \epsilon << 1$ as*

$$\Delta\theta(1/\epsilon) \equiv \theta_\epsilon(1/\epsilon) - \theta_\epsilon(0) - \int_0^{1/\epsilon} w(I_\epsilon(s), X(\epsilon s))ds$$

$$= \epsilon \int_0^{1/\epsilon} g(I_\epsilon(s), \theta_\epsilon(s), X(\epsilon s))ds$$

$$= \Delta\theta_0 + \epsilon\Delta\theta_1 + \epsilon^2\Delta\theta_2 + \ldots + \epsilon^n\Delta\theta_n + O(\epsilon^{n+1}).$$

Thus, the leading term in the asymptotic expansion for the adiabatic Hannay angle is the Hannay angle.

Remarks.

1. In general, our interest is in providing techniques capable of generating the full asymptotic expansion for the adiabatic Hannay angle. Of course, it is the leading term in the full expansion that is of particular interest and can often be given a geometrical interpretation separate from its dynamical one.

2. The existence of the limit defined above, along with the necessary technical assumptions, are discussed in Golin (1988) for single frequency systems, Golin and Marmi (1990) for more general systems. Geometric interpretations are given in Golin, Knauf, and Marmi (1989) and Montgomery (1988).   ◇

Example 5.6 (Restricted 3-vortex problem) Following Newton (1994), we compute the Hannay angle for a special restricted 3-vortex problem with two equal strength vortices Γ separated by distance $\tilde{D}$, and a passive tracer particle a distance $\tilde{r}$ from one of the point vortices. We let θ denote the angle the particle makes with the horizontal while ϕ denotes the angle the vortex makes with the horizontal. The equations of motion for the particle can be nondimensionalized by introducing the variables

$$r = \frac{\tilde{r}}{R_0}, \quad D = \frac{\tilde{D}}{R_0}, \quad t = \omega\tilde{t},$$

where R_0 denotes the radius of the circular particle orbit in the absence of the second vortex and $\omega = \Gamma/2\pi R_0^2$ its frequency. The nondimensional particle equations are then (see Newton (1994))

$$\dot{r} = -\frac{1}{D}\sin(\phi - \theta) + \frac{1}{D}\left[\frac{\sin(\phi - \theta)}{1 - (2r/D)\cos(\phi - \theta) + (r^2/D^2)}\right] \quad (5.3.3)$$

$$\dot{\theta} = \frac{1}{r^2} + \left(\frac{1}{rD}\right)\cos(\phi - \theta)$$

$$+ \left(\frac{1}{rD}\right)\left[\frac{(r/D) - \cos(\phi - \theta)}{1 - (2r/D)\cos(\phi - \theta) + (r^2/D^2)}\right], \quad (5.3.4)$$

with $\phi = 2t/D^2 + \phi(0)$, with general initial conditions $r(0)$, $\theta(0)$. We are interested in analyzing the particle orbit in the limit $D \gg 1$.

With this in mind, we expand the variables (r, θ) in inverse powers of D

$$r = \sum_{j=0}^{\infty}(\frac{1}{D^j})r_j$$

$$\theta = \sum_{j=0}^{\infty}(\frac{1}{D^j})\theta_j$$

and require

$$r_0(0) = r(0), \quad r_j(0) = 0, \ j \geq 1$$
$$\theta_0(0) = \theta(0), \quad \theta_j(0) = 0, \ j \geq 1.$$

Assuming the solution depends on both fast (t) and slow (τ) time, where

$$\tau = \frac{t}{D^2},$$

we use $d/dt \to \partial/\partial t + (1/D^2)\,(\partial/\partial\tau)$. This gives the following sequence of equations

$$O(1): \frac{\partial r_0}{\partial t} = 0$$

$$\frac{\partial \theta_0}{\partial t} = \frac{1}{r_0^2}$$

$$O(1/D): \frac{\partial r_1}{\partial t} = 0$$

$$\frac{\partial \theta_1}{\partial t} = -\frac{2r_1}{r_0^3}$$

$$O(1/D^2): \frac{\partial r_2}{\partial t} = -\frac{\partial r_0}{\partial \tau} + r_0 \sin 2(\phi - \theta_0)$$

$$\frac{\partial \theta_2}{\partial t} = -\frac{\partial \theta_0}{\partial \tau} + \frac{1}{r_0^2}\left(3\frac{r_1^2}{r_0^2} - 2\frac{r_2}{r_0}\right) - \cos 2(\phi - \theta_0)$$

$$O(1/D^3): \frac{\partial r_3}{\partial t} = -\frac{\partial r_1}{\partial \tau} + r_1 \sin 2(\phi - \theta_0) - 2r_0\theta_1 \cos 2(\phi - \theta_0)$$
$$+ r_0^2 \sin 3(\phi - \theta_0)$$

$$\frac{\partial \theta_3}{\partial t} = -\frac{\partial \theta_1}{\partial \tau} + \frac{2}{r_0^2}\left(-2\frac{r_1^3}{r_0^3} + \frac{3r_1r_2 - r_0r_3}{r_0^2}\right)$$
$$- 2\theta_1 \sin 2(\phi - \theta_0) - r_0 \cos 3(\phi - \theta_0).$$

Solving the system sequentially and imposing solvability conditions at each order gives

$$O(1): r_0 = r(0)$$

$$\theta_0 = \frac{t}{(r(0))^2} + \tilde{\theta}_0(\tau), \quad \tilde{\theta}_0(0) = \theta(0)$$

$$O(1/D): r_1 = 0$$

$$\theta_1 = \tilde{\theta}_1(\tau), \quad \tilde{\theta}_1(0) = 0$$

$$O(1/D^2): r_2 = \frac{(r(0))^3}{2}\cos 2(\phi - \theta_0) + \left[\tilde{r}_2(\tau) - \frac{(r(0))^3}{2}\cos 2(\phi(0) - \theta(0))\right]$$

$$\tilde{r}_2(0) = 0$$

$$\theta_2 = (r(0))^2 \sin 2(\phi - \theta_0) + \left[\tilde{\theta}_2(\tau) - (r(0))^2 \sin 2(\phi(0) - \theta(0))\right]$$

$$\tilde{\theta}_2(0) = 0.$$

Our goal is to derive the solvability condition for the phase correction term $\tilde{\theta}_0(\tau)$. The coupled phase-amplitude system to be solved is

$$\frac{d\tilde{\theta}_0}{d\tau} = \cos[2(\phi(0) - \theta(0))] - \frac{2}{(r(0))^3}\tilde{r}_2(\tau)$$

$$\frac{d\tilde{r}_2}{d\tau} = 0$$

where $\tilde{\theta}_0(0) = \theta(0)$ and $\tilde{r}_2(0) = 0$. The solution is

$$\tilde{\theta}_0(\tau) = \cos 2(\phi(0) - \theta(0)) + \theta(0)$$
$$\tilde{r}_2(\tau) = 0,$$

giving the leading order phase result

$$\theta_0(t, \tau) = \frac{t}{(r(0))^2} + \cos 2(\phi(0) - \theta(0))\tau + \theta(0).$$

Then, to compute the Hannay angle we take the difference between the perturbed and unperturbed angles,

$$\Delta\theta(t, \tau) \equiv \theta_0(t, \tau) - \left[\frac{t}{(r(0))^2} + \theta(0)\right]$$
$$= \cos[2(\phi(0) - \theta(0))]\tau$$

and evaluate it at the end of period T,

$$t = T = \pi D^2$$

to obtain the result

$$\Delta\theta = \cos[2(\phi(0) - \theta(0))] \cdot \pi. \qquad \lozenge$$

Remarks.

1. This result was first announced in Newton (1994) for the special case $\phi(0) = 0$, $\theta(0) = 0$, giving

$$\Delta\theta = \pi.$$

   Higher order corrections and more details are given in this paper as well.

2. The final result depends on the initial angle difference $(\phi(0) - \theta(0))$, and if averaged over 2π becomes zero. Discussions of the dependence of the Hannay angle on initial conditions are given in Berry and Morgan (1996) for the restricted three-body problem, where it is argued that its origin is due to nonsmoothness of the adiabatic variable at the initial $(t = 0)$ and final $(t = T)$ instant. Conditions on when the angle averages to zero (based on symmetries of the system) are also described in Golin and Marmi (1989).

3. For this configuration and all others in this chapter where we identify an adiabatic geometric phase, we nondimensionalize the system so that the "phase-object" (typically a fluid particle) is close to one of the vortices (the "parent vortex"), as compared with its distance to the "far-field vortices." This separation sets up a natural splitting of time scales in which the far-field is slowly varying compared to the phase object.   ◊

5.4   3-Vortex Problem

Our approach to calculating the geometric phase for the 3-vortex problem is based on separating time scales by placing two corotating vortices Γ_1 and Γ_2 close together, with a third vortex Γ_3 further away. The corotating pair (Γ_1, Γ_2) will be called the *primary pair* whose orientation we wish to track. The far-field vortex Γ_3 sees the other two as a single vortex of strength $\Gamma_1 + \Gamma_2$ (to leading order) and hence forms an effective 2-vortex system $(\Gamma_1 + \Gamma_2, \Gamma_3)$, moving on concentric circular orbits around the center of vorticity. The system was shown in Figure 1.6.

We know from Chapter 1, Example 1.8, that the primary pair, in the absence of Γ_3, rotates with period T_{12}, where

$$T_{12} = \frac{r^2}{(\Gamma_1 + \Gamma_2)}.$$

The effective pair $(\Gamma_1 + \Gamma_2, \Gamma_3)$ rotates with period T_{123}, where

$$T_{123} = \frac{D^2}{(\Gamma_1 + \Gamma_2 + \Gamma_3)},$$

around the center of vorticity. This effective pair forms the "averaged" system in which the high frequency motion of the primary pair is averaged out. Our goal is to compute the orientation $\theta(t)$ of the primary pair at the end of one period of the effective pair, hence we wish to compute $\theta(T_{123})$ and decompose it into a dynamic and geometric part,

$$\theta(T_{123}) = \theta_d + \theta_g.$$

We start with the equations of motion for the 3-vortex problem

$$\dot{z}_\alpha^* = (2\pi i)^{-1} \sum_{\beta=1}^{3} \Gamma_\beta (z_\alpha - z_\beta)^{-1},$$

where $z_\alpha = x_\alpha + iy_\alpha$, $\alpha = 1, 2, 3$, and (x_α, y_α) are the Cartesian coordinates of the vortices. For our purposes, it is more useful to write the equations in terms of their intervortex distances and relative orientations. Let $\hat{r}$ and $\hat{D}$ denote the dimensional distances of Γ_2 and Γ_3 from Γ_1, respectively, and θ and ϕ denote the angles that the lines joining Γ_2 and Γ_3 to Γ_1, respectively, make with the horizontal axis. The angles are measured clockwise from the negative x-axis. The change of variable is effected by letting

$$
\begin{aligned}
y_2 - y_1 &= \hat{r} \sin \theta \\
x_2 - x_1 &= -\hat{r} \cos \theta && (5.4.1) \\
y_3 - y_1 &= \hat{D} \sin \phi \\
x_3 - x_1 &= -\hat{D} \cos \phi && (5.4.2) \\
y_3 - y_2 &= \hat{D} \sin \phi - \hat{r} \sin \theta \\
x_3 - x_2 &= -\hat{D} \cos \phi + \hat{r} \cos \theta. && (5.4.3)
\end{aligned}
$$

The equations are made dimensionless as follows:

$$r = \frac{\hat{r}}{R_i}, \qquad D = \frac{\hat{D}}{D_i}, \qquad t = \omega \hat{t},$$

where R_i and D_i are the initial distances of Γ_2 and Γ_3 to Γ_1, while ω is taken to be the frequency of the primary pair (Γ_1, Γ_2), hence

$$\omega = \frac{(\Gamma_1 + \Gamma_2)}{R_i^2}.$$

We then define ϵ to be the ratio of R_i to D_i

$$\epsilon = \frac{R_i}{D_i}.$$

The system of dimensionless equations governing the relative positions (r, D) and orientations (θ, ϕ) is

$$\frac{dr}{dt} = \epsilon \left(\frac{\alpha_3}{2\pi D}\right) \sin(\phi - \theta) f(r, \theta, \phi, D; \epsilon) \tag{5.4.4}$$

$$\frac{d\theta}{dt} = \frac{\alpha_1 + \alpha_2}{2\pi r^2} + \epsilon \left(\frac{\alpha_3}{2\pi D}\right) \frac{1}{r} \cos(\theta - \phi) f(r, \theta, \phi, D; \epsilon)$$
$$+ \epsilon^2 \left(\frac{\alpha_3}{2\pi D^2}\right) (1 - f(r, \theta, \phi, D; \epsilon)) \tag{5.4.5}$$

$$\frac{dD}{dt} = \epsilon \left(\frac{\alpha_2}{2\pi r}\right) \sin(\phi - \theta) g(r, \theta, \phi, D; \epsilon) \tag{5.4.6}$$

$$\frac{d\phi}{dt} = \epsilon^2 \frac{\alpha_1 + \alpha_3}{2\pi D^2} + \epsilon \left(\frac{\alpha_2}{2\pi r D}\right) \cos(\phi - \theta) g(r, \theta, \phi, D; \epsilon)$$
$$+ \epsilon^2 \left(\frac{\alpha_2}{2\pi D^2}\right) f(r, \theta, \phi, D; \epsilon), \tag{5.4.7}$$

where

$$f(r, \theta, \phi, D; \epsilon) = 1 - \frac{1}{\left(1 - \frac{2\epsilon r}{D} \cos(\phi - \theta) + \frac{\epsilon^2 r^2}{D^2}\right)}$$

$$g(r, \theta, \phi, D; \epsilon) = 1 - \frac{\epsilon^2 r^2 / D^2}{\left(1 - \frac{2\epsilon r}{D} \cos(\phi - \theta) + \frac{\epsilon^2 r^2}{D^2}\right)}.$$

For initial conditions we give

$$r(0) = r_i, \quad \theta(0) = \theta_i, \quad D(0) = D_i, \quad \phi(0) = \phi_i,$$

and

$$\alpha_k = \frac{\Gamma_k}{(\Gamma_1 + \Gamma_2)}$$

are the dimensionless vortex strengths. Notice that the equations of motion have the following features:

1. The unperturbed problem $\epsilon = 0$ reduces to

$$\frac{dr}{dt} = 0$$
$$\frac{d\theta}{dt} = \frac{\alpha_1 + \alpha_2}{2\pi r^2}$$
$$\frac{dD}{dt} = 0$$
$$\frac{d\phi}{dt} = 0.$$

In this limit the primary pair (Γ_1, Γ_2) corotate in circular motion, with

$$r = r_i$$
$$\theta = \frac{t}{2\pi} + \theta_i.$$

2. The restricted 3-vortex problem, treated in Example 5.6, is obtained by setting $\alpha_2 = 0$ and $\Gamma_1 = \Gamma_3$.

3. When $0 < \epsilon << 1$, the far-field vortex perturbs the primary pair. The time-scales for this limit are the fast time scale t of the primary pair and the slow time scale $\tau = \epsilon^2 t$ of the far-field vortex period. This will form the basis for our multiscale analysis.

We now expand the variables as follows

$$r = \sum_{j=0}^{\infty} \epsilon^j r_j(t, \tau)$$

$$\theta = \sum_{j=0}^{\infty} \epsilon^j \theta_j(t, \tau)$$

$$D = \sum_{j=0}^{\infty} \epsilon^j D_j(t, \tau)$$

$$\phi = \sum_{j=0}^{\infty} \epsilon^j \phi_j(t, \tau),$$

with $\tau = \epsilon^2 t$ and initial conditions given by

$$r_0(0) = r_i, \quad \theta_0(0) = \theta_i, \quad D_0(0) = D_i, \quad \phi_0(0) = \phi_i$$
$$r_j(0) = 0, \quad D_j(0) = 0, \quad \theta_j(0) = 0, \quad \phi_j(0) = 0, \quad (j > 0).$$

Using the general formulation described in the previous section gives the following coupled system for the leading order phase correction $\tilde{\theta}_0$:

$$\frac{d\tilde{\theta}_0}{d\tau} = \frac{\alpha_3}{2\pi} \cos(2\theta_i - \phi_i) - \frac{\alpha_1 + \alpha_2}{\pi} \tilde{r}_2(\tau)$$

$$\frac{d\tilde{r}_2}{d\tau} = 0,$$

with

$$\tilde{\theta}_0(0) = \theta_i$$
$$\tilde{r}_2(0) = 0.$$

Solving for the slow phase correction gives

$$\tilde{\theta}_0 = \frac{\alpha_3 \tau}{2\pi} \cos 2(\theta_i - \phi_i) + \theta_i = \frac{\epsilon^2 \Gamma_3 t}{2\pi(\Gamma_1 + \Gamma_2)} \cos 2(\theta_i - \phi_i) + \theta_i.$$

To leading order, the long time period of the farfield vortex is

$$t = T \sim \frac{4\pi^2}{(\alpha_1 + \alpha_2 + \alpha_3)\epsilon^2} = \frac{4\pi^2(\Gamma_1 + \Gamma_2)}{\epsilon^2(\Gamma_1 + \Gamma_2 + \Gamma_3)}.$$

The Hannay angle formula, obtained from $\tilde{\theta}_0(T)$, can then be written

$$\theta_g = \left(\frac{\Gamma_3}{\Gamma_1 + \Gamma_2 + \Gamma_3} \right) 2\pi \cos 2(\theta_i - \phi_i)$$

which is the main result of this section. There are two important special cases to consider:

1. If $\Gamma_3 = 0$, there is only one time scale in the problem and there is no slowly varying background field — because of this, there is no geometric phase.

2. In the case of equal strength vortices $\Gamma_1 = \Gamma_2 = \Gamma_3$, the formula reduces to

$$\theta_g = \frac{2\pi}{3} \cos 2(\theta_i - \phi_i).$$

5.4.1 Geometric Interpretation

To understand why the geometric phase when it arises in the adiabatic limit of slow change of external parameters is geometric, one can view the problem in the following way. Consider the evolution of the phase variable $\theta(t, X(\epsilon t))$. Its change in value at the end of time period $T = \beta/\epsilon$ where $X(\beta) = X(0)$ is given by

$$\int_{\theta(0)}^{\theta(T)} d\theta = \int_0^T \frac{\partial \theta}{\partial t} dt + \oint \frac{\partial \theta(t, X)}{\partial X} \cdot dX.$$

The first integral measures the accumulated effect of local changes due to the instantaneous evolution of the system and is responsible for the dynamic phase contribution. The second integral, taken over the closed loop in parameter space, is written

$$\oint \frac{\partial \theta(t, X)}{\partial X} \cdot dX = \int_0^T \dot{X}(\epsilon t) \frac{\partial \theta(t, X(\epsilon t))}{\partial X(\epsilon t)} dt,$$

where $\dot{X}(\epsilon t)$ denotes the slow rate at which the parameters are changed. As $\epsilon \to 0$, $\dot{X}(\epsilon t) \to 0$, and $T \to \infty$. If $\partial \theta / \partial X$ is bounded for all times, then the integrand decreases as the integration range becomes larger. There is no a priori reason to believe that in the limit of vanishing ϵ the integral will be zero. The quantities can approach their limits in such a way as to give a nonzero result, which is precisely the geometric phase contribution. The angle change formula can be written as

$$\int_{\theta(0)}^{\theta(T)} d\theta(t, \tau) = \int_0^{T=\beta/\epsilon} \frac{\partial \theta(t, \tau)}{\partial t} dt + \int_0^\beta \frac{\partial \theta(t, \tau)}{\partial \tau} d\tau.$$

If we focus on the second integral and use the expansion

$$\theta(t, \tau) = \theta_0(t, \tau) + \epsilon\theta_1(t, \tau) + O(\epsilon^2),$$

we obtain

$$\int_0^\beta \frac{\partial\theta}{\partial\tau}d\tau = \int_0^\beta \frac{\partial\theta_0}{\partial\tau}d\tau + \epsilon\int_0^\beta \frac{\partial\theta_1}{\partial\tau}d\tau + O(\epsilon^2).$$

Then we make use of the leading order decomposition

$$\theta_0(t, \tau) = \theta_0(t) + \tilde{\theta}_0(\tau)$$

to obtain an expression at leading order

$$\int_0^\beta \frac{\partial\theta_0}{\partial\tau}d\tau = \int_0^\beta \frac{\partial\tilde{\theta}_0}{\partial\tau}d\tau.$$

If $\tilde{\theta}_0$ is linear in τ, i.e., $\tilde{\theta}_0 = \alpha\tau$, and assuming we can rewrite the orbit parameter τ in terms of some fixed Cartesian coordinates (X, Y), i.e., $\tau(X, Y)$, then

$$\int_0^\beta \frac{\partial\tilde{\theta}_0}{\partial\tau}d\tau = \alpha\int_0^\beta d\tau = \alpha\oint \frac{\partial\tau}{\partial X}dX + \frac{\partial\tau}{\partial Y}dY \equiv \oint \gamma,$$

where $\gamma \equiv \alpha\left[(\partial\tau/\partial X)dX + (\partial\tau/\partial Y)dY\right]$ is a 1-form defined in the (X, Y)-plane, and the final expression for the geometric phase is then

$$\theta_g = \oint \gamma. \tag{5.4.8}$$

For the 3-vortex configuration one considers the geometry of the closed circular orbit corresponding to the effective 2-vortex problem where Γ_3 co-orbits the combined vortex of strength $\Gamma_1 + \Gamma_2$ around the center of mass of the effective 2-vortex system $(\Gamma_3, \Gamma_1 + \Gamma_2)$ as shown in Figure 5.3. The angle $\phi(\tau)$ is given as

$$\phi(\tau) = \frac{(\alpha_1 + \alpha_2 + \alpha_3)\tau}{2\pi}.$$

Now consider a fixed Cartesian (X, Y) frame whose origin is at the center of vorticity of this 2-vortex problem. Then we can write

$$\tau(X, Y) = \left(\frac{2\pi}{\alpha_1 + \alpha_2 + \alpha_3}\right)\tan^{-1}\left(-\frac{Y}{X}\right)$$

and the geometric phase expression (5.4.8) is expressed in terms of the limiting 1-form

$$\gamma = \left(\frac{Y}{X^2 + Y^2}dX - \frac{X}{X^2 + Y^2}dY\right)\left(\frac{\Gamma_3}{\Gamma_1 + \Gamma_2 + \Gamma_3}\right)\cos 2(\theta_i - \phi_i).$$

To write this differently, let the radius of the circular orbit associated with Γ_3 be denoted R. Since $X^2 + Y^2 = R^2$ and

$$\int Y\,dX - X\,dY = 2A_v,$$

where $A_v = \pi R^2$ is the area swept out by the orbit of Γ_3, we obtain the following expression for the geometric phase

$$\theta_g = \oint \gamma = \frac{2A_v}{R^2} \left(\frac{\Gamma_3}{\Gamma_1 + \Gamma_2 + \Gamma_3} \right) \cos 2(\theta_i - \phi_i)$$

$$= \frac{8\pi^2 A_v}{\mathcal{L}^2} \left(\frac{\Gamma_3}{\Gamma_1 + \Gamma_2 + \Gamma_3} \right) \cos 2(\theta_i - \phi_i),$$

with $\mathcal{L} = 2\pi R$ being the circumference of the circle swept out by Γ_3. This final formula is similar to that given for other geometric phase problems, such as the bead-on-hoop example mentioned in Section 1.2.

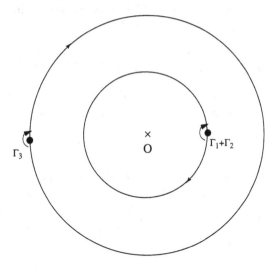

FIGURE 5.3. Effective 2-vortex limiting geometry.

5.4.2 Other Configurations

In this section we apply the previous formulas to the case of a vortex flow with boundaries and one with spatial symmetry. We outline the calculations in the following two examples and leave the details to be carried out as exercises.

Example 5.7 (Point vortex in a circle) In this example the geometric phase for a fluid particle is induced by the presence of a boundary. We track

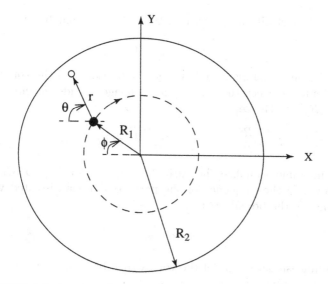

FIGURE 5.4. A point vortex and fluid particle in a circular domain.

the motion of a fluid particle orbiting a point vortex in a circular domain, whose basic configuration is shown in Figure 5.4. The circular trajectory of the point vortex is governed by the Hamiltonian

$$\mathcal{H}_v = \frac{\Gamma}{4\pi} \log \left[R_2^2 - (p^2 + q^2) \right],$$

with (p, q) being the canonical coordinates associated with the vortex of strength Γ in a Cartesian frame whose origin is located in the center of a circular boundary of radius R_2. We know that the vortex undergoes periodic motion on a circle of radius $R_1 = \sqrt{p^2 + q^2} < R_2$, with period

$$T = \frac{4\pi^2}{\Gamma} (R_2^2 - R_1^2).$$

For $R_1 \neq 0$ the Hamiltonian for the particle is

$$\mathcal{H}_p(x, y; t) = \frac{\Gamma}{4\pi} \frac{(x - p)^2 + (y - q)^2}{(x - ap)^2 + (y - aq)^2}, \qquad (5.4.9)$$

where $a \equiv (R_2/R_1)^2$ is the dimensionless ratio of the radii. To compute the phase of the particle after period T, we write the system in relative variables $(\hat{r}, \theta)$ defined as

$$x - p = -\hat{r} \cos \theta, \quad x - ap = D \cos \phi - \hat{r} \cos \theta$$
$$y - q = \hat{r} \sin \theta, \qquad y - aq = -D \sin \phi + \hat{r} \sin \theta,$$

(compare these with (5.4.1), (5.4.2), (5.4.3)) where

$$\phi = \frac{\Gamma}{2\pi} \frac{\hat{t}}{(R_2^2 - R_1^2)} = \frac{\Gamma}{2\pi} \frac{\hat{t}}{(a - 1)R_1^2}$$

is the angular velocity of the vortex. We take as the small parameter

$$\epsilon \equiv \frac{r_i}{D} << 1,$$

where r_i is the initial distance between the particle and the vortex and D is the distance between the vortex and its image outside the circle, located at radius R_2^2/R_1. Hence,

$$D \equiv \frac{R_2^2}{R_1} - R_1 = \frac{(R_2^2 - R_1^2)}{R_1} \equiv (a - 1)R_1.$$

We can nondimensionalize the variable $\hat{r}$ by r_i and $\hat{t}$ by ω, where ω is proportional to the frequency of the particle about an isolated vortex in the absence of the boundary, i.e.,

$$\omega \sim \frac{1}{r_i^2}.$$

The nondimensional period of the vortex then becomes

$$T \equiv \frac{4\pi^2}{\alpha(a - 1)\epsilon^2} = \frac{\beta}{\epsilon^2}$$

and the angle variable ϕ varies slowly, i.e.,

$$\tilde{\phi} = \frac{\alpha(a - 1)}{2\pi}(\epsilon^2 t)$$

where α is a nondimensional constant and the $\tilde{\phi}$ notation indicates that ϕ is a function only of the slow time scale. The nondimensionalized particle equations, after some straightforward alegbra, become

$$\frac{dr}{dt} = \frac{\alpha}{2\pi}\left[\epsilon \sin(\phi - \theta) - \frac{\epsilon \sin(\phi - \theta)}{1 - 2\epsilon r \cos(\phi - \theta) + \epsilon^2 r^2}\right] \qquad (5.4.10)$$

$$\frac{d\theta}{dt} = \frac{\alpha}{2\pi}\left[\frac{1}{r^2} + \frac{(\epsilon/r)\cos(\phi - \theta) - \epsilon^2}{1 - 2\epsilon r \cos(\phi - \theta) + \epsilon^2 r^2} - \frac{\epsilon}{r}\cos(\phi - \theta)\right], \qquad (5.4.11)$$

with initial conditions $r(0) = 1$, $\theta(0) = \theta_i$. As before, the slow time scale is $\tau = \epsilon^2 t$ and if we compare the particle equation (5.4.10) with the general form from equation (5.2.3), we find

$$f = \frac{\alpha}{2\pi} \frac{\epsilon r^2 \sin(\tilde{\phi} - \theta) - r \sin 2(\tilde{\phi} - \theta)}{1 - 2\epsilon r \cos(\tilde{\phi} - \theta) + \epsilon^2 r^2}$$

$$\Omega = \frac{\alpha}{2\pi}.$$

Hence,

$$f_0 = -\frac{\alpha}{2\pi}\sin 2(\tilde{\phi} - \frac{\alpha t}{2\pi} - \tilde{\theta}_0),$$

$$\int f_0 dt \mid_{t=0} = -\frac{1}{2}\cos 2\theta_i.$$

This gives

$$\theta_s \equiv \tilde{\theta}_0(\tau) = -\frac{\alpha}{2\pi}\tau \cos 2\theta_i + \theta_i,$$

with the Hannay–Berry phase expressed as

$$\theta_g = \int_0^\beta \frac{d\theta_s}{d\tau} d\tau$$

$$= -\frac{2\pi}{(a-1)} \cos 2\theta_i$$

$$= -\frac{2\pi}{(R_2/R_1)^2 - 1} \cos 2\theta_i.$$

We can interpret this term geometrically in the following way. From the discussion in Section 5.4.1, the angle can be written

$$\phi(\tau) = \frac{\alpha}{2\pi}(a-1)\tau,$$

and the relation $\tau(X, Y)$ takes the form

$$\tau(X, Y) = \frac{2\pi}{\alpha(a-1)} \tan^{-1}(-Y/X).$$

The 1-form γ is given by

$$\gamma = -\frac{\cos 2\theta_i}{(a-1)} \left(\frac{Y}{X^2 + Y^2} dX - \frac{X}{X^2 + Y^2} dY \right)$$

and the geometric phase is written in terms of this 1-form as

$$\theta_g = \oint \gamma = -\frac{2A_v}{(a-1)R_1^2} \cos 2\theta_i,$$

$$= -\frac{8A_v \pi^2}{(a-1)L^2} \cos 2\theta_i, \tag{5.4.12}$$

where $A_v = \pi R_1^2$ is the area of the vortex circle with circumference $L = 2\pi R_1$. The details of the calculations associated with this example will be carried out in an exercise. $\Diamond$

Example 5.8 (Shear layer model) In this example, we consider an infinite number of equally spaced point vortices of the same strength Γ, placed initially along the x-axis with spacing a between neighboring vortices. We denote the position of one of the vortices in the row

$$z_v = x_v + iy_v$$

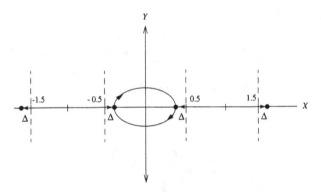

FIGURE 5.5. Shear layer model.

with the positions of all others generated by $z_v \mapsto z_v \pm na$ $(n = 1, 2, \ldots)$. If the symmetry about the vertical axis is broken by a subharmonic perturbation that pushes every adjacent pair of vortices toward each other by amount Δ, then by symmetry the vortices will pair up and perform periodic motion. The basic model for this is shown in Figure 5.5. The equations that govern the motion of the vortices are

$$\frac{dx_v}{dt} = \frac{\Gamma}{4a} \cdot \frac{\sinh(2\pi y_v/a)}{\cosh(2\pi y_v/a) - \cos(2\pi x_v/a)} \qquad (5.4.13)$$

$$\frac{dy_v}{dt} = -\frac{\Gamma}{4a} \cdot \frac{\sin(2\pi x_v/a)}{\cosh(2\pi y_v/a) - \cos(2\pi x_v/a)}, \qquad (5.4.14)$$

with governing Hamiltonian

$$\mathcal{H}_v = \frac{\Gamma}{8\pi} \log\left[\cosh(2\pi y_v/a) - \cos(2\pi x_v/a)\right].$$

With initial conditions given by

$$x_v(0) = \frac{a}{2} - \Delta$$
$$y_v(0) = 0,$$

it is possible to write down the exact solution for the periodic vortex motion in terms of Jacobi elliptic functions,

$$\tan(\pi x_v/a) = \cot(\pi\Delta/a) \cdot \mathrm{cn}\left(\frac{\pi\Gamma t}{4ka^2}; k\right) \qquad (5.4.15)$$

$$\tanh(\pi y_v/a) = -\frac{\cot(\pi\Delta/a)}{1 + \cot^2(\pi\Delta/a)} \cdot \mathrm{sn}\left(\frac{\pi\Gamma t}{4ka^2}; k\right). \qquad (5.4.16)$$

In these formulas, $k = \cos^2(\pi\Delta/a)$ is the modulus of the associated incomplete elliptic integral of the first kind. The time period of the vortex motion T_v can be expressed

$$T_v = \frac{16ka^2}{\pi\Gamma}K, \tag{5.4.17}$$

where K is the complete elliptic integral. The equations of motion for a particle in this flow located at (x, y) are

$$\frac{dx}{dt} = \frac{\Gamma}{4a} \cdot \frac{\sinh[\pi(y - y_v)/a]}{\cosh[\pi(y - y_v)/a] - \cos[\pi(x - x_v)/a]}$$

$$+ \frac{\Gamma}{4a} \cdot \frac{\sinh[\pi(y + y_v)/a]}{\cosh[\pi(y + y_v)/a] - \cos[\pi(x + x_v)/a]} \tag{5.4.18}$$

$$\frac{dy}{dt} = -\frac{\Gamma}{4a} \cdot \frac{\sin[\pi(x - x_v)/a]}{\cosh[\pi(y - y_v)/a] - \cos[\pi(x - x_v)/a]}$$

$$- \frac{\Gamma}{4a} \cdot \frac{\sin[\pi(x + x_v)/a]}{\cosh[\pi(y + y_v)/a] - \cos[\pi(x + x_v)/a]}. \tag{5.4.19}$$

We refer the reader to Shashikanth and Newton (1998), Meiburg and Newton (1991), and Meiburg, Newton, Raju, and Ruetsch (1995) for more details and references associated with this model. After appropriate nondimensionalization, [2] it is shown in Shashikanth and Newton (1999) that the particle equation is of the general form

$$\frac{dr}{dt} = \epsilon^2 f(r, \theta, D, \phi; \epsilon)$$

$$\frac{d\theta}{dt} = \frac{2k}{r^2} + \epsilon^2 g(r, \theta, D, \phi; \epsilon).$$

The slow phase term can be expressed

$$\theta_S \equiv \tilde{\theta}_0(\tau) = \left(\frac{k+3}{6}\right) \tau \cos 2\theta_i + \theta_i.$$

Hence, in the same way as before, the Hannay–Berry phase for this problem is given by

$$\theta_g = \int_0^{4K} \frac{d\theta_S}{d\tau} d\tau$$

$$= \left(\frac{2k+6}{3}\right) K \cos 2\theta_i. \tag{5.4.20}$$

In general terms, to obtain the expression geometrically, one writes the slow angle evolution as

$$\phi(\tau) = \tan^{-1}(-y_v/x_v).$$

---

[2]The small parameter ϵ in this problem is defined in the following way. We take the particle to lie close to one of the vortices in the pair, as compared to the distance between the two vortices in a pair, i.e., $\epsilon \equiv const \cdot r_i/D_i$. In this way, after nondimensionalizing, the pairing vortices act as a slowly varying background flow on the motion of the particle.

To arrive at a simple explicit 1-form for this problem we take the extreme case, where $\Delta \to a/2$, hence the vortices pair in (approximately) circular orbits since they are close together. The 1-form for this case can be written

$$\gamma = \frac{1}{2} \left(\frac{Y}{X^2 + Y^2} dX - \frac{X}{X^2 + Y^2} dY \right) \cos 2\theta_i$$

and as before, the geometric phase expressed as

$$\theta_g = \oint \gamma = \pi \cos 2\theta_i.$$

One can easily generalize this formula by allowing for arbitrary initial orientations of the pairing vortices, as denoted by the angle ϕ_i. Then the geometric phase formula reads

$$\theta_g = \left(\frac{2k + 6}{3} \right) K \cos 2(\theta_i - \phi_i). \qquad \Diamond$$

5.5   Applications

We turn now to a description of how these phase formulas arise in two applications.

5.5.1   Spiral Structures

The first is in the context of two-dimensional vortex-dominated flows where one is interested in calculating, among other things, growth rate formulas for the evolution of passive interfaces as they wrap locally around vortex maxima and deform and evolve globally. In particular, a frequently recurring theme in the fluid dynamics literature is the focus on "spiral structures" as important dynamical states in both two-dimensional and three-dimensional flows. These structures are especially prominent in vortex flows where they appear in the evolution of vorticity regions or interfaces of passive scalars such as concentration or temperature. Coherent vortices can acquire spiral-like structures near their vorticity maxima due to the winding up of variations in the initial distribution. As the flow evolves, further spirals are created through instabilities or collisions. This idea has been used more than once as the basis for phenomenological models of turbulent flows (see for example Lundgren (1982), Gilbert (1988), Everson and Sreenivasan (1992), or Moffatt (1993)). In astrophysical contexts, spiral structures and their formation and dynamics play a central role (see Lin (1976), Bertin and Lin (1996), or Nezlin and Snezhkin (1993) for a taste of this large literature).

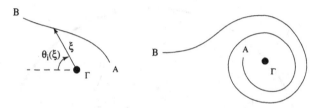

FIGURE 5.6. Spiral interface around an isolated point vortex.

To understand the evolution of a spiral interface under the influence of both the near-field vortex and the far-field, consider first a passive interface between two particles labeled A and B in the flowfield of an isolated point vortex, as shown in Figure 5.6. We view the passive interface as a smooth C^1 curve drawn in the flow domain, each point of which represents a passive particle at that location. We choose a coordinate frame centered at the vortex location and parametrize the interface by ξ, the distance from the vortex, as shown in the figure. Consider the evolution of an arbitrary particle $(r_0(t), \theta_0(t))$ on the interface, with

$$r_0(0) = \xi, \quad \xi_A \leq \xi \leq \xi_B$$
$$\theta(0) = \theta_i(\xi).$$

We know that the time evolution of such a particle is governed by

$$r_0(t) = \xi$$
$$\theta_0(t) = \frac{\Gamma t}{2\pi \xi^2} + \theta_i(\xi),$$

which implies that the interface wraps in a spiral around the vortex, as shown in Figure 5.6. The arc length $\mathcal{L}_0(t)$ is given by

$$\mathcal{L}_0(t) = \int_{\xi_A}^{\xi_B} \sqrt{(r_0(t)\frac{d\theta_0}{d\xi})^2 + (\frac{dr_0}{d\xi})^2} d\xi$$
$$= \int_{\xi_A}^{\xi_B} \sqrt{1 + \left(\xi\frac{d\theta_i}{d\xi} - \frac{\Gamma t}{\pi \xi^2}\right)^2} d\xi.$$

Now consider what happens to such an interface when subjected to an additional slowly varying background field (i.e., the far-field effects), hence allow the coordinates $(r(t,\tau), \theta(t,\tau))$ to depend on some slow time scale $\tau = \epsilon^2 t$. The perturbed interface length is

$$\mathcal{L}_\epsilon(t) = \int_{\xi_A}^{\xi_B} \sqrt{(r\frac{d\theta}{d\xi})^2 + (\frac{dr}{d\xi})^2} d\xi$$

and at any given time, we can form the difference between the unperturbed length and the perturbed length

$$\Delta \mathcal{L}(t) \equiv \mathcal{L}_\epsilon(t) - \mathcal{L}_0(t).$$

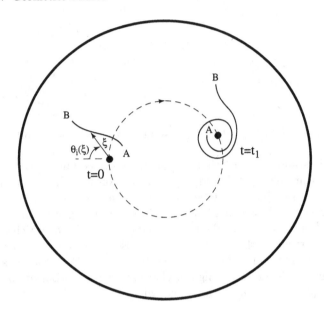

FIGURE 5.7. Spiral interfaces: Interface wrapping in a circular domain.

In addition, suppose the background field is periodic with period $T \sim 1/\epsilon^2$, then at time $t = T$ we have

$$\Delta \mathcal{L}(T) = \mathcal{L}_\epsilon(T) - \mathcal{L}_0(T).$$

We can then ask what $\Delta \mathcal{L}(T)$ is in the limit $\epsilon \to 0$, or equivalently as $T \to \infty$. If we define this limit, assuming it exists, as

$$\Delta \mathcal{L} \equiv \lim_{\epsilon \to 0} \Delta \mathcal{L}(T),$$

then it turns out that in certain flow configurations where geometric phases θ_g can be identified, we have the simple formula

$$\Delta \mathcal{L} = - \int_{\xi_A}^{\xi_B} d(\xi \theta_g).$$

In Shashikanth and Newton (1999) this formula is derived and its relevance is discussed in several canonical flow configurations, including flow in a circular boundary, as shown in Figure 5.7, and a model for the vortex pairing stage of the nonlinear evolution of a shear layer, shown in Figure 5.8. For this flow we have a simple formula for the geometric phase associated with a passive particle as the pairing process completes one period,

$$\theta_g = \left(\frac{2k + 6}{3} \right) K \cos 2(\theta_i - \phi_i),$$

where k is the modulus of the elliptic function solution of the periodic vortex motion and K is the complete elliptic integral associated with these solutions and related to the time period of the pairing process. The formula for $\Delta \mathcal{L}$ in this model becomes

$$
\begin{aligned}
\Delta \mathcal{L} &= -\int_{\xi_A}^{\xi_B} d(\xi \theta_g) \\
&= \int_{\xi_A}^{\xi_B} \left[-\left(\frac{2k+6}{3} \right) K \cos 2(\theta_i - \phi_i) \right. \\
&\quad \left. + \left(\frac{4k+12}{3} \right) K \xi \sin 2(\theta_i - \phi_i) \frac{d\theta_i}{d\xi} \right] d\xi.
\end{aligned}
$$

This result quantifies the additional stretching effect on an interface due to the pairing process in a simple model for shear layer evolution. It is known that the subharmonic pairing mechanism is a generic means by which shear layers evolve and grow (see the paper of Winant and Browand (1974), for example). It is hoped that results of the type outlined for pairing models could be useful for interpreting experiments on interfacial stretching in certain contexts. For more details, as well as applications of this formula to other canonical configurations, see Shashikanth and Newton (1999).

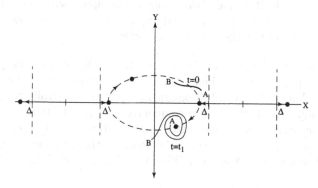

FIGURE 5.8. Spiral interfaces: Interface wrapping in a shear layer.

5.5.2 Pattern Evocation

The second interesting context of relevance to vortex flows, in which geometric phase ideas are currently being developed, is the use of special rotating reference frames in order to view trajectories that are quasiperiodic or chaotic. When the rotation rate is chosen appropriately, one can extract asymptotic patterns from the flowfield, patterns that would not be visible in the fixed frame — hence the phenomenon has been dubbed **pattern evocation** by Marsden and Scheurle (1995). The figures in Kunin et al (1992) show a beautiful example of the trajectories for a chaotic 4-vortex system that clearly exhibits a discrete symmetry. In the context of vortex problems, this idea of moving in special rotating reference frames was introduced by Kunin, Hussain, Zhou, and Kovich (1990), and Kunin, Hussain, Zhou, and Prishepionok (1992), while a more general theory has been developed in Marsden and Scheurle (1995), using the double spherical pendulum as their main example. In this work, a formula for the required angular velocity is derived. For periodic or quasiperiodic motions, as long as a certain KAM-type of condition holds, the special angular velocity can be chosen as constant and can be written as a combination of a dynamic phase and a geometric phase. In particular, it turns out that the angular velocity that evokes the patterns in the reduced space for the case of periodic orbits is the average of the so-called Tisserand frame velocity (the dynamic phase) minus the average velocity associated with the holonomy (the geometric phase). We refer the readers to the original papers for more details.

Remarks.

1. Spiral structures arise in other physical contexts, such as galaxy formation, and it is likely that phases associated with N-body Hamiltonians would be relevant and applicable in these contexts in much the same way as they are in the vortex flows.

2. Geometric phase expressions arise frequently in the control literature, since in applications such as robotic maneuvering (Leonard and Marsden (1997)), orbital control, and locomotion (Koiller, Ehlers, and Montgomery (1996)), it is frequently desirable to take advantage of phase anholonomies generated via manipulation of external control parameters. For an overview of some of these applications, see the review article of Marsden, O'Reilly, Wicklin, and Zombro (1991). ◊

5.6 Exercises

1. Consider the Foucault pendulum problem as described in Example 5.1.

   (a) Derive the Lagrangian for the system (5.1.1), (5.1.3), both in Cartesian and polar coordinates.

   (b) Derive the Hamiltonian for the system (5.1.2), (5.1.4), both in Cartesian and polar coordinates.

(c) Using the assumption $w/\Omega \equiv \epsilon \ll 1$, derive the asymptotic expansion for the angle variable at the end of one period and verify the expression (5.1.5) derived in the example.

2. Consider the generalized harmonic oscillator as described in Example 5.2.

   (a) Put the system in standard WKB form, as shown in (5.1.7), assuming that the coefficients (X, Y, Z) are slowly varying.

   (b) Use the WKB method to derive expressions (5.1.8), (5.1.9), for $Q(\tau)$ and $\theta(\tau)$.

   (c) Verify the geometric phase expression (5.1.10).

3. Go through the details of the multiscale formalism for the slowly varying oscillator described in Example 5.4.

4. Derive the nondimensional particle equations (5.3.3), (5.3.4) in Example 5.6 for the restricted 3-vortex problem and identify the vector field expressions (5.2.3), (5.2.4) for f and g.

5. Derive the nondimensional equations (5.4.4), (5.4.5), (5.4.6), (5.4.7) for the 3-vortex problem and verify the expression

$$\theta_g = \left(\frac{\Gamma_3}{\Gamma_1 + \Gamma_2 + \Gamma_3} \right) 2\pi \cos 2(\theta_i - \phi_i)$$

for the geometric phase.

6. For Example 5.7 of a point vortex and particle in a circular domain:

   (a) Derive the formula (5.4.9) for the particle Hamiltonian.

   (b) Derive the nondimensional particle equations (5.4.10), (5.4.11).

   (c) Derive the Hannay–Berry phase formula (5.4.12).

7. For the shear layer model of Example 5.8:

   (a) Verify that solutions (5.4.15), (5.4.16) satisfy the system (5.4.13), (5.4.14).

   (b) Derive expression (5.4.17) for the time period of the pairing process.

   (c) Derive equations (5.4.18), (5.4.19) for the motion of a particle in the field of the pairing vortices.

   (d) Derive the final expression (5.4.20) for the particle Hannay–Berry phase.

6

Statistical Point Vortex Theories

In a paper that initiated the modern treatment of "statistical hydrodynamics," Onsager (1949) described a point vortex-based theory of turbulence that has proven to be remarkably fruitful, if not all together perfect, and which continues to produce a lively and growing literature. Motivated largely by an attempt to explain the presence of large, isolated vortices in a wide range of turbulent flows (see the descriptive paper of McWilliams (1984))[1], Onsager's theory adopts three main assumptions:

- The theory is based on the inviscid Euler equations in two dimensions, thereby ignoring viscous dissipation (considered to be important at small scales) and vortex-stretching (considered to be important at all scales).

- The theory is based on a point vortex discretization of the Euler equations thereby producing a "vortex gas," which, while derived from the incompressible equations of motion, behaves in many respects like a compressible system with the ability of point vortices to cluster or expand.

- The theory makes use of equilibrium statistical mechanics to explain asymptotic properties of turbulent flows, typically thought to be more

---

[1]Onsager (1949) sought to explain the asymptotic formation of large structures by the clustering of point vortices — thus he sought an explanation for why like-signed vortices tended to attract over long time scales.

amenable to nonequilibrium techniques able to handle large flucuations far from equilibrium.

While many of the initial ideas of Onsager have now been modified, extended, and made rigorous, his primary contribution of positing the existence of **negative temperature states** remains one of the cornerstones of most present day studies. Other early statistical approaches include that of Synge and Lin (1943), Lee (1952), and Hopf (1952), as well as the more recent work of Novikov (1975).

As is now well understood, point vortex-based equilibrium theories can lead to ambiguities and some inconsistencies with the Euler equations. In particular, there are many different ways one can approximate a continuous vorticity distribution with a collection of point vortices, and it is known that different approximations lead to different statistical equilibria. For these and other reasons, equilibrium statistical theories have been developed that apply directly to the Euler equations. We refer the reader to the papers of Robert and Sommeria (1991) and Chavanis and Sommeria (1996) for further discussions , and Kraichnan and Chen (1989) for thoughts on the relevance of the statistical approach to the turbulence problem. A nice discussion on the relationship between deterministic and statistical theories can be found in Zaslavsky (1999). Statistical theories based directly on the full partial differential equations in Hamiltonian form have been developed for other systems as well, such as the nonlinear Schrödinger equation (Lebowitz, Rose, and Speer (1988)) and most recently for coupled Schrödinger equations governing the dynamics of nearly parallel vortex filaments (Lions and Majda (2000)).

After a brief introduction to several basic features of statistical physics and elements of ergodic theory in Section 6.1.1 and Section 6.1.2, we describe the physical meaning of the negative temperature states of Onsager in Section 6.1.3. In Section 6.2 we give an overview of the development of statistical equilibrium theories of planar "point vortex turbulence," starting with the Pointin–Lundgren approach in both the unbounded and bounded regions of the plane. We then describe in Section 6.3 an alternative approach based on maximizing a discrete entropy expression and taking the continuum limit, as described originally in Joyce and Montgomery (1973). These approaches all lead to "mean-field" equations of the general form

$$\Delta\Psi = F(\Psi, \alpha), \qquad (6.0.1)$$

where Ψ is an averaged streamfunction from which one obtains the macroscopic equilibrium vorticity distribution. Since these original papers, much of the work has been treated rigorously in the papers of Frohlich and Ruelle (1982), Caglioti, Lions, Marchioro, and Pulvirenti (1992, 1995), Kiessling (1993), and Eyink and Spohn (1993). In Section 6.4 we describe a nonequilibrium approach of Marmanis (1998) based on a kinetic theory formulation where a BBGKY hierarchy is derived. Section 6.5 discusses some of the more recent developments in the area. The interested reader can find a nice introductory treatment of many of these issues in Chorin (1994) or Marchioro and Pulvirenti (1994).

6.1 Basics of Statistical Physics

6.1.1 Probability Distribution Functions

The main goal of statistical physics is to predict the *macroscopic* properties of a system based solely on *microscopic* information, as well as some basic statistical assumptions. The macroscopic properties are obtained as averages over all the microscopic states. A starting point for this approach is the introduction of a probability distribution function (pdf) $f(x, v)$ of the position variables x and velocity components v, which defines a given microstate. In particular, the quantity $f(x, v)dxdv$ represents the probability that a "particle" is in the neighborhood dx of x, with velocity in the neighborhood dv of v. For most general mechanical systems the phase space (x, v) is unbounded, even if the physical system is constrained to move in a finite region of $\mathbb{R}^3$ (take, for example, an ideal gas in a closed container). For the case of point vortex motion, since the phase space is the physical plane, we can picture the situation as follows. Assume for the moment that all point vortices have the same strength. Then a particular configuration of point vortices, at some fixed energy, can be represented as points on a discrete grid, as is shown below in Figure 6.1. The probability that the collection of point vortices in a given configuration is then $f(\mathbf{x}, \mathbf{y})dxdy$ where $dx \equiv dx_1 dx_2 \cdots dx_N$, $dy \equiv dy_1 dy_2 \cdots dy_N$.

Each point vortex lies within some rectangle on the grid, which represents some uncertainty in its position. The vortices move on the fixed energy surface, where

$$H = -\frac{1}{4\pi} \sum_{\alpha \neq \beta}^{N} \Gamma_\alpha \Gamma_\beta \log(l_{\alpha\beta}) = E \qquad (6.1.1)$$

according to the canonical equations of motion

$$\Gamma_i \dot{x}_i = \frac{\partial H}{\partial y_i} \qquad (6.1.2)$$

$$\Gamma_i \dot{y}_i = -\frac{\partial H}{\partial x_i}. \qquad (6.1.3)$$

As is clear from the formula for the Hamiltonian (6.1.1), energies corresponding to $+\infty$ occur when vortices of the same sign collide, while those corresponding to $-\infty$ occur when vortices of opposite sign collide. Equivalently, for domains with boundaries, negative energy states indicate that

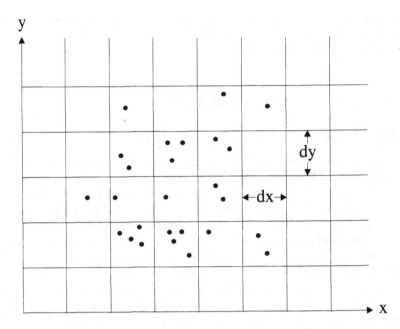

FIGURE 6.1. Point vortices of equal strength distributed randomly along a discrete grid.

vortices tend to congregate near boundaries where they are close to their images (recall that a vortex and its image have opposite signs), while positive energy states correspond to vortices of the same sign collecting in regions away from boundaries. If there are additional conserved quantities, the vortices move on a hypersurface made up of the intersecting surfaces due to each of the global conserved quantities.

Based on the incompressibility of the flow, we immediately have a **Liouville theorem**, which guarantees the invariance of the phase space volume, i.e.,

$$\frac{dH}{dt} \equiv \frac{\partial H}{\partial t} + \mathbf{u} \cdot \nabla H = 0$$

where

$$\mathbf{u} \equiv \left(\frac{dx_1}{dt}, \ldots, \frac{dx_N}{dt}; \frac{dy_1}{dt}, \ldots, \frac{dy_N}{dt} \right)$$

$$\nabla \equiv \left(\frac{\partial}{\partial x_1}, \ldots, \frac{\partial}{\partial x_N}, \frac{\partial}{\partial y_1}, \ldots, \frac{\partial}{\partial y_N} \right).$$

The assumption from statistical mechanics is that all microscopic states of a system that have equal energies are equally probable. The extent to which this is true for a given vortex configuration with *finite* N is an interesting and largely unresolved issue related to questions of ergodicity.

The probability that the system is in a particular microstate with energy E_k is denoted $P(E_k)$ and is written

$$P(E_k) = f(E_k)dxdy,$$

along with the normalization condition

$$\int_{-\infty}^{\infty} \cdots \int_{-\infty}^{\infty} f(E_k)dxdy = 1.$$

To obtain the proper functional form of $f(E_k)$, one need only assume that if the energies of two subsystems A and B are considered, the energy of the composite system is equal to the sum of the individual energies of each, i.e.,

$$P_{A+B}(E_{A+B}) = P_{A+B}(E_A + E_B).$$

Then, assuming that the events are independent, we obtain

$$P_{A+B}(E_A + E_B) = P_A(E_A)P_B(E_B).$$

We can differentiate the left hand side with respect to either E_A or E_B and since the derivatives must be equal, we have

$$P'_A(E_A)P_B(E_B) = P_A(E_A)P'_B(E_B),$$

which yields

$$\frac{P'_A(E_A)}{P_A(E_A)} = \frac{P'_B(E_B)}{P_B(E_B)} \equiv -\beta,$$

where $-\beta$ is a parameter that is introduced as a separation constant. Hence,

$$P'_A(E_A) + \beta P_A(E_A) = 0$$
$$P'_B(E_B) + \beta P_B(E_B) = 0,$$

from which we obtain the functional form

$$P_A(E_A) = C_1 \exp(-\beta E_A)$$
$$P_B(E_B) = C_2 \exp(-\beta E_B),$$

or more generally,

$$P(E) = C_1 \exp(-\beta E)$$
$$f(E) = C_2 \exp(-\beta E),$$

where C_1 and C_2 are normalization constants. The term $\exp(-\beta E)$ is called the **Gibbs factor**. Note that in this derivation, β is independent of the details of the particular system being considered, i.e., no assumptions have been made about the kinds of intermolecular forces being modeled. An important point should be made here regarding the probability of finding the system at any given energy state E. The fact that the Gibbs factor is maximized for $E = 0$ (assuming for the moment that $\beta > 0$, $E \geq 0$) does not imply that the state with zero energy is most probable. This is because there may be many different ways of arranging the system at any given energy, hence the probability of finding the system at a given energy level is equal to the product of the number of configurations at that energy, denoted $M(E)$ (sometimes referred to as the "degeneracy" of the system), with the probability of the system existing in that configuration. Thus, $M(E)P(E)$ is the probability of the system existing at energy level E. If $M(E)$ increases with E, i.e., if there are more states with higher energy, then the product $M(E)P(E)$ can certainly have a nonzero maximum $E = E_c$, despite the fact that $P(E)$ decreases exponentially, as is typically the case.

To obtain expressions for the normalization constants, we sum over all the microstates

$$\sum_k P(E_k) = \sum_k C_1 \exp(-\beta E_k) = C_1 \sum_k \exp(-\beta E_k) = 1,$$

or equivalently, for the continuum, we integrate

$$\int_{-\infty}^{\infty} \cdots \int_{-\infty}^{\infty} f(E)dx_1 \cdots dy_N = C_2 \int_{-\infty}^{\infty} \cdots \int_{-\infty}^{\infty} \exp(-\beta E)dx_1 \cdots dy_N = 1.$$

Then writing $C_1 = C_2 = Z^{-1}$ gives

$$P(E_k) = Z^{-1} \exp(-\beta E_k), \quad Z = \sum_k \exp(-\beta E_k)$$

and

$$f(E) = Z^{-1} \exp(-\beta E), \quad Z = \int_{-\infty}^{\infty} \cdots \int_{-\infty}^{\infty} \exp(-\beta E)dx_1 \cdots dy_N.$$

$$(6.1.4)$$

The function Z is called the **partition** function. Finally, to obtain the phase space average of a general function $g(\mathbf{x}, \mathbf{y})$ we write

$$< g > \equiv \int_{-\infty}^{\infty} \cdots \int_{-\infty}^{\infty} g(\mathbf{x}, \mathbf{y}) f(\mathbf{x}, \mathbf{y}) d\mathbf{x} d\mathbf{y}.$$

It is possible to relate the parameter β and the partition function Z to other standard thermodynamic quantities, such as the system's internal energy U:

$$U = Z^{-1} \sum_k E_k \exp(-\beta E_k).$$

From the fact that Z is a function of β, we obtain

$$\frac{\partial Z}{\partial \beta} = \sum_k \frac{\partial}{\partial \beta} \exp(-\beta E_k) = -\sum_k E_k \exp(-\beta E_k),$$

and

$$Z^{-1} \frac{\partial Z}{\partial \beta} \equiv \frac{\partial (\log Z)}{\partial \beta} = -\frac{\sum_k E_k \exp(-\beta E_k)}{\sum_k \exp(-\beta E_k)} = U.$$

Then, by evaluating the partition function Z for a particular system (the simplest being a perfect monatomic gas as discussed in Jackson (1968), but any system will suffice) and using it in the above formula with the corresponding representation of the internal energy U, we can conclude that

$$\beta = \frac{1}{kT},$$

where k is the Boltzmann constant and T is the temperature of the system — thus, β is inversely proportional to the temperature and the Gibbs factor becomes $\exp(-E/kT)$. For our purposes, from now on we scale the Boltzmann constant out of the problem, i.e., we take $k = 1$.

We are now in a position to define the entropy $S(E)$ and "temperature" T of the system at a given energy state.

Definition 6.1.1. *The **entropy** $S(E)$ of the system at energy level E is defined as*

$$S(E) \equiv \log \Lambda$$

where Λ is the total area of the energy surface $H = E$ on which the motion is constrained. The temperature T is inversely proportional to the rate at which the entropy varies with energy, i.e.,

$$T^{-1} \equiv \frac{dS}{dE}.$$

To obtain an appreciation of the meaning of entropy (and temperature) and its use in quantifying uncertainty in statistical systems, we introduce the concept of statistical uncertainty associated with the system by way of a simple example.

Example 6.1 (Fair and unfair dice) Consider an experiment in which a die is repeatedly rolled. In the first experiment the die is perfectly fair, hence each number appears with equal probability:

$$P_1 = P_2 = P_3 = P_4 = P_5 = P_6 = \frac{1}{6}.$$

In the second experiment the die is loaded so that the number 6 always appears:

$$P_1 = P_2 = P_3 = P_4 = P_5 = 0, P_6 = 1.$$

Clearly the outcome of this first experiment is far more uncertain than the second. The standard quantity that measures this statistical uncertainty is

$$\hat{H}(P_1, \dots, P_n) = -\sum_{i=1}^{n} P_i \log(P_i),$$

whose value in the first case of a perfectly fair die is

$$\hat{H}(\frac{1}{6}, \frac{1}{6}, \frac{1}{6}, \frac{1}{6}, \frac{1}{6}, \frac{1}{6}) = 6 \cdot \left(-\frac{1}{6} \log \frac{1}{6}\right)$$
$$= \log 6.$$

In the second case when the dice are loaded, we have

$$\hat{H}(0, 0, 0, 0, 0, 1) = -(0 + 0 + 0 + 0 + 0 + 1 \log 1) = 0.$$

Note that for $P_i = 0$ we evaluate the term $P_i \log(P_i)$ by taking the limit $P_i \to 0$. For a loaded die in which

$$P_1 = \frac{1}{2}, P_2 = 0, P_3 = 0, P_4 = 0, P_5 = 0, P_6 = \frac{1}{2},$$

we have

$$\hat{H}(\frac{1}{2}, 0, 0, 0, 0, \frac{1}{2}) = \log 2.$$

Since $0 < \log(2) < \log(6)$, we conclude that the largest uncertainty is associated with a perfectly fair die, while the smallest uncertainty occurs when the outcome is known for certain. All other events have uncertainties that fall between these two extremes.   ◊

We can now relate the statistical uncertainty to the entropy of the system.

Definition 6.1.2. *The **entropy** S of a (discrete) system is proportional to its statistical uncertainty and can be defined as*

$$S = \hat{H}(P_1, P_2, \dots, P_n)$$

where $\hat{H}(P_1, \dots, P_n) = -\sum_{i=1}^{n} P_i \log(P_i)$. For a continuous system this expression becomes

$$S = -\int_{-\infty}^{\infty} \cdots \int_{-\infty}^{\infty} f(E) \log(f(E)) d\mathbf{x} d\mathbf{y}.$$

Remark. If the statistical uncertainty increases with energy $(dS/dE > 0)$, the temperature is positive, while if it decreases with energy $(dS/dE < 0)$, the temperature is negative. $\Diamond$

From this definition, it is clear that the more uncertain we are about a particular state, the larger the value of its corresponding entropy. For an elementary yet lucid introduction to these ideas, see Jackson (1968). More advanced treatments can be found in Khinchin (1949), Sommerfeld (1956, Vol. V), or Landau and Lifshitz (1958). Chorin (1994) also has a concise description of many of these ideas.

6.1.2 Elements of Ergodic Theory

The essential mathematical ingredient of all statistical treatments of the Euler equations based on the point vortex discretization is ergodic theory, which is primarily concerned with the long-time averages of interacting systems, i.e., the macroscopic states associated with microscopic interactions. A nice discussion on the uses and techniques of ergodic theory in problems of classical mechanics can be found in Arnold and Avez (1989). One starts, in the simplest context, by considering a measure-preserving transformation $T_t : t \in \mathbb{R}$ acting on the measure space X, which maps to itself, hence $T : X \to X$, with the additional property that $T_{s+t} = T_s T_t$. The mapping T is assumed to be one-to-one and invertible. Then we can consider an orbit associated with the map, $T^n x : n \in \mathcal{Z}$, where $x \in X$ and $T^n x \equiv T^{n-1} \cdot Tx \equiv T^{n-2} \cdot T \cdot Tx \equiv \ldots$. Corresponding to any measurable function $f : X \to \mathbb{R}$ we can compute the image of each iterate, i.e., $f(x), f(Tx), f(T^2 x), \ldots$ and the associated finite-time and long-time averages as defined by

$$\tilde{f}(x) = \frac{1}{N} \sum_{k=0}^{N-1} f(T^k x)$$

$$\hat{f}(x) = \lim_{N \to \infty} \frac{1}{N} \sum_{k=0}^{N-1} f(T^k x).$$

Example 6.2 (Hamiltonian flows) For our purposes the evolution equations (6.1.2), (6.1.3) define a flow-map $T_t(\mathbf{x}, \mathbf{y})$ parametrized by time $t \in (-\infty, \infty)$ in the $2n$-dimensional phase space $\mathbb{R}^{2n}$. We know that the vector field

$$V(\mathbf{x}, \mathbf{y}) = \left(\frac{\partial H}{\partial x_1}, \ldots, \frac{\partial H}{\partial x_n}; -\frac{\partial H}{\partial y_1}, \ldots, -\frac{\partial H}{\partial y_n} \right)$$

is volume preserving, i.e., $\nabla \cdot V = 0$. If we denote the Jacobian of the flow-map T_t by $JT_t(\mathbf{x}, \mathbf{y})$, we have

$$\frac{\partial}{\partial t} JT_t(\mathbf{x}, \mathbf{y}) = 0.$$

If we then consider a measurable subset [2] $\mathcal{B} \subset \mathbb{R}^{2n}$ endowed with Lebesgue measure μ on $\mathbb{R}^{2n}$, we have

$$\mu(T_t \mathcal{B}) = \int_{\mathcal{B}} JT_t(\mathbf{x}, \mathbf{y}) d\mu(\mathbf{x}, \mathbf{y}).$$

From this we can prove that the flow-map preserves measure by calculating its time derivative

$$\frac{d}{dt} \mu(T_t \mathcal{B}) = \int_{\mathcal{B}} \frac{\partial}{\partial t} [JT_t(\mathbf{x}, \mathbf{y})] d\mu(\mathbf{x}, \mathbf{y}) = 0. \qquad \Diamond$$

Remark. In general, since $H = E = const$, the system cannot wander over the entire phase space, but is restricted to lie on a surface of constant energy. If there are additional conserved quantities, then it must be proven that the flow restricted to these subsurfaces also preserves measure. Good discussions of these and related issues can be found in Khinchin (1949).

$\Diamond$

The essential questions of ergodic theory, addressed in the 1930s and 1940s, had to do with convergence properties of the finite-time and long-time averages. In particular, von Neuman (1931) was the first to prove "mean-square" (i.e., L^2) convergence, while Birkhoff proved "almost everywhere" (i.e., pointwise) convergence. These results, known respectively as the **Mean Ergodic Theorem** and the **Pointwise Ergodic Theorem**, form the cornerstones of the modern theory, so we state them here without proofs, which can be found in Petersen (1983).

Theorem 6.1.1 (Mean Ergodic Theorem (von Neumann)). *Consider the measure space $(X, \mathcal{B}, \mu)$ along with the measure-preserving transformation $T : X \to X$ and square integrable function $f \in L^2(X, \mathcal{B}, \mu)$. Then there exists a function $\hat{f} \in L^2(X, \mathcal{B}, \mu)$ such that*

$$\lim_{n \to \infty} \|\frac{1}{n} \sum_{k=0}^{n-1} f \circ T^k - \hat{f}\|_2 = 0,$$

where $\| \cdot \|_2$ denotes the usual L^2 norm.

---

[2]Technically, one considers a σ-algebra of measurable subsets.

Theorem 6.1.2 (Pointwise Ergodic Theorem (Birkhoff)). *Consider the measure space $(X, \mathcal{B}, \mu)$ along with the measure-preserving transformation $T : X \to X$ and integrable function $f \in L^1(X, \mathcal{B}, \mu)$. Then:*

1. $\lim_{n\to\infty} \frac{1}{n} \sum_{k=0}^{n-1} f(T^k x) = \hat{f}(x)$ *exists almost everywhere (i.e., except on sets of measure zero).*

2. $\hat{f}(Tx) = \hat{f}(x)$ *almost everywhere.*

3. $\hat{f} \in L^1$ *and* $\|\hat{f}\|_1 \leq \|f\|_1$.

4. $\lim_{n\to\infty} \frac{1}{n} \sum_{k=0}^{n-1} f \circ T^k \to \hat{f}$ *in* L^1.

One can imagine that as a given trajectory evolves in phase space, there exist cases in which, after sufficiently long times, the orbit will have densely filled the entire phase space uniformly, so that equal subareas of the phase space are visited for equal lengths of time. In this case, one says that the temporal mean is equal to the spatial mean almost everywhere and the system is ***ergodic***.

Theorem 6.1.3. *The measure space $(X, \mathcal{B}, \mu, T)$ is ergodic if and only if for each $f \in L^1(X, \mathcal{B}, \mu)$, the temporal mean of f is equal to the spatial mean of f almost everywhere, i.e.,*

$$\hat{f}(x) = \lim_{n\to\infty} \frac{1}{n} \sum_{k=0}^{n-1} f(T^k x) = \int_X f \, d\mu.$$

Example 6.3 (Irrational rotations on the circle) One of the simplest examples of ergodic motion is the map corresponding to irrational rotations on the unit circle $\mathcal{K} = \{z \in \mathcal{C} : |z| = 1\}$, where we take the map T_α as

$$T_\alpha \exp(2\pi i \theta) = \exp[2\pi i (\theta + \alpha)].$$

For rational α the succession of points obtained under repeated applications of the mapping forms a closed, periodic orbit on the circle. However, when α is irrational, the orbit densely fills the unit circle and the orbit is ergodic.

$\Diamond$

Example 6.4 (Point vortex dipole motion on a flat torus) Consider a vortex dipole consisting of a pair of equal and opposite point vortices moving in a doubly periodic planar parallelogram. Configurations of this type were studied recently by Stremler and Aref (1999), who considered the motion of three vortices. The shape of the parallelogram is characterized by the complex numbers ω_1 and ω_2, and we choose the vertices of the parallelogram to lie at $0, 2\omega_1, 2\omega_1 + 2\omega_2, 2\omega_2$. We denote the parallelogram angle be θ, as shown in Figure 6.2. Doubly periodic domains of this type are topologically equivalent to a "flat torus." Since one of the conserved quantities is the distance between the two vortices, one can draw the line segment connecting them. The dipole then translates along a straight line, [3] called the dipole trajectory, which is the perpendicular bisector of this connecting line segment, and which moves at a fixed angle α with respect to the horizontal. Since the domain is doubly periodic, when the dipole trajectory hits one of the sides of the parallelogram we reflect it across the box and continue it with the same angle. The dipole orbit is the set of parallel lines obtained by following the trajectory over infinite times. Hence there are two cases to consider depending on whether the ratio α/θ is rational or irrational. For rational ratios the dipole orbit is closed and the vortices move in periodic motion. When the angle ratio is irrational, as shown in Figure 6.2, the dipole orbit never closes and the vortex motion is quasiperiodic, densely filling the parallelogram.   ◇

Remark. It should be noted that for systems of point vortices, due to the presence of periodic and quasiperiodic orbits, it is known that typical systems are not ergodic over the entire energy shell. Typically, the energy surface is decomposed into subregions, some of which support the ergodic hypothesis, some of which do not. It is believed that, generically, as $N \to \infty$ and $E \to \infty$, the measure of the nonergodic regions should shrink. Some discussion of these issues can be found in Weiss and McWilliams (1991), or Babiano, Boffetta, Provenzale, and Vulpiani (1994), but in general terms, proving ergodicity for vortex motion on general surfaces or closed domains is an open and interesting problem, one in which tools from billiard dynamics might eventually prove useful, such as those being developed in Gutkin, Smilansky, and Gutkin (1999).   ◇

6.1.3   Negative Temperature States

Certainly the main contribution of Onsager's (1949) paper was the prediction of **negative temperature states** ($dS/dE < 0$) for collections of point vortices occupying bounded phase space volumes, a concept not applicable to typical systems evolving through phase spaces that are unbounded. Since this original paper, the existence of these states have been rigorously proved by Eyink and Spohn (1993). Here we describe the original argument of Onsager (1949). With the volume element of phase space denoted

$$d\Omega = dx_1 dy_1 \cdots dx_n dy_n$$

---

[3]In a recent paper Kimura (1999) showed that vortex dipoles move along geodesics on surfaces of constant curvature.

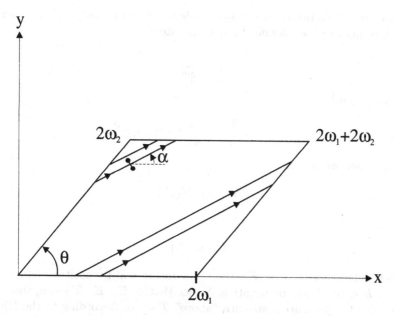

FIGURE 6.2. Motion of a dipole on a flat torus.

along with the fact that the total phase space is finite:

$$\int d\Omega = \left(\int d\mathbf{x} d\mathbf{y} \right)^n = A^n,$$

Onsager defines the phase volume $\Phi(E)$ as the volume of phase space that corresponds to energies less than E, i.e.,

$$\Phi(E) = \int_{H<E} d\Omega = \int_{-\infty}^{E} \Phi'(\tilde{E}) d\tilde{E},$$

where

$$\Phi(-\infty) = 0,$$
$$\Phi(\infty) = A^n.$$

The function $\Phi'(E)$ (called the **structure function** in Kinchin (1949)) is evidently positive for all energies,

$$\Phi'(E) > 0,$$

and since $\Phi(\infty)$ is finite, $\Phi'(E)$ must attain its maximum at some critical energy $E_c < \infty$, so that

$$\Phi''(E_c) = 0.$$

Since the temperature of the system is the rate of change of the energy with respect to the volume, [4] one can write

$$T \equiv \frac{\Phi'}{\Phi''},$$

or equivalently,

$$T^{-1} = \frac{\Phi''}{\Phi'} = \frac{d}{dE} \log[\Phi'(E)],$$

from which we see that

$$S \equiv \log[\Phi'(E)]$$

and

$$\Lambda \equiv \Phi'(E).$$

Then, when $-\infty < E < E_c$, the temperature will be positive, and when $E_c < E < \infty$, it will be negative. Notice that as $E \uparrow E_c$, $T \to \infty$, thus the negative temperature states are "above" $T = \infty$. According to the Gibbs factor $\exp(-E/T)$, if $T < 0$, high energy states have high probability, while if $T > 0$, low energy states have high probability — thus, negative temperature states refer to high energies where like-signed vortices tend to cluster to form large coherent structures — in this regime no spatially uniform macroscopic equilibrium states are expected (see Landau and Lifshitz (1958) for further discussion of this).

6.2 Statistical Equilibrium Theories

We now derive the equilibrium equations governing the reduced pdf's corresponding to these negative temperature states, or more generally, any macroscopic configuration of point vortices in statistical equilibrium.

6.2.1 Pointin–Lundgren Theory in the Unbounded Plane

One of the earliest systematic and straightforward treatments building on Onsager's basic approach is the work of Pointin and Lundgren (1976) and Lundgren and Pointin (1977a,b) where a statistical equilibrium theory based on point vortex motion in both bounded and unbounded planar regions is developed. For the bounded formulation, vortices of both

---

[4]One can find a discussion of this in any book on statistical physics, for example, Feynman (1998) or Landau and Lifshitz (1958).

signs are allowed, since the boundary constrains the phase space to finite volumes. However, as described in Chapter 3, the Hamiltonian is more complicated than for the unbounded case owing to the need to enforce the no-penetration boundary condition. By contrast, the formulation described in Lundgren and Pointin (1977b) for unbounded regions utilizes a simpler Hamiltonian, but only allows like-signed vortices so as to ensure that the phase space volume stays finite, (i.e., no vortices can pair up and translate out to infinity). Since the theory in the unbounded plane is simpler than that in closed boundaries, we describe it first and in greater detail. The theory for bounded planar regions will be summarized in Section 6.2.2. We start by assuming that all vortices have equal strength, hence $\Gamma_i \equiv \Gamma$, which gives the familiar equations of motion

$$\Gamma \frac{d\mathbf{x}_i}{dt} = -\hat{\mathbf{e}}_z \times \frac{\partial}{\partial \mathbf{x}_i} H(\mathbf{x}_1, \mathbf{x}_2, \ldots, \mathbf{x}_N),$$

where

$$H(\mathbf{x}_1, \mathbf{x}_2, \ldots, \mathbf{x}_N) = -\sum_{i>j}^{N} \frac{\Gamma^2}{2\pi} \log\left(\frac{|\mathbf{x}_i - \mathbf{x}_j|}{l_0}\right),$$

with $\hat{\mathbf{e}}_z$ being the unit normal vector to the plane of motion, and $\mathbf{x}_i = (x_i, y_i)$ denoting the coordinates of the ith vortex. The length parameter l_0 is inserted to make the argument of the log dimensionless, but otherwise plays no significant dynamical role.

We know from previous discussions in Chapter 2 that the conserved quantities are the Hamiltonian energy

$$H \equiv E,$$

the center of vorticity

$$\mathbf{R} = \frac{1}{N} \sum_{i=1}^{N} \mathbf{x}_i,$$

and the angular momentum around the center of vorticity

$$L^2 \equiv \frac{1}{N} \sum_{i=1}^{N} (\mathbf{x}_i - \mathbf{R})^2.$$

The main quantity of interest to us is the governing pdf $P_N(\mathbf{x}_1, \ldots, \mathbf{x}_N; t)$, whose evolution equation expresses the conservation of probability (Liouville equation):

$$\frac{dP_N}{dt} \equiv \frac{\partial P_N}{\partial t} + \{P_N, H\} \qquad (6.2.1)$$

$$= \frac{\partial P_N}{\partial t} + \sum_{i=1}^{N} \frac{\partial \mathbf{x}_i}{\partial t} \frac{\partial P_N}{\partial \mathbf{x}_i}$$

$$= \frac{\partial P_N}{\partial t} + \sum_{i=1}^{N} \frac{1}{\Gamma} \left(-\hat{\mathbf{e}}_z \times \frac{\partial H}{\partial \mathbf{x}_i} \right) \cdot \frac{\partial P_N}{\partial \mathbf{x}_i}$$

$$= 0,$$

where $\{,\}$ denotes the usual Poisson bracket. In statistical equilibrium P_N is a function only of the conserved integrals of motion $(H, \mathbf{R}, L^2)$, hence we can write the probability measure in terms of the microcanonical formulation

$$P_N \equiv \delta(H - E) \cdot \delta \left(\sum_{i=1}^{N} (\mathbf{x}_i - \mathbf{R})^2 - NL^2 \right) \cdot \delta(\sum_{i=1}^{N} \mathbf{x}_i - N\mathbf{R})/Q(E, L^2),$$

$$(6.2.2)$$

where $Q(E, L^2)$ is a normalizing factor used to enforce the condition

$$\int P_N(\mathbf{x}_1, \ldots, \mathbf{x}_N) d\mathbf{x}_1 \cdots d\mathbf{x}_N = 1.$$

Hence, we have

$$Q(E, L^2) = \int \delta(H - E) \cdot \delta \left(\sum_{i=1}^{N} (\mathbf{x}_i - \mathbf{R})^2 - NL^2 \right)$$

$$\cdot \delta \left(\sum_{i=1}^{N} \mathbf{x}_i - N\mathbf{R} \right) d\mathbf{x}_1 \cdots d\mathbf{x}_N.$$

Some approaches begin, alternatively, with the canonical Gibbs formulation

$$P_N = Q^{-1} \exp(-\beta H - \lambda_1 L^2 - \lambda_2 \mathbf{R}),$$

which can sometimes simplify some of the calculations — one can read about the various approaches in Chapter 7 of Marchioro and Pulvirenti (1994). Rigorous results are described in Caglioti, Lions, Marchioro, and Pulvirenti (1992, 1995), and Kiessling (1993) for more general interactions.

Remarks.

1. $Q(E, L^2)$ is called the ***density of states*** and, as defined above is clearly proportional to the area of the phase space determined by the constants of motion.

2. In the unbounded plane it is always possible to shift the origin to coincide with the center of vorticity, so there is no need to include the dependence of Q on $\mathbf{R}$.

3. The microcanonical formulation (6.2.2) reflects the assumption that equal areas of the intersection of the hypersurfaces defined by the conserved quantities are equally likely. In addition, an ergodic hypothesis is assumed, which for our purposes asserts that long-time averages over a single member of an ensemble is the same as the space average over all ensemble members.   ◊

In terms of $Q(E, L^2)$ we can define the system's entropy $S(E, L^2)$ and temperature T:

$$S(E, L^2) = \log[Q(E, L^2)]$$
$$\frac{1}{T} = \frac{\partial S}{\partial E} = \frac{1}{Q} \cdot \frac{\partial Q}{\partial E},$$

and in view of our previous definitions, we have

$$Q \equiv \Phi'(E) = \Lambda.$$

As described earlier, Onsager's idea that the entropy is maximized at some finite energy above which lie the negative temperature states, holds for bounded phase space volumes. In this case, owing to the extra angular momentum constraint $L^2 = \text{const}$, which effectively bounds the phase space volume, the same argument can be made.

The statistical equilibrium theory is formulated in terms of reduced pdf's, defined hierarchically as

$$P_1(\mathbf{x}_1) = \int P_N d\mathbf{x}_2 \cdots d\mathbf{x}_N$$

$$= Q(E, L^2)^{-1} \int \delta(H - E) \cdot \delta \left(\sum_{i=1}^N (\mathbf{x}_i - \mathbf{R})^2 - NL^2 \right)$$

$$\cdot \, \delta \left(\sum_{i=1}^N \mathbf{x}_i - N\mathbf{R} \right) d\mathbf{x}_2 \cdots d\mathbf{x}_N$$

$$P_2(\mathbf{x}_1, \mathbf{x}_2) = \int P_N d\mathbf{x}_3 \cdots d\mathbf{x}_N$$

and the goal is to derive the partial differential equation (or integral equation) for $P_1(\mathbf{x}_1)$, from which the spatial equilibrium structure of a large number ($N \to \infty$) of point vortices can be deduced.

If we denote the number density of vortices by $n(\mathbf{x})$, where

$$n(\mathbf{x}) = \sum_{i=1}^{N} \delta(\mathbf{x} - \mathbf{x}_i),$$

then the average density $< n(\mathbf{x}) >$ is given in terms of the reduced pdf $P_1(\mathbf{x})$,

$$< n(\mathbf{x}) >= \sum_{i=1}^{N} < \delta(\mathbf{x} - \mathbf{x}_i) >= NP_1(\mathbf{x}).$$

Likewise, the total vorticity of the system is

$$\omega(\mathbf{x}) = \Gamma n(\mathbf{x}) = \sum_{i=1}^{N} \Gamma \delta(\mathbf{x} - \mathbf{x}_i),$$

whose average is related to the average streamfunction

$$< \omega(\mathbf{x}) > = N\Gamma P_1(\mathbf{x}) \equiv -\nabla^2 < \psi(\mathbf{x}) > .$$

The velocity $\mathbf{u}$ is related to the streamfunction in the usual way as

$$\mathbf{u} \equiv -\hat{e}_z \times \nabla \psi,$$

and the average is given by

$$< \mathbf{u} > = \frac{N\Gamma}{2\pi} \int \left(\hat{e}_z \times \frac{\partial}{\partial \mathbf{x}} \log |\mathbf{x} - \mathbf{x}'| \right) P_1(\mathbf{x}') d\mathbf{x}'. \qquad (6.2.3)$$

The average energy and angular momentum can be shown to be

$$E =< H >$$
$$= -\frac{1}{2}(N^2 - N)\frac{\Gamma^2}{2\pi} \int \log \left(\frac{|\mathbf{x}_1 - \mathbf{x}_2|}{l_0} \right) P_2(\mathbf{x}_1, \mathbf{x}_2) d\mathbf{x}_1 d\mathbf{x}_2, \qquad (6.2.4)$$

$$NL^2 = < \sum_{i=1}^{N} (\mathbf{x}_i - \mathbf{R})^2 > = N \int (\mathbf{x}_1 - \mathbf{R})^2 P_1(\mathbf{x}_1) d\mathbf{x}_1. \qquad (6.2.5)$$

One of the tricky features common to all statistical formulations is that the equations for the reduced pdf's cannot be closed at any level without additional assumptions.[5] The two-point reduced pdf can be decomposed to the product of one-point reduced pdf's as

$$P_2(\mathbf{x}_1, \mathbf{x}_2) = P_1(\mathbf{x}_1) \cdot P_1(\mathbf{x}_2) + P_2'(\mathbf{x}_1, \mathbf{x}_2),$$

where the first term is interpreted as due to the mean flow, while the second term is caused by flucuations about the mean. $P_2'(\mathbf{x}_1, \mathbf{x}_2)$ must then satisfy the condition

$$\int P_2'(\mathbf{x}_1, \mathbf{x}_2) d\mathbf{x}_1 = \int P_2'(\mathbf{x}_1, \mathbf{x}_2) d\mathbf{x}_2 = 0.$$

The standard closure assumption typically invoked amounts to assuming that the flucuations are negligible compared to the mean, hence one replaces $P_2(\mathbf{x}_1, \mathbf{x}_2)$ with the product $P_1(\mathbf{x}_1) P_1(\mathbf{x}_2)$. This approximation follows from the expansion

$$P_2(\mathbf{x}_1, \mathbf{x}_2) - P_1(\mathbf{x}_1) \cdot P_1(\mathbf{x}_2) \equiv P_2'(\mathbf{x}_1, \mathbf{x}_2) = O(\lambda/N),$$

where $\lambda \equiv N\Gamma^2/8\pi T$ and we take the limit $\lambda/N \to 0$. Discussion and justifications of this closure assumption under various situations can be found in the literature (see Marchioro and Pulvirenti (1994) for a start).

We now proceed to derive a closed equation for $P_1(\mathbf{x}_1)$. We first differentiate the expression for $P_1(\mathbf{x}_1)$ to obtain

$$\frac{\partial P_1}{\partial \mathbf{x}_1} = \frac{(N-1)\Gamma^2}{2\pi} \int \left(\frac{\partial \log |\mathbf{x}_1 - \mathbf{x}_2|}{\partial \mathbf{x}_1} \right) \frac{1}{Q} \cdot \frac{\partial}{\partial E} Q P_2(\mathbf{x}_1, \mathbf{x}_2) d\mathbf{x}_2$$
$$- 2(\mathbf{x}_1 - \mathbf{R}) \frac{1}{Q} \cdot \frac{\partial}{\partial NL^2} Q P_1(\mathbf{x}_1) - \frac{1}{Q} \cdot \frac{\partial}{\partial N\mathbf{R}} Q P_1(\mathbf{x}_1). \qquad (6.2.6)$$

Notice that the right hand side involves the higher order pdf $P_2(\mathbf{x}_1, \mathbf{x}_2)$ and one therefore must use a closure assumption to generate a scalar equation for $P_1(\mathbf{x}_1)$. Taking the limit $N \to \infty$, we can show that (6.2.6) simplifies to

$$\frac{\partial P_1(\mathbf{x}_1)}{\partial \mathbf{x}_1} = 4\lambda \int \left(\frac{\partial}{\partial \mathbf{x}_1} \log |\mathbf{x}_2 - \mathbf{x}_1| \right) P_1(\mathbf{x}_2) \cdot P_1(\mathbf{x}_1) d\mathbf{x}_2$$
$$- 2(\mathbf{x}_1 - \mathbf{R}) \left(\frac{1+\lambda}{L^2} \right) P_1(\mathbf{x}_1), \qquad (6.2.7)$$

which we solve, subject to the normalization condition

$$\int P_1(\mathbf{x}_1) d\mathbf{x}_1 = 1.$$

---

[5] How one deals with the closure problem is, of course, one of the central themes in the turbulence literature. One can read more about these and related issues in the review paper of Kraichnan and Montgomery (1980) or in Orszag (1970).

As in the original paper of Lundgren and Pointin (1977b), we scale the variables

$$\tilde{P}_1(\boldsymbol{\eta}_1) \equiv L^2 P_1(\mathbf{x}_1), \quad \boldsymbol{\eta}_i \equiv \frac{(\mathbf{x}_i - \mathbf{R})}{L},$$

from which we obtain the cleaner equation

$$\frac{\partial}{\partial \boldsymbol{\eta}_1} \log(\tilde{P}_1) = 4\lambda \int \left(\frac{\partial}{\partial \boldsymbol{\eta}_1} \log |\boldsymbol{\eta}_1 - \boldsymbol{\eta}_2| \right) \tilde{P}_1(\boldsymbol{\eta}_2) d\boldsymbol{\eta}_2 - 2(1 + \lambda)\boldsymbol{\eta}_1.$$

$$(6.2.8)$$

We next take the divergence of (6.2.7) using the fact that

$$\Delta \log \|\boldsymbol{\eta}\| = 2\pi\delta\|\boldsymbol{\eta}\|$$

to arrive at the nonlinear ordinary differential equation for $\log(\tilde{P}_1)$ as a function of the radial variable $\eta \equiv \|\boldsymbol{\eta}_1\|$:

$$\frac{d^2 \log(\tilde{P}_1)}{d\eta^2} + \frac{1}{\eta} \frac{d \log(\tilde{P}_1)}{d\eta} = -4(1 + \lambda) + 8\pi\lambda \exp[\log(\tilde{P}_1)]. \quad (6.2.9)$$

This equation, whose solutions determine the statistical equilibrium vorticity distribution, is the main result of the Pointin–Lundgren theory. It is a radially symmetric version of the general mean-field equation (6.0.1) and is studied numerically for various values of λ, in Lundgren and Pointin (1977b). The main features are as follows:

1. For large η, the asymptotic profile of $\tilde{P}_1$ is

   $$\tilde{P}_1 \sim C_1 \exp[-(1 + \lambda)\eta^2]\eta^{4\lambda},$$

   from which it immediately follows that $\lambda \geq -1$ are the only allowable values of λ.

2. For $\lambda = 0$, one obtains the Gaussian form (after normalization)

   $$\tilde{P}_1(\eta) = \frac{1}{\pi} \exp(-\eta^2).$$

3. For $\lambda > 0$, which correspond to positive temperature states, the vortex density $\tilde{P}_1$ as a function of the radial variable is flat near the center and decays to zero at infinity.

4. As $\lambda \to -1$, the density profile is sharper and more highly peaked at the center, with approximate form (near $\lambda \sim -1$)

   $$\tilde{P}_1 = \frac{A}{(1 - \pi\lambda A\eta^2)^2} \exp[-(1 + \lambda)\eta^2],$$

where A satisfies the formula

$$A + \frac{1}{\pi} \log A = -\frac{1}{\pi} \log \pi - \frac{\gamma}{\pi} - \frac{1}{\pi} \log(1 + \lambda),$$

with γ being the Euler constant.

Remarks.

1. It is also possible to cast the main equation (6.2.8) in a different but still useful form by integrating it once and writing it as an integral equation:

$$\tilde{P}_1(\boldsymbol{\eta}_1) = C_1 \exp\left[-(1+\lambda)\boldsymbol{\eta}_1^2 + 4\lambda \int \log|\boldsymbol{\eta}_1 - \boldsymbol{\eta}_2|\tilde{P}_1(\boldsymbol{\eta}_2)d\boldsymbol{\eta}_2\right].$$

From this formula, one immediately sees the Gaussian form when $\lambda = 0$.

2. The paper of Lundgren and Pointin (1977b) also compares theory with numerical simulations, based on decompositions with $N \sim 60$. In general, there is good agreement between the predicted equilibrium distributions and the ensemble averaged simulations. However, in certain cases flucuations from equilibrium were large and convergence rates were slow. Most current simulations could easily use far more vortices and do many more runs (see the more recent work of Takaki and Utsumi (1992), for example). The paper also develops a theory for point vortex motion in a weak external velocity field, with governing equations

$$\Gamma \frac{d\mathbf{x}_i}{dt} = -\hat{e}_z \times \frac{\partial H}{\partial \mathbf{x}_i} + \Gamma \mathbf{v}(\mathbf{x}_i, t),$$

where $\mathbf{v}(\mathbf{x}_i, t) \equiv -\hat{e}_z \times \nabla\psi(\mathbf{x}_i, t)$ is a prescribed velocity field. As an example they choose a model that introduces plane strain and solid body rotation,

$$\psi = a(\frac{3}{2}x^2 + \frac{1}{2}y^2).$$

The interesting principal conclusion is that due to the external field, clusters of point vortices tend to *grow* in size, hence lead to nonequilibrium behavior.

3. Mezić and Min (1997) have generalized the approach of Lundgren and Pointin (1997b) to allow for vortices of differing strengths $\Gamma_i > 0, i = 1, \ldots, N$. For this case it is necessary that the reduced pdf's be indexed as

$$P_1^{(i)}(\mathbf{x}_i) = \int P_N dr_1 \cdots dr_{i-1} dr_{i+1} \cdots dr_N,$$

since now

$$P_1^{(i)}(\mathbf{x}_i) \neq P_1^{(j)}(\mathbf{x}_j) \quad (i \neq j).$$

The main result of Mezić and Min (1997) is the derivation of a power law formula relating the scaled reduced pdf's for the different vortices

$$p_1^{(i)}(\eta) = C \left(p_1^{(j)}(\eta) \right)^{\Gamma_i/\Gamma_j},$$

where

$$p_1^{(i)}(\boldsymbol{\eta}_i) \equiv L^2 P_1^{(i)}(\mathbf{x}_i), \qquad \boldsymbol{\eta}_i \equiv \frac{(\mathbf{x}_i - \mathbf{R})}{L},$$

and

$$\eta \equiv \frac{\|\mathbf{x} - \mathbf{R}\|}{L}.$$

This formula says that, on average, stronger vortices spend more time closer to the center of vorticity than do weaker ones.   ◊

6.2.2 *Pointin–Lundgren Theory in Bounded Regions*

The earlier paper of Pointin and Lundgren (1976) develops the equilibrium statistical theory for planar regions that are enclosed by solid boundaries. Since the boundary will constrain opposite-signed vortices from traveling to infinity, the theory allows for vortices of differing sign — this is the principal generalization of the paper. The Hamiltonian $H(\mathbf{x}_1, \mathbf{x}_2, \dots, \mathbf{x}_N)$ is given by

$$H(\mathbf{x}_1, \mathbf{x}_2, \dots, \mathbf{x}_N) = -\sum_{i>j=1}^{N} \Gamma_i \Gamma_j G(\mathbf{x}_i, \mathbf{x}_j) - \sum_{i=1}^{N} \frac{\Gamma_i^2}{2} g(\mathbf{x}_i, \mathbf{x}_j),$$

following the discussion of Chapter 3, with

$$g(\mathbf{x}, \mathbf{x}_0) = G(\mathbf{x}, \mathbf{x}_0) - \frac{1}{2\pi} \log |\mathbf{x} - \mathbf{x}_0|$$

and $G(\mathbf{x}, \mathbf{x}_0)$ is the Green's function for the bounded domain.

For the general case where the domain has no symmetries the only conserved quantity is the energy E, and the N-point pdf finds its expression as

$$P_N(\mathbf{x}_1, \mathbf{x}_2, \dots, \mathbf{x}_N) = \frac{\delta(H - E)}{Q(E)}$$

where

$$\int_V P_N(\mathbf{x}_1, \mathbf{x}_2, \cdots, \mathbf{x}_N) d\mathbf{x}_1 \ldots d\mathbf{x}_N = 1.$$

For domains with symmetries there are additional conserved quantities. In circular domains, for example, the angular momentum L^2 is also conserved,

$$L^2 = \frac{1}{N\Gamma} \sum_{i=1}^{N} \Gamma_i \mathbf{x}_i^2.$$

Here, Γ is a positive reference circulation used for dimensional consistency, and the N-point pdf becomes

$$P_N(\mathbf{x}_1, \ldots, \mathbf{x}_N) = \delta(H - E) \cdot \delta\left(\sum \Gamma_i \mathbf{x}_i^2 - N\Gamma L^2\right) / Q(E, L^2).$$

Pointin and Lundgren (1976) consider mixed populations of vortices of strength $\Gamma_i \equiv \Gamma > 0$ and $\Gamma_j \equiv -\Gamma$, with $n^+ N$ vortices of strength Γ and $n^- N$ vortices of strength $-\Gamma$, where $n^+ + n^- = 1$, with $n^+ \geq 0$, $n^- \leq 1$. The reduced single-point pdf's for the two populations are written

$$P_1^+(\mathbf{x}_1) = \int P_N(\mathbf{x}_1, \ldots, \mathbf{x}_N) d\mathbf{x}_2 \cdots d\mathbf{x}_N$$

$$P_1^+(\mathbf{x}_2) = \int P_N(\mathbf{x}_1, \ldots, \mathbf{x}_N) d\mathbf{x}_1 d\mathbf{x}_3 \cdots d\mathbf{x}_N$$

$$.$$
$$.$$
$$.$$

$$P_1^+(\mathbf{x}_{n+N}) = \int P_N(\mathbf{x}_1, \ldots, \mathbf{x}_N) d\mathbf{x}_1 \cdots d\mathbf{x}_{n+N-1} d\mathbf{x}_{n+N+1} \cdots d\mathbf{x}_N$$

$$P_1^-(\mathbf{x}_{n+N+1}) = \int P_N(\mathbf{x}_1, \ldots, \mathbf{x}_N) d\mathbf{x}_1 \cdots d\mathbf{x}_{n+N} d\mathbf{x}_{n+N+2} \cdots d\mathbf{x}_N$$

$$.$$
$$.$$
$$.$$

$$P_1^-(\mathbf{x}_N) = \int P_N(\mathbf{x}_1, \ldots, \mathbf{x}_N) d\mathbf{x}_1 \cdots d\mathbf{x}_{N-1}.$$

The two-point pdf's are defined similarly; hence, for example,

$$P_2^{++}(\mathbf{x}_1, \mathbf{x}_2) = \int P_N(\mathbf{x}_1, \ldots, \mathbf{x}_N) d\mathbf{x}_3 \cdots d\mathbf{x}_N.$$

Basic symmetries also apply:

$$P_2^{+-}(\mathbf{x}_2, \mathbf{x}_1) = P_2^{-+}(\mathbf{x}_1, \mathbf{x}_2).$$

From these quantities, the mean vorticity and streamfunctions can be defined:

$$< \omega(\mathbf{x}) > \equiv \int \left[\sum_{i=1}^{N} \Gamma_i \delta(\mathbf{x} - \mathbf{x}_i) \right] P_N d\mathbf{x}_1 \cdots d\mathbf{x}_N$$

$$= N\Gamma \left[n^+ P_1^+(\mathbf{x}) - n^- P_1^-(\mathbf{x}) \right]$$

$$\equiv -\nabla^2 < \psi(\mathbf{x}) >$$

with the average streamfunction

$$< \psi(\mathbf{x}) > = -N\Gamma \int G(\mathbf{x}, \xi) \left[n^+ P_1^+(\xi) - n^- P_1^-(\xi) \right] d\xi,$$

and the mean energy of the system is obtained from the Hamiltonian:

$$< E > \equiv < H >$$

$$= -\frac{N^2 \Gamma^2}{2} \int G(\mathbf{x}_1, \mathbf{x}_2) \left[n^+ P_1^+(\mathbf{x}_1) - n^- P_1^-(\mathbf{x}_1) \right]$$

$$\cdot \left[n^+ P_1^+(\mathbf{x}_2) - n^- P_1^-(\mathbf{x}_2) \right] d\mathbf{x}_1 d\mathbf{x}_2$$

$$- \frac{N^2 \Gamma^2}{2} \int G(\mathbf{x}_1, \mathbf{x}_2) \left[(n^+)^2 P_2^{\prime++}(\mathbf{x}_1, \mathbf{x}_2) - 2n^+ n^- P_2^{\prime+-}(\mathbf{x}_1, \mathbf{x}_2) \right]$$

$$+ \left[(n^-)^2 P_2^{\prime--}(\mathbf{x}_1, \mathbf{x}_2) \right] d\mathbf{x}_1 d\mathbf{x}_2$$

$$+ \frac{N\Gamma^2}{2} \int G(\mathbf{x}_1, \mathbf{x}_2) \left[n^+ P_2^{++}(\mathbf{x}_1, \mathbf{x}_2) + n^- P_2^{--}(\mathbf{x}_1, \mathbf{x}_2) \right] d\mathbf{x}_2 d\mathbf{x}_2$$

$$- \frac{N\Gamma^2}{2} \int g(\mathbf{x}_1, \mathbf{x}_2) \left[n^+ P_1^+(\mathbf{x}_1) + n^- P_1^-(\mathbf{x}_1) \right] d\mathbf{x}_1.$$

Remark. From the energy formula the following scaling laws hold. Assuming the first term in the above formula does not vanish, then $E \sim N^2 \Gamma^2$, i.e., the energy scales with the square of the number of vortices — called the high energy scaling limit. If, however, the first two terms vanish, then $E \sim N\Gamma^2$, i.e., the energy scales with N — called the low energy scaling limit. Equilibrium theories for these two limits behave quite differently. $\Diamond$

We now proceed to derive equations for the equilibrium distributions. Since the calculations closely follow those in the previous section, we skip most of the details in deriving the equation governing the reduced pdf's, which in this case are

$$\frac{\partial P_1^+}{\partial \mathbf{x}_1} = \int \frac{\partial G(\mathbf{x}_1, \mathbf{x}_2)}{\partial \mathbf{x}_1} \left(8\pi\lambda[(n^+ - N^{-1})P_2^{++}(\mathbf{x}_1, \mathbf{x}_2) - n^- P_2^{+-}(\mathbf{x}_1, \mathbf{x}_2)] \right.$$

$$+ N^{-1} \left[(n^+ - N^{-1}) \frac{\partial P_2^{++}(\mathbf{x}_1, \mathbf{x}_2)}{\partial \tilde{E}} - n^- \frac{\partial P_2^{+-}(\mathbf{x}_1, \mathbf{x}_2)}{\partial \tilde{E}} \right] \right) dx_2$$

$$+ \frac{1}{2N} \frac{\partial g(\mathbf{x}_1, \mathbf{x}_2)}{\partial \mathbf{x}_1} \left[8\pi\lambda P_1^+(\mathbf{x}_1) + N^{-1} \frac{\partial P_1^+(\mathbf{x}_1)}{\partial \tilde{E}} \right]$$

where the normalized reciprocal temperature λ and the scaled energy $\tilde{E}$ are defined by

$$\lambda = \frac{N\Gamma^2}{8\pi T}, \qquad \tilde{E} = \frac{E}{N^2\Gamma^2}.$$

One can see from this equation (and the general equation for the s-point pdf found in Pointin and Lundgren (1976)) that a closure assumption must be made, which we take, as usual, to be

$$P_2^{++}(\mathbf{x}_1, \mathbf{x}_2) \sim P_1^+(\mathbf{x}_1) \cdot P_1^+(\mathbf{x}_2).$$

If we then take the limit $N \to \infty$, thereby neglecting terms that are $O(1/N)$ or smaller, we arrive at the simpler equation

$$\frac{\partial P^+(\mathbf{x}_1)}{\partial \mathbf{x}_1} = 8\pi\lambda P^+(\mathbf{x}_1) \int \frac{\partial G(\mathbf{x}_1, \mathbf{x}_2)}{\partial \mathbf{x}_1} [n^+ P^+(\mathbf{x}_2) - n^- P^-(\mathbf{x}_2)] dx_1$$

$$(6.2.10)$$

which should be compared with (6.2.7) in the previous section. Taking the divergence of (6.2.10), one obtains

$$\nabla_{\mathbf{x}_1}^2 \log \frac{P^+(\mathbf{x}_1)}{P_0} = 8\pi\lambda[n^+ P^+(\mathbf{x}_1) - n^- P^-(\mathbf{x}_1)], \qquad (6.2.11)$$

where the scaling constant P_0 is defined so that

$$\int P_0 dx_1 = 1.$$

A simpler and more convenient scaled equation can be written as

$$\Delta\Phi = \alpha[\exp(\Phi) - C\exp(-\Phi)], \qquad (6.2.12)$$

where

$$C = \frac{n^- \int \exp(\Phi)dx}{[n^+ \int \exp(-\Phi)dx]},$$

with $\Phi(\mathbf{x}) = 0$ on the domain boundary, and

$$\alpha = 8\pi n^+ C^+ \lambda,$$
$$\Phi = \log \frac{P^+(\mathbf{x}_1)}{C^+}.$$

Finally, (6.2.12) can also be written

$$\Delta \tilde{\Phi} = 2\tilde{\alpha} \sinh \tilde{\Phi} \tag{6.2.13}$$

via the simple change of variables

$$\tilde{\Phi} = \Phi - \frac{1}{2} \log C$$
$$\tilde{\alpha} = \alpha \sqrt{C}.$$

The sinh-Poisson equation (6.2.13) and equations similar to it in structure are the key equations governing statistical equilibrium distributions for multi-species point vortex gases. A thorough study and discussion of the analytic aspects of the sinh-Poisson equation and some of its solutions can be found in Ting, Chen, and Lee (1987). Recent papers of Chavanis and Sommeria (1996, 1998) are also quite nice for these purposes, in particular in their discussions of the multiple solutions inherent to these theories. The central role played by this equation is further heightened by more recent work of Montgomery et al (1992), Brands et al (1997), and Mikelic and Robert (1998), which shows that decaying two-dimensional turbulent simulations of the Navier–Stokes equations "relax," after a sufficiently long time, to sinh-Poisson maximum entropy solutions.

The equilibrium pdf's can be recovered from the relations

$$C^+ = [\int \exp(\Phi)d\mathbf{x}]^{-1}, \quad C^- = [\int \exp(-\Phi)d\mathbf{x}]^{-1}$$
$$P^+ = C^+ \exp(\Phi), \qquad P^- = C^- \exp(-\Phi)$$
$$\lambda = \frac{\alpha}{8\pi n^+ C^+}.$$

Based on this formulation, Pointin and Lundgren (1976) carry out numerical simulations in bounded regions that have rectangular and circular shapes, both for neutral ($n^+ = n^-$) and non-neutral ($n^+ \neq n^-$) vortex gases. Their basic conclusions, in general, are that the system spends its time near (most likely) maximum entropy states, whose spatial structures are generally the most simple — these are states that have single minima or maxima within the domain and correspond to high energy, spatially nonuniform configurations. From the entropy expression

$$S = -\int \left[n^+ P^+(\mathbf{x}_1) \log \left(\frac{P^+(\mathbf{x}_1)}{P_0} \right) + n^- P^-(\mathbf{x}_1) \log \left(\frac{P^-(\mathbf{x}_1)}{P_0} \right) \right] d\mathbf{x}_1,$$

they conclude that the states with highest entropy for a given energy are the most relevant. Hence, one might then try to develop a theory based directly on maximizing the entropy expression subject to certain constraints (see, for example, Erikson and Smith (1988)). We describe this approach in the next section.

6.3   Maximum Entropy Theories

A fundamentally different but nonetheless complementary approach is based on maximizing the entropy of the given system of point vortices subject to the constraints dictated by the conservation of energy and whatever other conserved quantities exist. Maximum entropy approaches are used in a wide variety of contexts, and interesting descriptions can be found in the collection edited by Levine and Tribus (1979), particularly the article of Jaynes (1979) and Robertson (1979). For vortex systems the approach was used by Joyce and Montgomery (1973), Montgomery and Joyce (1974), Kida (1975), and, more recently, directly for the Euler equations by Miller (1990), Miller, Weichman, and Cross (1992), Robert (1991), and Robert and Sommeria (1991). We first describe the main ideas in some generality, following part of the discussion given in Chorin (1994, 1996).

As shown in Figure 6.1, we first subsection the $(\mathbf{x}, \mathbf{y})$-plane into discrete cells. The spatial vorticity distribution is then obtained by identifying the number of point vortices in each cell. As long as the vortices stay within their given cell or, if any two vortices of the same strength in two different cells are exchanged, the spatial vorticity distribution remains invariant. The coarse-grained distribution is simply the product of the total area of the cells in which the vortices are located, with the number of different ways one can allocate a given number of vortices to each of the cells. This phase space volume then corresponds to the probability of realizing a given spatial distribution, hence the distribution that corresponds to maximal volume (subject to all the relevant constraints) is also the most probable.

6.3.1   Joyce–Montgomery Approach

To be more specific, we now consider a collection of N–point vortices occupying a finite region of the plane of area A. Assume for the moment that we have N^+ vortices of strength $\Gamma = +1$ and N^- of strength $\Gamma = -1$, with $N^+ + N^- \equiv N$. The net vorticity of the system is then $N^+ - N^-$ and if $N^+ = N^-$, we have a "neutral" vortex gas.

To obtain a combinatorial formula for the probability of achieving a given spatial distribution, we partition the area A into M squares of area h^2, hence

$$h^2 \cdot M \equiv A,$$

and we denote n_i^+ (n_i^-) the number of positive (negative) vortices in square i ($i = 1, \ldots, M$). Since there are

$$\frac{N^+!}{n_1^+! n_2^+! \cdots n_M^+!}$$

ways of arranging N^+ vortices into M squares, and

$$\frac{N^-!}{n_1^-! n_2^-! \cdots n_M^-!}$$

ways of arranging N^- vortices into M squares, the probability W associated with a given distribution is

$$W = \left(\frac{N^+!}{n_1^+! n_2^+! \cdots n_M^+!} \right) \cdot \left(\frac{N^-!}{n_1^-! n_2^-! \cdots n_M^-!} \right) \cdot h^{2N}.$$

Since this expression also corresponds to the phase space volume occupied by the vortices, the entropy S can be obtained via

$$S \sim \log W.$$

We then must maximize this discrete expression (using the Lagrange multiplier method) subject to the constraints

$$\sum_{i=1}^{M} n_i^+ = N^+, \quad \sum_{i=1}^{M} n_i^- = N^-,$$

with

$$E = \frac{1}{2} \sum_{i \neq j} (n_i^+ - n_i^-) G_{ij} (n_j^+ - n_j^-) = \text{const},$$

where

$$G_{ij} = -\frac{1}{2} \log |\mathbf{x}_i - \mathbf{x}_j| + \text{const},$$

with $\mathbf{x}_i$ being a point in the ith box and $\mathbf{x}_j$ being a point in the jth box. Upon maximizing S, one obtains the equilibrium system

$$\log(n_i^+) + \alpha^+ + \beta \sum_j G_{ij} (n_j^+ - n_j^-) = 0$$

$$\log(n_i^-) + \alpha^- - \beta \sum_j G_{ij} (n_j^+ - n_j^-) = 0,$$

with Lagrange multipliers α^+, α^-, β. This system can be further manipulated to obtain an expression for the discrete vorticity distribution

$$n_i^+ - n_i^- = \exp[-\alpha^+ - \beta \sum_j G_{ij}(n_j^+ - n_j^-)]$$

$$- \exp[-\alpha^- + \beta \sum_j G_{ij}(n_j^+ - n_j^-)],$$

where $i = 1, \ldots, M$.

After this we pass to the continuum limit defined by

$$h \to 0,$$
$$(n_i^+ - n_i^-) \to \omega(\mathbf{x})h^2 = \omega(\mathbf{x})dx,$$

which leads directly to the nonlinear equation

$$-\Delta\psi(\mathbf{x}) = \omega(\mathbf{x})$$
$$= d_+ \exp(-\beta\psi) - d_- \exp(\beta\psi),$$

where ψ is the equilibrium streamfunction, β is a Lagrange multiplier, and d_+ and d_- are normalization constants; this equation is often referred to as the Joyce–Montgomery equation (1973) for statistical equilibrium. Under certain conditions, such as the case of a periodic domain where $\psi = 0$ on the boundary and $N^+ = N^- = N/2$, $d_+ = d_- \equiv d$, the equation simplifies to the sinh-Poisson equation

$$-\Delta\psi(\mathbf{x}) = d\sinh[\beta\psi(\mathbf{x})].$$

Remarks.

1. In a very thorough and detailed treatment, Kida (1975) follows essentially the same ideas as those in Joyce and Montgomery (1973) by maximizing the discrete entropy for a two-species vortex gas in a closed circular container. This leads him to a nonlinear integral equation analogous to the Pointin–Lundgren equation (6.2.10).

2. The developments of Robert (1991), to a large extent, represent a rigorous treatment of the maximum entropy approaches developed in Joyce and Montgomery (1973) and Kida (1975). Robert (1991) introduces the idea of "Young measures" to characterize the small-scale oscillations associated with the vorticity fields, applied directly to the Euler equations, without the use of the point vortex approximation. From this point of view his approach follows more closely ideas developed in Miller (1990). Discussions of these methods can be found in DiPerna and Majda (1987) and more recently in Michel and Robert (1994).

3. One of the central issues in all of the statistical treatments is the number of conserved quantities that are automatically built in to the theory. The theory of Kraichnan (1975), for example, relies only on conservation of energy and entropy, while the theories of Miller (1990), Miller, Weichman, and Cross (1992), Robert (1991), and Robert and Sommeria (1991) are based on an infinite set of constraints. Other more recent approaches such as that in Turkington (1998) use only a small number of constraints — the paper of Majda and Holen (1997) shows that the "few constraint" theories are in fact statistically sharp macroscopic states associated with the infinite constraint theories. The paper of Miller, Weichman and Cross (1992) is particularly complete and is recommended as a starting point.

4. As an alternative to maximum entropy theories, one can consider macroscopic states that correspond to vorticity fields with minimum enstrophy, an idea that originated in Leith (1984). Maximum entropy approaches are based on inviscid dynamics (either discrete or continuous), whereas minimum enstrophy theories are based essentially on an inviscid limit $\nu \to 0$. A connection between the two theories is discussed in Chavanis and Sommeria (1996, 1998) in the limit of "strong mixing," where the maximization of entropy becomes equivalent to the minimization of the coarse-grained enstrophy. These ideas are further discussed in Brands et al (1999).

5. Recent work of Lim (1999) introduces, for the first time, the use of the Bogoliubov–Feynman inequalities in the derivation of the sinh-Poisson and related equations and thus represents a significant new direction. In addition, the work of Lim (1998c) and Newton and Mezić (1999) develops a statistical theory for spherical surfaces, thus making it potentially applicable to atmospheric flows. Mathematically, spherical domains are attractive because they offer the advantages of compact, rotationally symmetric domains, without the disadvantages of boundaries.

6. The statistical mechanics of the two-dimensional Euler equation is very similar to the theory of "violent relaxation" developed by Lynden-Bell (1967) for the Vlasov equation describing collisionless stellar systems (e.g., elliptical galaxies). The analogy between stellar systems and two-dimensional vortices has recently been investigated by Chavanis (1996, 1998). This analogy concerns not only equilibrium states, but also relaxation toward equilibrium (Chavanis et al (1996), Chavanis (1998)) and the statistics of flucuations (Chavanis and Sire (2000a,b)). ◊

6.4 Nonequilibrium Theories

We turn next to a time-dependent treatment capable of incorporating nonequilibrium aspects of vortex interactions, based on a "kinetic theory" approach. There are not many results in this direction, so we only outline the approach, based on a recent paper of Marmanis (1998). The general goal is to derive the so-called BBGKY hierarchy for a system of point vortices in the unbounded plane, which gives the evolution of the reduced pdf's and allows us, under certain assumptions that we make explicit, to determine the nonequilibrium evolution of these variables. A general introduction to the BBGKY hierarchy and the kinetic theory approach to fluids can be found in Cole (1967).

Our starting point is the Liouville equation (6.2.1) governing $P_N(\mathbf{q}, \mathbf{p}; t)$ with the usual Hamiltonian (6.1.1). Because of the large number of independent variables it contains, this equation is impractical to use and we will use it only as the starting point to derive a simpler set of "kinetic" equations that govern the reduced pdf's. For this, we follow Marmanis (1998) and denote the reduced distribution function $P_{mn}(\mathbf{q}, \mathbf{p}; t)$,

$$
P_{mn}(\mathbf{q}, \mathbf{p}; t) = A^{m+n} \int P_N(\mathbf{q}, \mathbf{p}; t) \prod_{i=m+1}^{N^+} dq_i dp_i \prod_{j=N^++n+1}^{N^++N^-} dq_j dp_j,
$$

where we have assumed N positive vortices (the $+$ superscript denotes positive circulation) and N negative vortices (the $-$ superscript denotes negative circulation) of equal and opposite strength. A is the area that encloses all the vortices. The expression

$$
\frac{1}{A^{m+n}} P_{mn}(\mathbf{q}, \mathbf{p}; t) \prod_{i=1}^{m} dq_i dp_i \prod_{j=N^++1}^{N^++n} dq_j dp_j
$$

is the probability that the group of m positive and n negative vortices lie within

$$
dq_1 \cdots dq_m dp_1 \cdots dp_m
$$

of (q_i, p_i), $i = 1, \ldots, m$ and within

$$
dq_1 \cdots dq_n dp_1 \cdots dp_n
$$

of (q_i, p_i), $i = N+1, \ldots, N+n$. In terms of P_N, we can write the average velocity expressions as

$$
\begin{aligned}
<u> &\equiv \int \left(\sum_{i=1}^{2N} \frac{\Gamma_i}{2\pi} \frac{y - y_i}{r_i^2} \right) P_N d\mathbf{q} d\mathbf{p} \\
&= \sum_{i=1}^{N^+} \frac{(\Gamma_i)^{3/2}}{2\pi} \int \frac{(\sqrt{\Gamma_i} y - p_i)}{l_i^2} P_{10}(q_i, p_i; t) dq_i dp_i \\
&\quad - \sum_{i=1}^{N^-} \frac{|\Gamma_i|^{3/2}}{2\pi} \int \frac{(\sqrt{|\Gamma_i|} y - q_i)}{l_i^2} P_{01}(q_i, p_i; t) dq_i dp_i
\end{aligned}
$$

$$
\begin{aligned}
<v> &\equiv \int \left(-\sum_{i=1}^{2N} \frac{\Gamma_i}{2\pi} \cdot \frac{x - x_i}{r_i^2} \right) P_N d\mathbf{q} d\mathbf{p} \\
&= -\sum_{i=1}^{N^+} \frac{(\Gamma_i)^{3/2}}{2\pi} \int \frac{(\sqrt{\Gamma_i} x - q_i)}{l_i^2} P_{10}(q_i, p_i; t) dq_i dp_i \\
&\quad + \sum_{i=1}^{N^-} \frac{|\Gamma_i|^{3/2}}{2\pi} \int \frac{(\sqrt{|\Gamma_i|} x - p_i)}{l_i^2} P_{01}(q_i, p_i; t) dq_i dp_i,
\end{aligned}
$$

where

$$
\begin{aligned}
r_i^2 &= (x - x_i)^2 + (y - y_i)^2 \\
l_i^2 &= (\sqrt{|\Gamma_i|} x - q_i)^2 + (\sqrt{|\Gamma_i|} y - p_i)^2.
\end{aligned}
$$

The average vorticity becomes

$$
\begin{aligned}
<\omega> &= \int \left(\sum_{i=1}^{2N} \Gamma_i \delta(x - x_i) \delta(y - y_i) \right) P_N d\mathbf{q} d\mathbf{p} \\
&= \sum_{i=1}^{N^+} \Gamma_i \int \delta(x - x_i) \delta(y - y_i) P_{10}(q_i, p_i; t) dq_i dp_i \\
&\quad - \sum_{i=1}^{N^-} |\Gamma_i| \int \delta(x - x_i) \delta(y - y_i) P_{01}(q_i, p_i; t) dq_i dp_i \\
&= \sum_{i=1}^{N^+} \Gamma_i^2 P_{10}(\sqrt{\Gamma_i} x, \sqrt{\Gamma_i} y; t) - \sum_{i=1}^{N^-} |\Gamma_i|^2 P_{01}(\sqrt{|\Gamma_i|} y, \sqrt{|\Gamma_i|} x; t).
\end{aligned}
$$

For notational convenience Marmanis (1998) sets $A = 1$ with the proviso that whenever needed, the correct expressions can be deduced from the "scaled" ones by dimensional considerations. This will be clear as we proceed.

We now define the following integral reduction operators

$$\hat{\Theta}^m \equiv \int_{-\infty}^{\infty} \cdots \int_{-\infty}^{\infty} d\mathbf{x}_{m+1} \cdots d\mathbf{x}_N$$

$$\hat{A}^n \equiv \int_{-\infty}^{\infty} \cdots \int_{-\infty}^{\infty} d\mathbf{x}_{N+n+1} \cdots d\mathbf{x}_{2N},$$

where $\mathbf{x}_i \equiv (q_i, p_i)$ for $1 \leq i \leq 2N$. This notation simplifies the representation of the reduced pdf P_{mn} to

$$P_{mn}(\mathbf{x}_1, \ldots, \mathbf{x}_m, \mathbf{x}_{N+1}, \ldots, \mathbf{x}_{N+n}; t) = \hat{\Theta}^m \hat{A}^n P_N(\mathbf{x}; t).$$

Our goal is to derive the simplified evolution equations for P_{mn} starting from the Liouville equation (6.2.1). Using the fact that H is a function only of the relative distance between vortex pairs, we can write

$$H \equiv \sum_{1 \leq i < j \leq N} H(|\mathbf{x}_i - \mathbf{x}_j|) + \sum_{N+1 \leq i < j \leq 2N} H(|\mathbf{x}_i - \mathbf{x}_j|)$$

$$+ \sum_{1 \leq i \leq N} \sum_{N+1 \leq j \leq 2N} H(|q_i - p_j|, |p_i - q_j|),$$

whose various terms represent the interactions between all possible combinations of positive and negative vortices in pairs. Then, we can operate with $\hat{\Theta}^m$ and $\hat{A}^n$ on (6.2.1) to obtain

$$\frac{\partial P_{mn}}{\partial t} = \sum_{1 \leq i < j \leq N} \hat{\Theta}^m \hat{A}^n \{H(|\mathbf{x}_i - \mathbf{x}_j|), P_N\}$$

$$+ \sum_{N+1 \leq i < j \leq 2N} \hat{\Theta}^m \hat{A}^n \{H(|\mathbf{x}_i - \mathbf{x}_j|), P_N\}$$

$$+ \sum_{1 \leq i \leq N} \sum_{N+1 \leq j \leq 2N} \hat{\Theta}^m \hat{A}^n \{H(|q_i - p_j|, |p_i - q_j|), P_N\}.$$

Upon simplifying this equation (see Marmanis (1998) for details), we arrive at the hierarchical system

$$
\frac{\partial P_{mn}}{\partial t} = \sum_{1 \le i < j \le m} \{H(|\mathbf{x}_i - \mathbf{x}_j|), P_{mn}\} + \sum_{N+1 \le i < j \le N+n} \{H(|\mathbf{x}_i - \mathbf{x}_j|), P_{mn}\}
$$

$$
+ \sum_{1 \le i \le m} \sum_{N+1 \le j \le N+n} \{H(|q_i - p_j|, |p_i - q_j|), P_{mn}\}
$$

$$
+ \frac{(N-m)}{A} \sum_{1 \le i \le m} \int \{H(|\mathbf{x}_i - \mathbf{x}_{m+1}|), P_{(m+1)n}\} d\mathbf{x}_{m+1}
$$

$$
+ \frac{(N-n)}{A} \sum_{N+1 \le i \le N+n} \int \{H(|\mathbf{x}_i - \mathbf{x}_{N+n+1}|), P_{m(n+1)}\} d\mathbf{x}_{N+n+1}
$$

$$
+ \frac{(N-m)}{A} \times
$$

$$
\sum_{N+1 \le j \le N+n} \int \{H(|q_{m+1} - p_j|, |p_{m+1} - q_j|), P_{(m+1)n}\} dq_{m+1} dp_{m+1}
$$

$$
+ \frac{(N-n)}{A} \times
$$

$$
\sum_{1 \le i \le m} \int \{H(|q_i - p_{N+n+1}|, |p_i - q_{N+n+1}|), P_{m(n+1)}\} dq_{N+n+1} dp_{N+m+1}.
$$

$$(6.4.1)$$

From this equation we see that the evolution of P_{mn} depends on terms involving $P_{(m+1)n}$, $P_{m(n+1)}$, $\ldots$; hence, to close the system, we need further assumptions. The evolution of the first three reduced pdf's, P_{01}, P_{10}, P_{11}, for example, are given by

$$
\frac{\partial P_{01}}{\partial t} = \frac{N}{A} \int \{H(|q_1 - p_{N+1}|, |p_1 - q_{N+1}|), P_{11}\} dq_1 dp_1
$$

$$
+ \frac{(N-1)}{A} \int \{H(|\mathbf{x}_{N+1} - \mathbf{x}_{N+2}|), P_{02}\} d\mathbf{x}_{N+2}
$$

$$
\frac{\partial P_{10}}{\partial t} = \frac{N}{A} \int \{H(|q_1 - p_{N+1}|, |p_1 - q_{N+1}|), P_{11}\} dq_{N+1} dp_{N+1}
$$

$$
+ \frac{(N-1)}{A} \int \{H(|\mathbf{x}_1 - \mathbf{x}_2|), P_{20}\} d\mathbf{x}_2
$$

$$\frac{\partial P_{11}}{\partial t} = \{H(|\mathbf{x}_1 - \mathbf{x}_{N+1}|), P_{11}\}$$

$$+ \frac{(N-1)}{A} \int \{H(|\mathbf{x}_1 - \mathbf{x}_2|), P_{21}\} d\mathbf{x}_2$$

$$+ \frac{(N-1)}{A} \int \{H(|\mathbf{x}_{N+1} - \mathbf{x}_{N+2}|), P_{12}\} d\mathbf{x}_{N+2}$$

$$+ \frac{(N-1)}{A} \int \{H(|q_2 - p_{N+1}|, |p_2 - q_{N+1}|), P_{21}\} dq_2 dp_2$$

$$+ \frac{(N-1)}{A} \int \{H(|q_1 - p_{N+2}|, |p_1 - q_{N+2}|), P_{12}\} dq_{N+2} dp_{N+2}$$

$$= \{H(|\mathbf{x}_1 - \mathbf{x}_{N+1}|), P_{11}\}$$

$$+ \frac{(N-1)}{A} \times$$

$$\int \{H(|\mathbf{x}_1 - \mathbf{x}_2|) + H(|q_2 - p_{N+1}|, |p_2 - q_{N+1}|)), P_{21}\} d\mathbf{x}_2$$

$$+ \frac{(N-1)}{A} \times$$

$$\int \{(H(|\mathbf{x}_{N+1} - \mathbf{x}_{N+2}|) + H(|q_1 - p_{N+2}|, |p_1 - q_{N+2}|)), P_{12}\} d\mathbf{x}_{N+2}$$

$$= \{H(|\mathbf{x}_1 - \mathbf{x}_{N+1}|), P_{11}\} + \frac{(N-1)}{A} \int \{\mathcal{H}, P_{22}\} d\mathbf{x}_2 d\mathbf{x}_{N+2},$$

where $\mathcal{H}$ in the last term stands for the interaction between the vortex at $\mathbf{x}_1$ with those at $\mathbf{x}_2$ and $\mathbf{x}_{N+2}$, and the vortex at $\mathbf{x}_{N+1}$ with those at $\mathbf{x}_2$ and $\mathbf{x}_{N+2}$. Since we have assumed a population of vortices of the same or opposite strength, this last expression can be thought of as governing interactions of a gas of vortex dipoles (pairs of equal and opposite vortices) — the second term above can be thought of as a collision operator governing dipole interactions.

We would now like to derive a closed evolution equation of the form

$$\frac{\partial P_{mn}}{\partial t} = G(\mathbf{x}, P_{mn}).$$

To achieve this, assume

$$P_{22}(\mathbf{x}_1, \mathbf{x}_2, \mathbf{x}_{N+1}, \mathbf{x}_{N+2}) = P_{11}(\mathbf{x}_1, \mathbf{x}_{N+1}) P_{11}(\mathbf{x}_2, \mathbf{x}_{N+2}),$$

which Marmanis (1998) calls the *vortex-dipole-chaos* (VDC) assumption and which amounts to the same closure assumption adopted by Lundgren and Pointin (1977b). It assumes that colliding dipoles come from different regions of phase space and are uncorrelated before colliding and scattering.

The closed system governing P_{11} then becomes

$$\frac{\partial P_{11}}{\partial t} + (\mathbf{u}_1 \cdot \nabla_*^{(1)}) P_{11}$$
$$= \frac{N-1}{A} \int [H, P_{11}(\mathbf{x}_1, \mathbf{x}_{N+1}) P_{11}(\mathbf{x}_2, \mathbf{x}_{N+2})] d\mathbf{x}_2 d\mathbf{x}_{N+2} \qquad (6.4.2)$$

where

$$\nabla_*^{(1)} \equiv \left((\frac{\partial}{\partial q_1^+} + \frac{\partial}{\partial p_1^-}), (\frac{\partial}{\partial p_1^+} + \frac{\partial}{\partial q_1^-}) \right)$$

with $\mathbf{u}_1$ being the velocity of the dipole with subscript 1. This is the basic kinetic equation, whose left hand side represents the change in P_{11} due to evolution and convection and whose right hand side represents the change due to collisions.

Remarks.

1. Marmanis (1998) simplifies the equation further for the special case of a statistically homogeneous and isotropic dipole vortex gas by introducing polar variables and assuming no angular dependence upon interactions. Equation (6.4.2) is similar in structure to the equation for molecular evolution in an ordinary gas using the Boltzmann approach. With this analogy each dipole is thought of as a fundamental building block making up the gas via interactions. Whether such an approach ultimately will prove useful remains to be seen, but the approach seems overly restrictive since it is known that a dipole in a sea of vortices is not a robust or stable structure. In Newton and Mezić (2000) a more general approach is taken, based on the interaction of monopoles, dipoles, and tripoles as the basic ballistic elements.

2. A different approach is suggested in the work of Chavanis (1998), where the relaxation of point vortices towards statistical equilibrium is governed by a Fokker–Planck equation involving both diffusion and drift, where the drift coefficient is a generalization of the Einstein formula in the theory of Brownian motion and is always perpendicular to the mean-field velocity of the vortices. The Fokker-Planck approach is consistent with a maximum entropy production principle, as discussed in Robert and Sommeria (1992). These methods can be extended, in principle, to the case of continuous vorticity fields, making it possible to derive an H-theorem for the mixing entropy (see Chavanis (2000a,b)).

3. Decaying two-dimensional turbulence is a nonequilibrium process exhibiting many interesting features. It is thought to be driven by the

merging of like-signed vortices when they approach a critical distance. Therefore, the density n of vortices decreases with time, while their characteristic radii increase. To understand this decay process, a "puncuated" Hamiltonian model has been introduced by Weiss and McWilliams (1993) in which point vortices follow the usual Hamiltonian dynamics but merge via emperical rules derived from the conservation laws. These types of models were investigated recently by Sire and Chavanis (2000) using a renormalization group procedure, which allows for much larger simulation times than could otherwise be achieved. In this work an effective 3-vortex interaction theory shows that the decay of the total area occupied by the vortices results in a physical process by which merging occurs principally via 3-vortex collisions involving vortices of different signs. A kinetic theory based on this 3-body process leads to a decay of the vortex density like $n \sim t^{-\xi}$, with $\xi = 1$. See also the recent numerical simulations of Laval et al (2000).

6.5   Exercises

1. Consider a point vortex dipole made up of equal and opposite point vortices of strength Γ and $-\Gamma$, separated by distance d.

   (a) In the unbounded plane prove that they move along a straight line that bisects in a perpendicular way the line segment joining them.

   (b) Consider their motion in a doubly periodic parallelogram, as discussed in Example 6.4. As shown in Figure 6.2, show that they move along a straight line that bisects in a perpendicular way the line segment joining them.

2. Go through the details of the derivation leading to (6.2.9).

3. Derive the sinh-Poisson equation (6.2.13) starting from (6.2.11).

4. Derive the evolution equation (6.4.1) for P_{mn} and verify the equations for the special cases P_{10}, P_{01}, P_{11}.

5. Plot $\Phi(E)$, $\Phi'(E)$, $\Phi''(E)$, and $T \equiv \Phi'/\Phi''$.

6. Derive formulas (6.2.3), (6.2.4), (6.2.5) for the average velocity, energy, and angular momentum.

7
Vortex Patch Models

In this chapter we relax the assumption that the vorticity is distributed singularly and we describe solutions to the two-dimensional Euler equations in which it is uniformly distributed over compact planar regions. Since elliptical regions of constant vorticity play a particularly important role in the literature for reasons that are both physical and mathematical, much of the focus is on solutions and models based on this shape. We start by describing Kida's (1981) exact solution of an unsteady elliptical region of constant vorticity, which persists in a background flowfield including strain and additional uniform vorticity. This solution generalizes the classical one described in Lamb (1932) of an isolated elliptical region undergoing self-induced rotation — a so-called Kirchhoff ellipse. It is also a time-dependent generalization of the steady state solutions discovered by Moore and Saffman (1971). The Kida vortex represents one of the few known exact, nonlinear, time-dependent solutions to the Euler equations, where vorticity is not distributed singularly. Our presentation combines the original approach of Kida (1981) with the more general approach of Neu (1984) that allows for out-of-plane strain and makes use of a Hamiltonian formulation. We then use the Melnikov method to prove that in the presence of a three-dimensional *time-periodic* strain field, the elliptical vortex parameters oscillate chaotically, as was shown originally in the work of Bertozzi (1988) and more recently analyzed in Ide and Wiggins (1995). In Section 7.4 we describe the *moment model* of Melander, Zabusky, and Styczek (1986) (see also Melander, Syczek, and Zabusky (1984) for the original reference), which allows for the interaction of neighboring vortex patches under the simplifying assumption that the regions of uniform vorticity are well-separated. We describe a hierarchical formulation in which a self-consistent Hamiltonian system of ordinary differential equations governing the evolution of the physical space moments of the individual patches can be derived. The truncated second order moment model is of particular interest since it can be viewed as a generalization of the point vortex system of equations that allows for interacting elliptically shaped regions of uniform vorticity. This model is then used as the starting point for the calculation of a geometric phase associated with two corotating elliptical patches, following the work of Shashikanth and Newton (2000). We then describe a shear layer model introduced in Meiburg and Newton (1991), which includes the effects of viscous diffusion, based on the exact Oseen solution of a viscously decaying isolated point vortex. It is interesting that the system retains its Hamiltonian structure (although it is nonautonomous) even though viscous effects are included. In the final section we summarize other work, including estimates on the growth of compact vortex patches, as well as results on their nonlinear stability.

7.1   Introduction to Vortex Patches

A vortex patch is a desingularization of a point vortex in which the vorticity is a bounded function over a finite area of the plane — see Lamb (1932) and Saffman (1992) for an introduction to the classical theory. It can be viewed as the perpendicular section of an infinitely long rectilinear vortex tube of area A, whose vorticity distribution is invariant along the length of the tube. The induced velocity field is obtained by integrating over the area of the patch. For a patch with vorticity distribution $\omega(x,y)$, the streamfunction is

$$\psi(x,y) = -\frac{1}{2\pi} \int_A \omega(\xi,\eta) \log\left[(x-\xi)^2 + (y-\eta)^2\right] d\xi d\eta,$$

and the velocity components (u,v) at a point (x,y) are related to the streamfunction in the usual fashion: $u = \partial\psi/\partial y$ and $v = -\partial\psi/\partial x$. The boundary of each patch is advected by the flowfield according to the basic vorticity evolution equation

$$\omega_t + \mathbf{u} \cdot \nabla\omega = 0,$$

where, in general, $\mathbf{u}$ is the total velocity field due to all of the patches.

Example 7.1   (Kirchhoff ellipse) A simple example of an isolated uniform vortex patch is the Kirchhoff elliptical vortex described in Lamb (1932). The vorticity is distributed uniformly ($\omega = $ const) over an elliptical region with constant aspect ratio $\lambda = a/b$, where a is the major axis and b is the minor axis. The patch rotates about the center of the ellipse with constant angular velocity given by

$$\dot{\theta} = \frac{\omega\lambda}{(1+\lambda)^2} = \frac{\Gamma\lambda}{A(1+\lambda)^2}, \tag{7.1.1}$$

in which $\Gamma = \omega A$ is the strength of the patch. As shown in Lamb (1932), where details of the derivation can be found, a fluid particle on the patch moves in a circular orbit with *twice* this frequency. Other classical work includes the paper of Lord Kelvin (1880) and that of Love (1893), who showed that the patch is neutrally stable if and only if $a/b < 3$. Also see the paper of Wan (1986) for results on stability.   ◊

For a system of N disjoint patches, the velocity field is obtained by linear superposition of the velocity field due to each patch, just as in the case of N-point vortices. However, the evolution of a system of N interacting patches is, in general, more complicated than a system of N interacting point vortices. There is a large body of numerical work documenting various complex processes associated with the evolution of vortex patches, such as merger and filamentation; a good overview can be found in Chapter 9 of Saffman (1992). The numerical procedure used to simulate these processes goes under the name of "contour dynamics." One can find discussions of it starting with the paper of Zabusky, Hughes, and Roberts (1979). There are, however, certain special and interesting configurations that can be treated analytically in which the patch boundary evolves in such a way as to retain its initial shape. We describe such a case in the following section.

7.2   The Kida-Neu Vortex

In this section we formulate the equations of motion for an elliptical patch of constant vorticity under the influence of an external flowfield. Our description combines the analysis of Kida (1981) and Neu (1984) in the sense that Kida's analysis only allows for a symmetric planar external strain field, with additional background vorticity, while Neu's does not include additional background vorticity, but allows the planar strain to be asymmetric, and includes out of plane stretching. Consider an elliptical region in the plane, whose boundary is defined by the level curve

$$\begin{aligned} f(x, y; t) &\equiv \frac{[x\cos\theta(t) + y\sin\theta(t)]^2}{[a(t)]^2} \\ &+ \frac{[-x\sin\theta(t) + y\cos\theta(t)]^2}{[b(t)]^2} - 1 \\ &= 0. \end{aligned} \tag{7.2.1}$$

The principal axes of the ellipse are denoted $a(t)$, $b(t)$, with the angle between the x-axis and the axis aligned with $a(t)$ denoted by $\theta(t)$. The kinematic evolution equation for the elliptical boundary is

$$\frac{Df}{Dt} \equiv \frac{\partial f}{\partial t} + \mathbf{u} \cdot \nabla f = 0, \tag{7.2.2}$$

where $\mathbf{u}$ denotes the total velocity field. Hence, we write $\mathbf{u}$ as

$$\mathbf{u} = \mathbf{u}_k + \mathbf{u}_1 + \mathbf{u}_2 + \mathbf{u}_r, \tag{7.2.3}$$

where $\mathbf{u}_k$ is the self-induced velocity field due to the elliptical patch (as obtained from the isolated Kirchhoff ellipse described in Example 7.1), $\mathbf{u}_1$ is that due to the in-plane irrotational strain field, which we take to be

$$\mathbf{u}_1 = (\gamma_1 x, -\gamma_2 y, 0),$$

$\mathbf{u}_2$ is due to the out-of-plane irrotational strain

$$\mathbf{u}_2 = (0, 0, \gamma_3 z),$$

and $\mathbf{u}_r$ is due to the planar background rotation

$$\mathbf{u}_r = (-\gamma_4 y, \gamma_4 x, 0).$$

The incompressibility condition requires that

$$\gamma_1 - \gamma_2 + \gamma_3 = 0.$$

To make direct comparisons with the analysis of Neu (1984), we require $\gamma_1 = \gamma'$, $\gamma_2 = \gamma$, $\gamma_3 = \gamma''$, $\gamma_4 = 0$, while comparisons with Kida (1981) require $\gamma_1 = e$, $\gamma_2 = e$, $\gamma_3 = 0$, $\gamma_4 = \gamma$.

The evolution equations for $(a(t), b(t), \theta(t))$ are then obtained by using the representation for the elliptic boundary (7.2.1) in the evolution equation (7.2.2), using the total velocity field (7.2.3). One then obtains the closed system

$$\dot{a} = (\gamma_1 \cos^2 \theta - \gamma_2 \sin^2 \theta)a \tag{7.2.4}$$

$$\dot{b} = (\gamma_1 \sin^2 \theta - \gamma_2 \cos^2 \theta)b \tag{7.2.5}$$

$$\dot{\theta} = \frac{\omega a b}{(a+b)^2} - \frac{1}{2}(\gamma_1 + \gamma_2)\frac{a^2 + b^2}{a^2 - b^2}\sin 2\theta + \gamma_4. \tag{7.2.6}$$

Notice that γ_3 enters only through the incompressibility condition. Since Kida's analysis requires that $\gamma_1 = \gamma_2$, then necessarily $\gamma_3 = 0$, i.e., there is no out-of-plane strain component.

One can easily obtain the evolution equation for the cross-sectional area of the ellipse $A = \pi a b$ by multiplying (7.2.4) by πb and (7.2.5) by πa and adding

$$\dot{A} = (\gamma_1 - \gamma_2)A \Rightarrow A(t) = A_0 \exp(\gamma_1 - \gamma_2)t.$$

Since the vorticity and cross-sectional area are related by

$$w(t) = \frac{\Gamma}{A(t)},$$

we have an expression for the vorticity evolution in the elliptical region

$$w(t) = w_0 \exp(\gamma_2 - \gamma_1)t,$$

which shows that unless $\gamma_1 = \gamma_2$, the vorticity is time-dependent and the system of equations (7.2.4), (7.2.5), (7.2.6) is nonautonomous.

Taking $\gamma_4 = 0$ to follow the analysis of Neu (1984), it is useful to rewrite the system (7.2.4), (7.2.5), (7.2.6) in Hamiltonian form, using the aspect ratio $\lambda = a/b > 1$ and orientation angle θ as dependent variables. Using the rescaled time

$$\tau = w \left(\frac{\lambda^2}{\lambda^2 - 1} \right) t,$$

we obtain the canonical equations

$$\dot{\lambda} = -\frac{\partial \mathcal{H}}{\partial \theta}, \qquad \dot{\theta} = \frac{\partial \mathcal{H}}{\partial \lambda},$$

$$\mathcal{H}(\lambda, \theta; \tau) = \log \frac{(1 + \lambda)^2}{\lambda} - \frac{1}{2} \frac{\gamma_1 + \gamma_2}{w} \left(\lambda - \frac{1}{\lambda} \right) \sin 2\theta. \qquad (7.2.7)$$

By plotting level curves of the Hamiltonian function (7.2.7), one obtains a good understanding of the full range of solutions for the autonomous case $\gamma_1 = \gamma_2$. The phase portraits, obtained by plotting level curves of the Hamiltonian in the (x, y)-plane, with

$$x = \lambda \cos \theta, \qquad y = \lambda \sin \theta,$$

are shown in Figure 7.1 in a series of figures with increasing parameter values in the range $0 \leq \gamma_1/w \equiv \gamma_2/w \leq 0.25$, as depicted originally in Neu (1984). Each increase in the parameter value represents an increase in the strength of the external strain imposed on the elliptical vortex region.

The first case, shown is for the parameter value $\gamma_1/w = 0$, in which case the elliptical vortex patch rotates uniformly about its center, subject only to its self-induced velocity field; this corresponds to the Kirchhoff elliptical vortex patch of Example 7.1. As the strain field increases in strength, as shown for the cases $0 < \gamma_1/w < 0.15$, there exist four critical points, symmetrically placed along the line with slope $\pi/4$. The two closest to the origin are centers, while the outer two are saddles. These four solutions represent the steady elliptical patches discovered by Moore and Saffman (1971). The centers correspond to patches that are neutrally stable with respect to elliptical perturbations, while the saddles correspond to unstable patches. The regions which enclose the centers contain a continuous family of closed orbits, each one representing an ellipse that oscillates symmetrically about either $\theta = \pi/4$ or $\theta = 5\pi/4$. The shaded regions enclosed by the separatrices connecting the two saddles contain a family of closed orbits representing rotating elliptical patches. The case $\gamma_1/\omega = 0.1227$ is significant in that the shaded region disappears for this particular parameter value, hence it represents a bifurcation point. The next important bifurcation is shown for the case $\gamma_1/\omega = 0.15$, where the centers and saddles coalesce. For parameter values larger than 0.15, there are no critical points in the region $\lambda > 1$, as shown in the original paper of Neu (1984). In this regime, the ellipse is subject to extreme strain, and it undergoes irreversible elongation as the major axis aligns itself with the x-axis.

We next consider briefly the nonautonomous case in which there is out of plane strain, i.e., $\gamma_3 > 0$. For simplicity, we restrict ourselves to the adiabatic regime in which $0 < (\gamma_2 - \gamma_1)/\omega << 1$, hence the vorticity can be regarded as slowly varying. We know from discussions in Chapter 1 that the area enclosed by a trajectory in the phase plane is an adiabatic invariant. In particular, if we assume that at $t = 0$ the phase point orbits on or along a closed trajectory of area α, then as ω evolves slowly, the phase point will continue along a trajectory that continues to enclose the same area. As a simple example, consider a phase point initially lying at one of the centers, the corresponding enclosed area being zero. Then, as ω evolves, the phase point is forced to track the center (in order to enclose zero area) while it slowly drifts, so long as the center continues to exist. Since each center represents a static elliptical configuration of Moore and Saffman (1971), this analysis suggests that as $\omega \to \infty$, these elliptical patches remain in their stable configurations even as they undergo slow stretching. The reader is referred to the discussions in Neu (1984) for more details.

7.3   Time-Dependent Strain

We now consider the situation in which the Kida vortex is subject to a time-periodic strain field, as analyzed by Bertozzi (1988) and Ide and Wiggins (1995). The discussion here follows that in Bertozzi (1988). We are interested in the parameter regime $0 < \gamma_1/\omega < 0.15$, where we let $\gamma_3 = \epsilon g(t)$ ($0 < \epsilon << 1$), with $g(t)$ being a T-periodic function of time. For our purpose we rewrite the equations in real (instead of scaled) time,

$$\frac{d\theta}{dt} = \frac{\omega\lambda}{(1+\lambda)^2} - \frac{1}{2}(\gamma_1 + \gamma_2)\frac{\lambda^2 + 1}{\lambda^2 - 1}\sin 2\theta,$$

$$\frac{d\lambda}{dt} = (\gamma_1 + \gamma_2)\lambda\cos 2\theta.$$

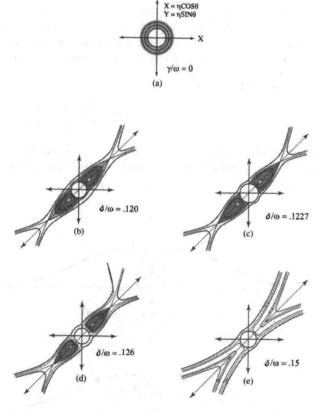

FIGURE 7.1. Phase portraits corresponding to time-dependent Kida–Neu vortices.

Notice that states corresponding to values (λ, θ) and $(\lambda^{-1}, \theta + \pi/2)$ represent the same ellipse, and hence it is preferable to use new coordinates in which there is a one-to-one correspondence between the physical configuration and the coordinate values. For this purpose, it is convenient to introduce the coordinates

$$r = \log \lambda, \quad \phi = 2\theta,$$

in which case the new evolution equations become

$$\dot{r} = (\gamma_1 + \gamma_2) \cos \phi \tag{7.3.1}$$

$$\dot{\phi} = \frac{2\omega e^r}{(e^r + 1)^2} - (\gamma_1 + \gamma_2) \frac{e^{2r} + 1}{e^{2r} - 1} \sin \phi. \tag{7.3.2}$$

The Hamiltonian associated with this system is given by

$$\mathcal{H} = \log \left[\frac{(1 + e^r)^2}{e^r} \right] - \frac{\gamma_1 + \gamma_2}{2\omega} (e^r - e^{-r}) \sin \phi. \tag{7.3.3}$$

We show in Figure 7.2 the phase portrait from Bertozzi (1988) in the Cartesian plane $(x, y) = (r \cos \phi, r \sin \phi)$. Illustrated in these figures are the cases $\gamma_1/\omega = 0.1, 0.1227, 0.1429$. For the parameter range under consideration, the unperturbed system has a hyperbolic fixed point p_0 at values $\phi = \pi/2$, $r = r_0$, where r_0 is the largest real root of the transendental equation

$$e^{3r} + e^{2r}(1 - B) + e^r(1 + B) + 1 = 0$$

$$B = \left(\frac{\gamma_1 + \gamma_2}{2\omega} \right)^{-1}.$$

The homoclinic saddle connection is denoted Γ_0 and our focus will be on the break-up of this structure under perturbations, as discussed in Chapter 1, Section 1.2.2.

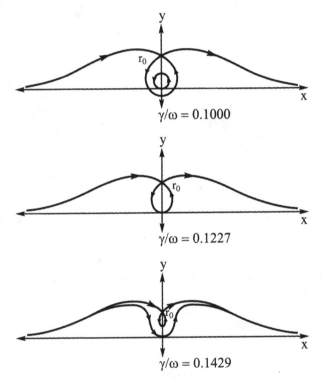

FIGURE 7.2. Phase portrait of a Kida vortex in a time-periodic strain field for three different parameter values.

The perturbed system takes on the general form

$$\dot{r} = C_0 \cos \phi + \epsilon g_1(r, \phi, t)$$

$$\dot{\phi} = \frac{2\omega_0 e^r}{(e^r + 1)^2} - C_0 \frac{e^{2r} + 1}{e^{2r} - 1} \sin \phi + \epsilon g_2(r, \phi, t),$$

with g_1 and g_2 being periodic functions of time and $C_0 = \gamma_1(0) + \gamma_2(0)$, $\omega_0 = \omega(0)$. In particular, with $\gamma_3 = \epsilon g(t)$, the perturbations become

$$g_1 = C_1(t) \cos \phi$$
$$g_2 = \frac{2C_2(t)e^r}{(e^r + 1)^2} - C_1(t) \frac{e^{2r} + 1}{e^{2r} - 1} \sin \phi,$$

where now $C_1(t)$ and $C_2(t)$ are T-periodic functions of time. If we restrict ourselves to the symmetric case where the incompressibility condition is maintained for all time, i.e.,

$$\gamma_1 - \gamma_2 + \gamma_3 = \gamma_1 - \gamma_2 + \epsilon g(t)$$
$$= 0,$$

and in addtion, $\gamma_1 + \gamma_2 = const$, then $C_1(t)$ must vanish, so that

$$g_1 \equiv 0, \quad g_2(t) = \frac{2C_2(t)}{(e^r + 1)^2} e^r.$$

We present the Melnikov calculation, as detailed in Bertozzi (1988), in the (r, ϕ)-plane only. This requires us to avoid the special value $r = 0$, a case that is treated separately in Bertozzi (1988). Our unperturbed system takes on the form

$$\dot{r} = f_1(r, \phi)$$
$$\dot{\phi} = f_2(r, \phi),$$

where in the neighborhood of the homoclinic saddle connection Γ_0 we have

$$f_1 = C_0 \cos \phi,$$
$$f_2 = \frac{2\omega_0 e^r}{(e^r + 1)^2} - C_0 \sin \phi \frac{(e^{2r} + 1)}{(e^{2r} - 1)}.$$

To compute the Melnikov function, we need an expression for

$$\exp \left(\int_0^t \mathrm{tr} Df(\Gamma_0(s)) ds \right).$$

Using

$$Df \equiv \begin{pmatrix} \frac{\partial f_1}{\partial r} & \frac{\partial f_1}{\partial \phi} \\ \frac{\partial f_2}{\partial r} & \frac{\partial f_2}{\partial \phi} \end{pmatrix}$$

one has

$$\mathrm{tr} Df = -C_0 \frac{e^{2r} + 1}{e^{2r} - 1} \cos \phi = -\frac{\dot{r}(e^{2r} + 1)}{(e^{2r} - 1)},$$

from which it follows that

$$\exp\left(\int_0^t \mathrm{tr}Df(\Gamma_0(s))ds\right) = \frac{e^{r(t)}(e^{r_0} - e^{-r_0})}{e^{2r(t)} - 1}.$$

This then gives rise to a Melnikov function of the form (see Bertozzi (1988) for details)

$$M(t_0) = C_3 \int_{-\infty}^{\infty} \frac{e^{2r}\cos\phi}{(e^r + 1)^2(e^{2r} - 1)} C_2(t + t_0)dt \qquad (7.3.4)$$

$$C_3 = C_0(e^{r_0} - e^{-r_0}). \qquad (7.3.5)$$

We now make the specific choice $C_2 = \cos kt$ to obtain the expression

$$M(t_0) = \sin kt_0 \cdot M_0(k),$$

with

$$M_0(k) \equiv C_3 \int_{-\infty}^{\infty} \frac{e^{2r}\cos\phi}{(e^r + 1)^2(e^{2r} - 1)} \sin(kt)dt.$$

It now remains only to prove that there exist some values of k for which $M_0(k) \neq 0$. Then, since $\sin(kt_0)$ premultiplies $M_0(k)$, $M_0(t)$ must have a simple zero. The conclusion that there exists an interval for which $M_0(k)$ is not zero follows from a simple argument based on the fact that the expression is in the form of a Fourier sine transform of an L^1 function, hence is a uniformly continuous function of k, not identically zero. From this, it follows that there exist values of k such that $M_0(k) \neq 0$ in a window, say $k_1 \leq k \leq k_2$. We refer the reader to Bertozzi (1988) for precise details of the argument.

Remark.   There are several other nice papers that discuss the chaotic dynamics in and around a Kida vortex. Ide and Wiggins (1995) analyze the same general set-up as Bertozzi (1988), whereas the works of Polvani and Wisdom (1990a,b) and Dahleh (1992) numerically documents the chaotic mixing of passive scalars in the presence of linear shear. The more recent work of Kawakami and Funakoshi (1999) takes this analysis one step further by studying the mixing as a function of the strain rate, using mostly analytical methods. In particular, a Melnikov analysis is carried out in this paper using the external strain rate as a small parameter. See also the work of Dhanak and Marshall (1993). $\Diamond$

7.4   Melander–Zabusky–Styczek Model

From the discussion of Section 7.2 it should be clear that elliptically shaped vortex patches are highly robust structures. In isolation they remain intact, and when subject to simple strain fields they also retain their elliptical shape. In addition, there is numerical evidence that elliptical vorticity regions form dynamically under a wide range of initial conditions.[1] One can imagine using interacting elliptically shaped vortex patches, instead of point vortices, as the basis for a more accurate model for the evolution of vorticity in the Euler equations, which is what we describe in this section, following the work of Melander, Styczek, and Zabusky (1984), and Melander, Zabusky, and Styczek (1986), referred to as the MZS model.

The model, whose derivation we outline in this section, is given by a set of equations governing the shape λ_k orientation ϕ_k, and centroid location, (x_k, y_k), of N interacting vortex patches

$$\Gamma_k \dot{y}_k = \frac{\partial \mathcal{H}}{\partial x_k} \tag{7.4.1}$$

$$\Gamma_k \dot{x}_k = -\frac{\partial \mathcal{H}}{\partial y_k} \tag{7.4.2}$$

$$\frac{\Gamma_k A_k}{8\pi} \cdot \frac{1 - \lambda_k^2}{\lambda_k^2} \dot{\lambda}_k = \frac{\partial \mathcal{H}}{\partial \phi_k} \tag{7.4.3}$$

$$\frac{\Gamma_k A_k}{8\pi} \cdot \frac{1 - \lambda_k^2}{\lambda_k^2} \dot{\phi}_k = -\frac{\partial \mathcal{H}}{\partial \lambda_k}. \tag{7.4.4}$$

The Hamiltonian representing the total energy of the system of N elliptical patches is given by

$$\mathcal{H} = \mathcal{H}_1 + \mathcal{H}_2 + \mathcal{H}_3, \tag{7.4.5}$$

with

$$\mathcal{H}_1 = \sum_{k=1}^{N} \frac{\Gamma_k^2}{8\pi} \log \frac{(1 + \lambda_k)^2}{4\lambda_k} - \frac{\Gamma_k^2}{16\pi} \tag{7.4.6}$$

$$\mathcal{H}_2 = \sum_{k=1}^{N} \sum_{j \neq k}^{N} \frac{\Gamma_j \Gamma_k}{4\pi} \log R_{kj} \tag{7.4.7}$$

$$\mathcal{H}_3 = \sum_{k=1}^{N} \sum_{j \neq k}^{N} \frac{\Gamma_j \Gamma_k}{32\pi^2 R_{kj}^2} \tag{7.4.8}$$

$$\times \left[A_j \frac{1 - \lambda_j^2}{\lambda_j} \cos 2(\theta_{kj} - \phi_j) + A_k \frac{1 - \lambda_k^2}{\lambda_k} \cos 2(\theta_{kj} - \phi_k) \right].$$

---

[1] In a two-dimensional numerical simulation of McWilliams (1984), it was found that elliptical vorticity regions emerged from initial conditions with power law spectra and random phases, while Melander, McWilliams, and Zabusky (1986) demonstrated that nearly elliptical vortices emerged from two interacting circular initial states.

$\mathcal{H}_1$ is the self-energy of an elliptical patch, $\mathcal{H}_2$ accounts partially for patch interactions based only on intercentroid distances (as in the point vortex case), while $\mathcal{H}_3$ is also an interaction energy term involving the patch orientations. It can readily be verified that

$$\frac{d\mathcal{H}}{dt} = 0,$$

and one can put (7.4.3), (7.4.4) in canonical form by using the variables

$$\left(\frac{\Gamma_k A_k}{16\pi} \cdot \frac{(\lambda_k - 1)^2}{\lambda_k}, 2\phi_k \right).$$

Here, as in the Kida solution, $\lambda_k = a_k/b_k$ denotes the aspect ratio of the kth patch, while ϕ_k denotes its orientation with respect to the x-axis. The variables R_{kj} and θ_{kj} measure the relative positions of the patch centroids,

$$\mathbf{x}_k = \int_{\mathcal{D}_k} \mathbf{x} \, \frac{d\sigma}{A_k}$$

defined by

$$\mathbf{x}_k - \mathbf{x}_j = R_{kj}(\cos\theta_{kj}, \sin\theta_{kj}). \tag{7.4.9}$$

This set of equations is derived in the paper of Melander, Zabusky, and Styczek (1986) as a special case of a more general moment model whose main features we now describe.

Start by considering N arbitrarily shaped domains $\mathcal{D}_k$ of constant vorticity ω_k, with respective areas A_k. Within each domain we can assign a local coordinate system $\boldsymbol{\xi}_k = (\xi_k, \eta_k)$, whose origin is located at the centroid of the domain and whose axes are parallel to the global inertial axis $\mathbf{x} = (x, y)$, located at the global center of vorticity of the system of k patches, as long as it is not at infinity. We can now define the local geometric moments associated with a given patch:

Definition 7.4.1. *The **local geometric moment** of order (m, n) associated with patch $\mathcal{D}_k$ is defined as*

$$J_k^{(m,n)} \equiv \int_{\mathcal{D}_k} \xi_k^m \eta_k^n \, d\sigma. \tag{7.4.10}$$

Remark. Note that the lowest order moment $J_k^{(0,0)}$ is the area of patch $\mathcal{D}_k$, which is invariant as a result of Kelvin's theorem, i.e.,

$$\dot{J}_k^{(0,0)} = 0. \qquad\qquad \Diamond$$

Our goal is to derive a system of evolution equations for the local geometric moments and to show under what conditions one obtains the simpler system (7.4.1), (7.4.2), (7.4.3), (7.4.4). The basic assumption is that the patches are small and well-separated. In particular, the following hypotheses are made (which are subject to verification in any subsequent analysis based on the model):

- The maximum diameter of each patch is much smaller than the minimum distance between any two centroids associated with a patch. In particular, this gives rise to the small parameter $\epsilon = $ maximum patch diameter/ minimum intercentroid distance.

- The centroid of each patch stays within the patch as the system evolves.

The total velocity field generated by the system of patches, denoted

$$\mathbf{u} = (u, v) = (\psi_y, -\psi_x),$$

can be broken down into a near-field generated by the kth patch and the far-field due to the remaining $N - 1$ patches. In particular,

$$\mathbf{u} = \mathbf{u}_k + \mathbf{U}_k,$$

where

$$\mathbf{u}_k = \left(\frac{\partial \psi_k}{\partial y}, -\frac{\partial \psi_k}{\partial x} \right)$$

$$\mathbf{U}_k = \left(\frac{\partial \Psi_k}{\partial y}, -\frac{\partial \Psi_k}{\partial x} \right).$$

The near-field and far-field streamfunctions are given by

$$\psi_k(\mathbf{x}) = -\frac{\omega_k}{2\pi} \int_{D_k} \log |\mathbf{x}_k + \boldsymbol{\xi}_k - \mathbf{x}| d\sigma$$

$$\Psi_k(\mathbf{x}) = -\sum_{j \neq k}^{N} \frac{\omega_k}{2\pi} \int_{D_j} \log |\mathbf{x}_j + \boldsymbol{\xi}_j - \mathbf{x}| d\sigma.$$

The centroid evolution equation is obtained by differentiating the expression for the centroid,

$$\dot{\mathbf{x}}_k = \frac{d}{dt} \int_{D_k} \mathbf{x} d\sigma = \int_{D_k} \frac{D}{Dt} \mathbf{x} d\sigma$$

$$= \int_{D_k} \mathbf{u} d\sigma = \int_{D_k} (\mathbf{u}_k + \mathbf{U}_k) d\sigma = \int_{D_k} \mathbf{U}_k d\sigma,$$

which shows that only the far-field influences the evolution of a given centroid. The evolution equations for the higher order moments are obtained similarly:

$$
\begin{aligned}
j_k^{(m,n)} &= \frac{d}{dt} \int_{D_k} \xi_k^m \eta_k^n d\sigma \\
&= \int_{D_k} \frac{D}{Dt} (\xi_k^m \eta_k^n) d\sigma \\
&= \int_{D_k} \left[m\eta_k^n \xi_k^{m-1} \frac{D\xi_k}{Dt} + n\eta_k^{n-1} \xi_k^m \frac{D\eta_k}{Dt} \right] d\sigma \\
&= \int_{D_k} \eta_k^{n-1} \xi_k^{m-1} \left[m\eta_k \frac{D}{Dt}(x - x_k) + n\xi_k \frac{D}{Dt}(y - y_k) \right] d\sigma \\
&= \int_{D_k} \eta_k^{n-1} \xi_k^{m-1} \left[m\eta_k(u - \dot{x}_k) + n\xi_k(v - \dot{y}_k) \right] d\sigma \\
&= j_{*k}^{(m,n)} \\
&\quad + \int_{D_k} \eta_k^{n-1} \xi_k^{m-1} [m\eta_k(U_k - \dot{x}_k) + n\xi_k(V_k - \dot{y}_k)] d\sigma, \qquad (7.4.11)
\end{aligned}
$$

where

$$
j_{*k}^{(m,n)} \equiv \int_{D_k} \eta_k^{n-1} \xi_k^{m-1} (m\eta_k u_k + n\xi_k v_k) d\sigma. \qquad (7.4.12)
$$

The integral (7.4.12) represents the self-interaction effect of each patch on its moment evolution, while the second integral in (7.4.11) captures the combined influence of all the other patches, and at this stage no approximations have been invoked.

The next step in the model is to use the assumptions that the patches are well-separated, which allows us to expand U_k and u_k around x_k and thus express the right hand side of (7.4.11) entirely in terms of the local geometric moments. This process is carried out in detail in Melander, Zabusky, and Styczek (1986) and leads to an infinite system of ordinary differential equations for the geometric moments $J_k^{(m,n)}(t)$. The model can be truncated at any order. For example, if second and higher moments are dropped, we obtain the point vortex system. Improving on this, we can keep the second order moments, which implicitly assumes that each patch retains its elliptical shape, hence we arrive at the so-called second order moment model. For this model we can use the exact angular velocity associated with an isolated Kirchhoff ellipse (7.1.1) to evaluate self-interaction terms such as (7.4.12), and hence this truncation is particularly appealing.

The moments of inertia for an ellipse are given explicitly as

$$J^{(2,0)} = \frac{A^2}{4\pi\lambda}[\lambda^2 + (1 - \lambda^2)\sin^2\phi]$$

$$J^{(0,2)} = \frac{A^2}{4\pi\lambda}[\lambda^2 + (1 - \lambda^2)\cos^2\phi]$$

$$J^{(1,1)} = -\frac{A^2}{8\pi\lambda}(1 - \lambda^2)\sin 2\phi.$$

Note that the area and moments of inertia of an ellipse are related via the expression

$$\frac{A^4}{16\pi^2} = J^{(2,0)}J^{(0,2)} - J^{(1,1)}J^{(1,1)}. \tag{7.4.13}$$

The self-interaction moment evolution terms can be written

$$\dot{J}_*^{(2,0)} = \frac{\omega A^2}{4\pi} \cdot \frac{1 - \lambda}{1 + \lambda}\sin 2\phi$$

$$\dot{J}_*^{(0,2)} = -\dot{J}_*^{(2,0)}$$

$$\dot{J}_*^{(1,1)} = -\frac{\omega A^2}{4\pi} \cdot \frac{1 - \lambda}{1 + \lambda}\cos 2\phi.$$

By manipulating these equations further (see Melander, Zabusky, and Styczek (1986) for details), one arrives at a system that generalizes the Kida–Neu system by taking into account the leading order interaction of the elliptical patches

$$\dot{\lambda}_k = \lambda_k \sum_{j \neq k}^{N} \frac{\omega_j}{\pi} \cdot \frac{A_j}{R_{kj}^2}\sin 2(\theta_{kj} - \phi_k) \tag{7.4.14}$$

$$\dot{\phi}_k = \omega_k \frac{\lambda_k}{(1 + \lambda_k)^2} + \frac{1 + \lambda_k^2}{1 - \lambda_k^2}\sum_{j \neq k}^{N} \frac{\omega_j}{2\pi} \cdot \frac{A_j}{R_{kj}^2}\cos 2(\theta_{kj} - \phi_k). \tag{7.4.15}$$

The centroid motion, which is coupled to the above, is governed by

$$\dot{x}_k = \sum_{\substack{j \neq k}}^{N} -\frac{\omega_j}{2\pi} \cdot \frac{A_j}{R_{kj}} \sin \theta_{kj}$$

$$+ \frac{\omega_j}{8\pi^2} \cdot \frac{A_j^2}{R_{kj}^3} \frac{1 - \lambda_j^2}{\lambda_j} \sin(3\theta_{kj} - 2\phi_j)$$

$$+ \frac{\omega_j}{8\pi^2} \cdot \frac{A_j A_k}{R_{kj}^3} \cdot \frac{1 - \lambda_k^2}{\lambda_k} \sin(3\theta_{kj} - 2\phi_k) \qquad (7.4.16)$$

$$\dot{y}_k = \sum_{\substack{j \neq k}}^{N} \frac{\omega_j}{2\pi} \cdot \frac{A_j}{R_{kj}} \cos \theta_{kj}$$

$$- \frac{\omega_j}{8\pi^2} \cdot \frac{A_j^2}{R_{kj}^3} \cdot \frac{1 - \lambda_j^2}{\lambda_j} \cos(3\theta_{kj} - 2\phi_j)$$

$$- \frac{\omega_j}{8\pi^2} \cdot \frac{A_j A_k}{R_{kj}^3} \cdot \frac{1 - \lambda_k^2}{\lambda_k} \cos(3\theta_{kj} - 2\phi_k). \qquad (7.4.17)$$

The system (7.4.1), (7.4.2), (7.4.3), (7.4.4) is the Hamiltonian representation of this model.

Remarks.

1. One might wonder what are the conserved quantities associated with the truncated moment model. It turns out, as shown in Melander, Zabusky, and Styczek (1986), that three quantities are conserved for the second order model: The conservation of any function of the vorticity I_1; the conservation of global centroid I_2; and the conservation of angular impulse I_3, where

$$I_1 = \int_{\mathcal{R}^2} F(\omega) d\sigma \qquad (7.4.18)$$

$$I_2 = \int_{\mathcal{R}^2} \mathbf{x}\omega d\sigma \qquad (7.4.19)$$

$$I_3 = \int_{\mathcal{R}^2} \mathbf{x} \cdot \mathbf{x}\omega d\sigma. \qquad (7.4.20)$$

2. For the second order moment model, the angular impulse can be written explicitly in terms of the familiar point vortex angular impulse, plus an elliptical correction term

$$I_3 = \sum_{k=1}^{N} \Gamma_k \left[\mathbf{x}_k \cdot \mathbf{x}_k + \frac{A_k}{4\pi} \cdot \frac{1 + \lambda_k^2}{\lambda_k} \right]. \qquad \Diamond$$

It is worthwhile to explicitly work out some of the consequences of the second order moment model for comparison with their corresponding point vortex configurations.

Example 7.2 (Symmetric corotating states) Consider two elliptical vortex patches of area A and vorticity ω, placed symmetrically at (x_1, y_1) and $(-x_1, -y_1)$. The separation of their centroids, nondimensionalized, is

$$\mu = R_{12} \left(\frac{\pi}{A}\right)^{1/2}.$$

We can now look for steady state solutions of the system (7.4.14), (7.4.15), (7.4.16), (7.4.17), by letting

$$\dot{\lambda} = 0,$$

which from (7.4.15) gives

$$\theta = \phi + \frac{n\pi}{2}.$$

From (7.4.15) we have that $\dot{\phi} = \dot{\theta} = const$, and from (7.4.16), (7.4.17) we obtain

$$R_{12} = const,$$

$$\dot{\theta} = \dot{\phi} = \frac{\omega}{\mu^2} \left[1 - \frac{1}{2\mu^2} \cdot \frac{1 - \lambda^2}{\lambda}\right]. \tag{7.4.21}$$

The first term in this last expression for angular velocity can be written

$$\frac{\omega}{\mu^2} = \frac{A\omega}{\pi R_{12}^2},$$

which agrees with the corresponding formula for corotating point vortex pairs,

$$\frac{\Gamma}{\pi D^2},$$

as worked out in Example 1.8. The governing equation for the aspect ratio λ is

$$\frac{1}{\mu^2} \left[1 - \frac{1}{2\mu^2} \cdot \frac{1 - \lambda^2}{\lambda}\right] = \frac{\lambda}{(1 + \lambda)^2} + \frac{1 + \lambda^2}{1 - \lambda^2} \cdot \frac{1}{2\mu^2},$$

which can be rewritten as a quintic in λ and solved numerically (see Melander, Zabusky, and Styczek (1986) for further detail). ◊

Example 7.3 (Translating states) We next consider symmetrically placed translating elliptical patches of area A and aspect ratio λ, with vorticity ω and $-\omega$, symmetrically placed on opposite sides of the x-axis. Again, we let

$$\dot{\lambda} = 0$$
$$\theta = \phi + \frac{n\pi}{2},$$

which then implies that $\dot{\theta} = \dot{\phi} = 0$. The pair propagates forward along the x-axis, with linear velocity given by the formula

$$\dot{x} = \frac{\omega A}{2\pi R_{12}} \left[1 - \frac{A}{2\pi R_{12}^2} \cdot \frac{\lambda^2 - 1}{\lambda} \right]. \qquad (7.4.22)$$

The first term of this formula agrees with the corresponding one for translating point vortex pairs, whose velocity is given by

$$\frac{\Gamma}{2\pi D},$$

as shown in Example 1.8. The aspect ratio satisfies the equation

$$0 = \frac{\lambda}{(1+\lambda)^2} + \frac{1+\lambda^2}{1-\lambda^2} \cdot \frac{1}{2\mu^2},$$

where $\mu = R_{12}(\pi/A)^{1/2}$, and the lengths are normalized so that

$$1 = \left(\frac{A}{\pi\lambda} \right)^{1/2} + \frac{1}{2} R_{12}.$$

The equation for the aspect ratio reduces to a cubic equation

$$0 = \lambda^3 + (1 - 2\mu^2)\lambda^2 + (1 + 2\mu^2)\lambda + 1.$$

Pierrehumbert (1980) also considers configurations of this type, computing a family of steadily translating pairs.   ◇

7.5 Geometric Phase for Corotating Patches

As an extension to the geometric phase calculations for point vortex configurations of Newton (1994), and Shashikanth and Newton (1998, 1999) described in Section 5.4, we now ask if adiabatic phases can arise in the case where the vorticity is uniformly distributed. In particular, in this section we calculate the geometric phase associated with the motion of corotating vortex patches, as modeled by the second order moment model described in the previous section. Hence, we take two patches which are of elliptical shape, with time-varying eccentricities and nonuniform rates of rotation around their respective centroids, which move, to leading order, like point vortices. Hence, the patches can be viewed as Kirchhoff ellipses with moving centroids, and time-varying aspect ratios and angular velocities. Geometric phases are calculated in the adiabatic setting in which there are two length scales, due to our assumptions about the interpatch separation and patch size. The time scales represent the "slow" point vortex-like motion of the centroids and the "fast" rotation of the Kirchhoff elliptical patches. We ask what the change is (to leading order) in the orientation angle of each patch at the end of one appropriately defined time period of centroid motion. Can this angle change be expressed explicitly as the sum of a "dynamic" part plus a "geometric" part?

Consider the configuration shown in Figure 7.3 in which two like-signed elliptical patches of respective areas A_1 and A_2 co-rotate. We will denote the patch strengths Γ_1 and Γ_2, and we use the second order moment equations, as described in the previous section, appropriately scaled. Using the angular velocity formula (7.1.1) associated with an isolated ellipse, we can associate a (short) time period to each patch as

$$T_s \sim A_i.$$

There is also a (longer) time period associated with the corotation scale, given as

$$T_l \sim D^2,$$

where D is a typical intercentroid distance. Defining a small dimensionless parameter ϵ as

$$\epsilon^2 = \frac{(A_1 + A_2)}{D_i^2},$$

with D_i being the initial intercentroid distance, we would like to compute the orientation of each patch after one period T_l. Since $\epsilon \sim T_s/T_l$, small values of ϵ define an adiabatic process just as in the point vortex problems discussed in Chapter 5 — there exist two time scales, one associated with the rotation of the patch and the other with the (relatively) slow revolution of the patch centroids. As $\epsilon \to 0$, we will compute the $O(1)$ contribution to the angle change of the patches at the end of time period T_l.

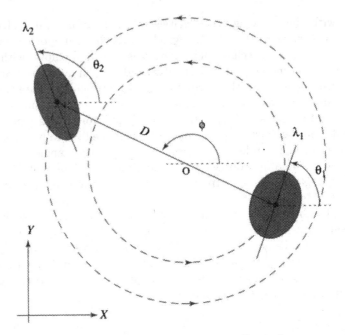

FIGURE 7.3. Motion of two well-separated uniform elliptical vortex patches of the same sign in the MZS model.

The equations governing the motion of the variables $(\hat{D}(t), \phi(t))$ and $(\lambda_1(t), \theta_1(t))$, $(\lambda_2(t), \theta_2(t))$ are given by

$$\frac{d\hat{D}}{dt} = \frac{\Gamma_1 + \Gamma_2}{8\pi^2 \hat{D}^3} \sum_{k=1}^{2} A_k \frac{1 - \lambda_k^2}{\lambda_k} \sin 2(\phi - \theta_k)$$

$$\frac{d\hat{\phi}}{dt} = \frac{\Gamma_1 + \Gamma_2}{2\pi \hat{D}^2} - \frac{\Gamma_1 + \Gamma_2}{8\pi^2 \hat{D}^4} \sum_{k=1}^{2} A_k \frac{1 - \lambda_k^2}{\lambda_k} \cos 2(\phi - \theta_k)$$

$$\frac{d\lambda_1}{dt} = \lambda_1 \frac{\Gamma_2}{\pi \hat{D}^2} \sin 2(\phi - \theta_1)$$

$$\frac{d\theta_1}{dt} = \frac{\omega_1 \lambda_1}{(1 + \lambda_1)^2} + \frac{1 + \lambda_1^2}{1 - \lambda_1^2} \cdot \frac{\Gamma_2}{2\pi \hat{D}^2} \cos 2(\phi - \theta_1)$$

$$\frac{d\lambda_2}{dt} = \lambda_2 \frac{\Gamma_1}{\pi \hat{D}^2} \sin 2(\phi - \theta_2)$$

$$\frac{d\theta_2}{dt} = \frac{\omega_2 \lambda_2}{(1 + \lambda_2)^2} + \frac{1 + \lambda_1^2}{1 - \lambda_1^2} \cdot \frac{\Gamma_1}{2\pi \hat{D}^2} \cos 2(\phi - \theta_2),$$

where $\hat{D}$ represents the dimensional intercentroid distance. The Hamiltonian for this system is written

$$\mathcal{H} = \sum_{k=1}^{2} \frac{\Gamma_k^2}{8\pi} \log \frac{(1+\lambda_k)^2}{4\lambda_k} + \frac{\Gamma_1 \Gamma_2}{2\pi} \log \hat{D}$$

$$+ \frac{\Gamma_1 \Gamma_2}{16\pi^2 \hat{D}^2} \sum_{k=1}^{2} A_k \frac{(1-\lambda_k^2)}{\lambda_k} \cos 2(\phi - \theta_k).$$

Remarks.

1. These equations show that for large $\hat{D}$ or for small A_1, A_2, the centroids move like two point vortices of strengths Γ_1 and Γ_2.

2. The equations for the ellipse parameters $(\lambda_1, \theta_1, \lambda_2, \theta_2)$ are (7.4.14), (7.4.15), which imply that for large values of $\hat{D}$, or small A_1, A_2 (equivalently, large ω_1, ω_2), the patches rotate like uncoupled Kirchhoff ellipses. The coupling arises only through the $(\hat{D}, \phi)$ variables.

3. As noted in the paper of Melander, Zabusky, and Styczek (1986), the equations for θ_1 and θ_2 are not defined when the patches are perfectly circular, i.e., $\lambda = 1$. Thus, in their paper they introduce a rescaled version of the governing equations to rectify this, since they are interested in patches that are nearly circular. We find it more convenient to use the original unscaled variables. $\Diamond$

We next nondimensionalize the system by introducing the variables

$$D = \frac{\hat{D}}{D_i}, \qquad t = \Omega \hat{t},$$

where Ω is the mean strength over the mean area of the patches,

$$\Omega \equiv \frac{(\Gamma_1 + \Gamma_2)}{(A_1 + A_2)}.$$

The nondimensional equations then become

$$\frac{dD}{dt} = \frac{\epsilon^4}{8\pi^2 D^3} \sum_{k=1}^{2} \sigma_k f(\lambda_k) \sin[2(\phi - \theta_k)]$$

$$\frac{d\phi}{dt} = \frac{\epsilon^2}{2\pi D^2} - \frac{\epsilon^4}{8\pi^2 D^4} \sum_{k=1}^{2} \sigma_k f(\lambda_k) \cos[2(\phi - \theta_k)]$$

$$\frac{d\lambda_1}{dt} = \epsilon^2 \lambda_1 \frac{\alpha_2}{\pi D^2} \sin[2(\phi - \theta_1)]$$

$$\frac{d\theta_1}{dt} = \frac{\alpha_1 g(\lambda_1)}{A_1} + \epsilon^2 \frac{\alpha_2 h(\lambda_1)}{2\pi D^2} \cos[2(\phi - \theta_1)]$$

$$\frac{d\lambda_2}{dt} = \epsilon^2 \lambda_2 \frac{\alpha_1}{\pi D^2} \sin[2(\phi - \theta_2)]$$

$$\frac{d\theta_2}{dt} = \frac{\alpha_2 g(\lambda_2)}{A_2} + \epsilon^2 \frac{\alpha_1 h(\lambda_2)}{2\pi D^2} \cos[2(\phi - \theta_2)],$$

where

$$\sigma_k = \frac{A_k}{(A_1 + A_2)}, \qquad \alpha_k = \frac{\Gamma_k}{(\Gamma_1 + \Gamma_2)}$$

$$f(\lambda) = \frac{(1 - \lambda^2)}{\lambda} \qquad g(\lambda) = \frac{\lambda}{(1 + \lambda)^2} \qquad h(\lambda) = \frac{(1 + \lambda^2)}{(1 - \lambda^2)}.$$

Our initial conditions for the nodimensional system are $D(0) = 1$, $\phi(0)$, $\theta_1(0)$, $\theta_2(0)$, $\lambda_1(0)$, $\lambda_2(0)$.

The main result, described in Shashikanth and Newton (2000), is based on a multiscale perturbation analysis using the slow time scale $\tau \equiv \epsilon^2 t$ associated with the corotation period. The leading order solutions for the angle variables ϕ, θ_k are given by

$$\phi = \frac{\tau}{2\pi} + \phi(0) + O(\epsilon), \tag{7.5.1}$$

$$\theta_k = \frac{\alpha_k}{\sigma_k} g(\lambda_k(0)) t - \frac{\alpha_{k'}}{2\pi} \frac{1 - \lambda_k(0)}{1 + \lambda_k(0)} \cos[2(\phi(0) - \theta_k(0))] \cdot \tau + \theta_k(0)$$
$$+ O(\epsilon), \tag{7.5.2}$$

where $k' = 2$ if $k = 1$, and $k' = 1$ if $k = 2$. At the end of one period of corotation, $T = 4\pi^2$ (to leading order, hence $t = 4\pi^2/\epsilon^2$), the change in the patch orientation angles, which we denote $\Delta\theta_k$, is

$$\Delta\theta_k = \frac{\alpha_k}{\sigma_k} g(\lambda_k(0)) \frac{4\pi^2}{\epsilon^2} - \alpha_{k'} \frac{1 - \lambda_k(0)}{1 + \lambda_k(0)} 2\pi \cos[2(\phi(0) - \theta_k(0))] + O(\epsilon^2).$$

From this we obtain the following result, which is discussed in more detail in Shashikanth and Newton (2000).

Theorem 7.5.1 (Shashikanth, Newton (2000)). *The adiabatic geometric phase for a system of two elliptical corotating vortex patches, evolving according to the equations of the second order truncated MZS model, are*

$$(\theta_g)_1 = -\frac{\Gamma_2}{\Gamma_1 + \Gamma_2} \cdot \frac{1 - \lambda_1(0)}{1 + \lambda_1(0)} 2\pi \cos[2(\phi(0) - \theta_1(0))]$$

$$(\theta_g)_2 = -\frac{\Gamma_1}{\Gamma_1 + \Gamma_2} \cdot \frac{1 - \lambda_2(0)}{1 + \lambda_2(0)} 2\pi \cos[2(\phi(0) - \theta_2(0))].$$

For a geometric interpretation of this result and more details of the derivation, we refer the reader to the original paper of Shashikanth and Newton (2000).

Remark. In Shashikanth and Newton (2000) we compare the geometric phase formulas above with a simpler configuration made up of four single-point vortices in order to highlight the effects of allowing for elliptical deformations. ◇

7.6 Viscous Shear Layer Model

We now describe a model of a planar shear layer of the type considered in Section 5.5, but with the important additional feature that viscous diffusion is included. To understand the effect that viscous diffusion has on vorticity, we start by recalling the example from Chapter 1 of the viscous decay of an isolated point vortex, called the Lamb–Oseen vortex. The azimuthal and radial components of velocity are

$$u_\theta(r, \theta) = \frac{\Gamma}{2\pi r}[1 - \exp\left(-\frac{r^2}{4\nu t}\right)]$$

$$u_r = 0,$$

while the vorticity distribution is given by

$$\omega = \frac{\Gamma}{4\pi\nu t} \exp\left(-\frac{r^2}{4\nu t}\right).$$

If we introduce the Reynolds number R,

$$R \equiv \frac{\Gamma}{\nu},$$

and normalize the flowfield so that $\Gamma = 1$, then the solution corresponds to a viscously decaying point vortex, whose core grows according to the growth law

$$\sigma(t) = (4\nu t)^{1/2} = \left(\frac{4t}{R}\right)^{1/2}. \tag{7.6.1}$$

Note that this is simply the solution of the radially symmetric heat equation.

If, however, there is more than one point vortex in the flow, or if there are boundaries or some additional background flow that breaks the radial symmetry, then the vortex will no longer remain radially symmetric for all time. For sufficiently weak symmetry-breaking effects, one can assume that as long as t/R is sufficiently small, each vortex will diffuse according to the isolated core growth formula (7.6.1). This idea is the basis for the viscous shear layer model introduced and analyzed in Meiburg and Newton (1991), whose main features we now describe.

Our model is based on Stuart's (1967) one parameter family of solutions to the time-independent Euler equations, whose normalized streamfunction is given by

$$\psi(x, y) = \log[(\cosh 2\pi y) - (\rho \cos 2\pi x)]. \tag{7.6.2}$$

Here, the parameter $0 \leq \rho \leq 1$ determines the concentration of vorticity. For $\rho = 1$, the solution corresponds to a periodic row of point vortices, evenly spaced along the x-axis at points $x = x_n = \pm n$. For $\rho = 0$, the flow corresponds to parallel shear flow along the x-axis. For all intermediate values of the parameter ρ, the flow corresponds to a row of vortices with distributed vorticity. We illustrate in Figure 7.4 (reprinted with permission from Meiburg and Newton (1991)) the level curves of the streamfunction (7.6.2) for various values of the parameter ρ. Shown is the familiar cat's eye streamline pattern, in one periodic window, which gets increasingly thin as $\rho \to 0$.

To incorporate the core growth law (7.6.1) into our model, we solve for the y-coordinate of the separating streamline along the x-axis, which is the cat's eye half-width; hence

$$y_w = \frac{1}{2\pi} \cosh^{-1}(1 + 2\rho).$$

We then relate the parameter ρ to a core size, which evolves according to (7.6.1). To achieve this, consider the u-velocity profiles of the Stuart vortex, as shown in Figure 1 of Meiburg and Newton (1991). One sees that there is a unique value along the y-axis, which we label $y(u = u_{max})$, that maximizes the u velocity, u_{max}. One can then define a core size σ as the vertical distance from the origin to $y(u = u_{max})$. Hence, consider the streamfunction evaluated along the y-axis,

$$\psi(x = 0, y) = \log[(\cosh 2\pi y) - \rho].$$

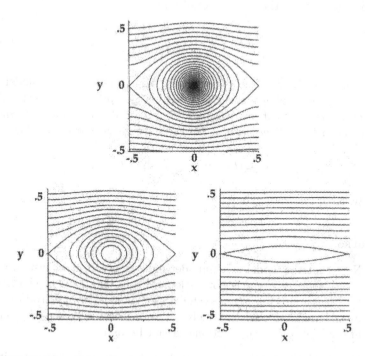

FIGURE 7.4. Streamline level curves of the Stuart family of solutions, for increasing parameter values ρ.

This gives the formula for the x-component of velocity:

$$u(x = 0, y) = \frac{\partial \psi}{\partial y}(x = 0, y)$$

$$= \frac{2\pi \sinh 2\pi y}{\cosh 2\pi y - \rho}.$$

From this we can compute

$$\frac{\partial u}{\partial y}(x = 0, y) = \frac{4\pi^2(1 - \rho \cosh 2\pi y)}{[(\cosh 2\pi y) - \rho]^2},$$

whereupon we set the numerator to zero and solve for $y(u = u_{max})$ to obtain

$$y(u = u_{max}) \equiv \sigma = \frac{1}{2\pi} \cosh^{-1}\left(\frac{1}{\rho}\right). \tag{7.6.3}$$

The final step in formulating our model is to use the core growth law (7.6.1) in (7.6.3) and invert the formula to obtain

$$\rho(t) = \frac{1}{\cosh 4\pi \left(\frac{t}{R}\right)^{1/2}}. \tag{7.6.4}$$

The equations (7.6.2), (7.6.4) constitute our time-dependent model of a viscously decaying planar shear layer. Note that $\rho(0) = 1$ corresponds to a periodic row of point vortices, while $\rho(\infty) = 0$ corresponds to smooth parallel shear flow. The width of the cat's eye y_w goes from a value of

$$y_w(t = 0) = \frac{1}{2\pi} \cosh^{-1} 3$$

initially, to $y_w(t \to \infty) = 0$.

Remarks.

1. Note that the model is a time-dependent Hamiltonian system where the time-dependence is not periodic. From this point of view, many of the dynamical systems techniques that have been introduced and used on other models that rely on periodicity are not useful here — in particular Poincaré sections cannot be used.

2. One of the features of the model is that particles starting initially within the cat's eye eventually cross the separatrix and end up in the free stream. Mixing and stretching due to separatrix crossings is one of the focuses of Meiburg and Newton (1991), and is discussed in different contexts in Section 1.2.3.

3. A careful analysis of the errors associated with this model has not been carried out, and in view of the observations in Greengard (1985), one should be careful when making conclusions about solutions of the Euler equations based solely on the model. With this in mind, the model seems useful nonetheless, since detailed analytical results can be derived from it and in the future could form a basis of comparison with numerical simulations of the Euler equations. $\Diamond$

We now summarize several of the main conclusions based on the model, all of which are discussed more thoroughly in Meiburg and Newton (1991).

- As a result of the viscous core growth, a particle initially trapped within the separatrix region, oscillating around the vortex center, will eventually escape via a separatrix-crossing event. For a schematic

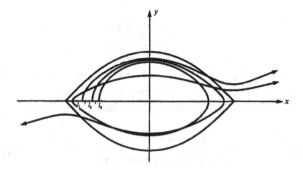

FIGURE 7.5. Schematic diagram of particles escaping from cat's eye which shrinks due to viscous diffusion.

diagram of this process, see Figure 7.5. Via this process, one can define an infinite number of intervals along the x-axis inside the cat's eye, which we label l_n. These intervals characterize the initial particle positions according to whether a particle will escape at the top half of the cat's eye (n odd) or the bottom half (n even). Particles on the interval boundaries evolve to one of the two stagnation points as $t \to \infty$, hence never escape. The intervals are indexed so that particles starting initially in the interval l_n will escape on the nth half-cycle around the vortex center, or equivalently, after crossing the x-axis $(n-1)$ times. Since the intervals become increasingly thin as $n \to \infty$, in such a way that

$$\sum_{n=1}^{\infty} l_n = \frac{1}{2},$$

one can find (near the vortex center) an arbitrarily large number of particles, which start infinitesimally close to each other initially, but which ultimately separate in alternating directions, one above the mixing layer, the other below, etc.

- It is shown in Meiburg and Newton (1991) that the interval widths scale like

$$l_n \sim R^{-1/2}(n^{1/2} - (n-1)^{1/2})$$

for $n \gg 1$.

- A second quantity of interest in characterizing the mixing process due to viscous diffusion is the flux of fluid $F(t/R)$ across the separatrix as a function of time. If we use the scaled time $\hat{t} \equiv t/R$, we define the flux as

$$F(\hat{t}) \equiv \frac{dA/d\hat{t}}{A(0)},$$

where $A(\hat{t})$ is the area of the cat's eye at time $\hat{t}$. Hence, the flux is the rate of change of the cat's eye area, normalized by its initial value. In Meiburg and Newton (1991) we obtain the expression

$$F(\hat{t}) = -\frac{2\pi \sinh 4\pi \hat{t}^{1/2}}{\hat{t}^{1/2} \int_0^{1/2} f(0,x)dx \cosh^2 4\pi \hat{t}^{1/2}} \int_0^{1/2} \frac{1 + \cos 2\pi x}{\sinh[f(\hat{t},x)]} dx,$$

where

$$f(\hat{t},x) \equiv \cosh^{-1}\left[1 + \frac{1 + \cos 2\pi x}{\cosh 4\pi \hat{t}^{1/2}}\right].$$

See Figure 10(b) of Meiburg and Newton (1991) for a plot of the flux function.

- We also quantify the mixing process by characterizing how the fluid within the cat's eye mixes with fluid outside it, and how fluid initially within the top half mixes with fluid in the bottom half. The key to quantifying this process is an estimate on the interfacial area being generated. This, in turn, depends on two different mechanisms. First, on the rate at which new lobes of fluid emerging from the cat's eye are being formed (corresponding to the escape layers previously discussed). Second, on the rate at which the interface of each lobe is stretched. The local rate of interfacial stretching is linked to the escape time of the particles that form the lobe's interface. We define the time of emergence of lobe i as the time at which the first particle from the interval l_i crosses the separatrix. For this we arrive at the governing equation for the escape time t_{esc} in the first lobe,

$$\tanh(2\pi^2 t_{esc}) = \frac{1}{(\cosh 2\pi (4t_{esc}/R)^{1/2})^{1/2}}.$$

Escape times for the lobes with higher index are computed numerically and shown in Meiburg and Newton (1991).

- An interesting feature of the model, which we prove in Meiburg and Newton (1991), is that near the interval boundaries the escape time blows up in a logarithmic, self-similar fashion, dictating the lobe formation process. In Figure 7.6 we show plots taken from Meiburg and Newton (1991), which depict this process, as fluid is squeezed out of the cat's eye. In these figures the inner region, where no fluid has yet emerged, is cross-hatched and not considered. We track the stripping of the outer region only, seeing that the formation of thin viscous filaments, which separate in alternating directions, are subsequently elongated by the free stream. One of the main conclusions of the study is that, due to the lobe formation process, viscosity fundamentally alters the mixing rates. In general, viscous effects lead to an additional factor, which is linearly proportional to time for interfacial growth rates.

- In Newton and Meiburg (1991) we take the model one step further by including the effects of Brownian diffusion on the particle motion. This is achieved by adding a random walk component to the particle motion at each timestep, thereby allowing a mechanism for particles to cross the separatrices even without the cat's eye shrinking due to core diffusion. For this it is necessary to introduce a separate dimensionless parameter, the Peclet number $P_e = \Gamma/D$, where Γ is the vortex strength and D is the diffusion coefficient. In this paper, scaling and numerical results are obtained for the escape intervals as a function of both P_e and R.

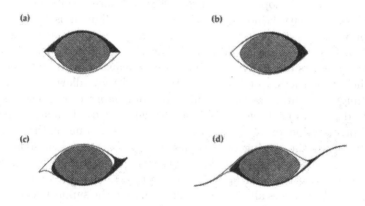

FIGURE 7.6. Viscous lobe formation process. Time increases from (a)–(d), squeezing liquid out into filaments.

7.7 Bibliographic Notes

We mention briefly in this section other interesting work involving the evolution of more generally shaped vortex patches. Despite the fact that it is known through numerical simulation that the detailed evolution of a vortex patch can be extremely complex, involving filamentation, elongation, deformation, and tearing (see, for example, Christiansen and Zabusky (1973), Overman and Zabusky (1982), Melander, McWilliams, and Zabusky (1987), Dritschel (1985), and the review paper of Saffman and Baker (1979)), some gross features associated with a patch can sometimes be estimated analytically. For example, a question of interest is how fast does an isolated vortex patch grow, subject only to its own self-induced dynamics? The first result along these lines is that due to Marchioro (1994), where it is proven that the diameter of the support of a nonnegative initial vorticity patch grows no faster than $O(t^{1/3})$. This result has subsequently been improved upon by Iftimie, Sideris, and Gamblin (1999), who prove the following theorem.

Theorem 7.7.1 (Iftimie, Sideris, Gamblin (1999)). *Let $\omega(t, \mathbf{x})$ be the solution of the two-dimensional Euler equations with positive, compactly supported initial vorticity $\omega_0 \in L^\infty(\mathcal{R}^2)$. There exists a constant C_0 such that for every time $t \geq 0$, the support $\omega(t, \cdot)$ is contained within the ball*

$$\|\mathbf{x}\| < 4d_0 + C_0[t\log(2 + t)]^{1/4}.$$

See also the work of Lopes and Lopes (1998) for more results along these lines. If the vorticity is not required to have one sign, it is known that the support grows at most linearly in t, a result that is sharp, since an example is given in Iftimie, Sideris, and Gamblin (1999) that achieves this rate. Sharp results of this type are important because they do not depend on explicit construction of known solutions, and they allow us to better understand the limitations and ramifications of making specific choices for vorticity distributions. In general, these estimates yield information on the rate of propagation of vorticity, which, since this is one of the driving mechanisms for the mixing of particles, should be related in some way to mixing rates of passive particles, a quantity of both fundamental and applied interest. We should remark that for the case of point vortices we know that if the vortices all have the same sign, then the support is bounded by a constant, whereas if the vortices are allowed to have opposite sign, the support grows linearly.

There is a large and growing literature documenting various nonlinear stability and instability results for vortex patches, starting with the classical works of Arnold (1965, 1966a, 1966b, 1969) introducing variational-type methods for stationary flows. These methods are thoroughly developed and explained in Holm, Marsden, Ratiu, and Weinstein (1985). Specific results on stability of circular vortex patches can be found in Wan and Pulvirenti (1985), Marchioro and Pulvirenti (1985), Constantin and Titi (1988), with more general rotating patches in Wan (1986) and Dritschel (1988). For results on global regularity see Chemin (1991), Bertozzi and Constantin (1993), and the paper of DiPerna and Majda (1987). Good sources for learning abstract techniques associated with the evolution of vorticity include the book of Marchioro and Pulvirenti (1994), the books of Lions (1996), or Chemin (1998), or the review paper of Majda (1986). We finish this chapter by mentioning recent work of Shkoller (1999), and Oliver and Shkoller (1999), in which "vortex blob" solutions to the Euler-α equations are derived. These equations, as discussed in Holm, Marsden, and Ratiu (1998), and Marsden, Ratiu, and Shkoller (1999), can be thought of as suitably regularized Euler equations, where spatial scales smaller than a cut-off parameter α are smoothed out by an averaging procedure. Because of this, the singularity associated with a point vortex is smeared out and one avoids some of the technicalities associated with these measure-valued solutions.

7.8   Exercises

1. Prove that the Kirchhoff elliptical patch rotates with frequency given by the formula (7.1.1), and that a particle located inside the patch rotates in a circular orbit of twice this frequency.

2. Derive the evolution equations for the boundary of the Kida vortex, as given by (7.2.4), (7.2.5), (7.2.6).

3. Derive the Hamiltonian system for the Kida vortex, as given by (7.2.7).

4. Derive the formula for the Melnikov function (7.3.5), for the Kida vortex in a time-dependent strain field.

5. Put the equations (7.4.3), (7.4.4) in standard canonical form.

6. Using formulas (7.4.6), (7.4.7), (7.4.9) for $\mathcal{H}_1, \mathcal{H}_2, \mathcal{H}_3$ in the MZS model, prove that

$$\frac{d\mathcal{H}}{dt} = 0$$

   for $\mathcal{H} = \mathcal{H}_1 + \mathcal{H}_2 + \mathcal{H}_3$.

7. Prove that the area of an ellipse is related to its moments according to the formula (7.4.13).

8. Use the MZS second order moment model to prove that the quantities (7.4.18), (7.4.19), (7.4.20) are invariant.

9. Derive the angular velocity formula (7.4.21) of Example 7.2 and the linear velocity formula (7.4.22) of Example 7.3 governing a translating pair of elliptical patches.

10. Using the multiscale method:

   (a) Derive the angle formulas (7.5.1), (7.5.2) for corotating elliptical
       patches.

   (b) Prove Theorem 7.5.1 regarding the geometric phase for corotat-
       ing elliptical patches.

11. Using the decaying shear layer model (7.6.2), (7.6.4), derive an expres-
    sion for the error by plugging the expression into the Euler equation
    and computing the remainder.

8

Vortex Filament Models

In this chapter we discuss the evolution of vorticity in three dimensions, whose equation is given by

$$\frac{D\omega}{Dt} = \omega \cdot \nabla \mathbf{u}.$$

As mentioned in Chapter 1, the fundamental complication arising in three dimensions, but not present in two dimensions, is the "vortex-stretching" term $\omega \cdot \nabla \mathbf{u}$. The velocity field can be obtained from the vorticity distribution by invoking the Biot–Savart formula (1.1.10)

$$\mathbf{u}(\mathbf{x}) = -\frac{1}{4\pi} \int \frac{(\mathbf{x} - \mathbf{z}) \times \omega(\mathbf{z}) d\mathbf{z}}{\|\mathbf{x} - \mathbf{z}\|^3}.$$

Our basic goal in this chapter is to describe some of the analytical models governing vorticity evolution in three dimensions, based primarily on what are commonly called "vortex filaments." In Section 8.1 we define a vortex filament and derive the localized self-induction equation (LIE), first obtained by DaRios (1906), then rederived by Arms and Hama (1965), governing the approximate motion of an isolated filament. As a special case we show that the equations for filaments that remain perfectly parallel and whose vorticity is given by delta function distributions correspond exactly to the point vortex equations of motion. We also show the classical result (Lamb (1932)) — that a perfectly circular vortex ring retains its shape and moves with uniform speed proportional to its curvature in the direction perpendicular to the ring's plane. This result heuristically motivates the form of the evolution equation for a more general noncircular filament under the localized induction approximation (LIA). In Section 8.2 we describe the intrinsic equations of Betchov (1965) and DaRios (1906), which govern the general evolution of the curvature and torsion of a filament under LIA as functions of time and arc length along the filament. We describe the coordinate transformation due to Hasimoto (1972) in Section 8.3, which takes the familiar Serret–Frenet formulas governing the geometry of a curved filament and the localized self-induction equation governing its dynamics into the nonlinear Schrödinger equation, a completely integrable, infinite-dimensional, Hamiltonian system about which much is known — see, for example, the recent book of Sulem and Sulem (1999) or the books of Lamb (1980) or Newell (1985). Using this knowledge, we can write down a special solution, which corresponds to the motion of a soliton loop traveling on the filament. Section 8.4 describes the invariant quantities associated with the localized self-induction approximation. A shortcoming of the localized self-induction approximation is that, on its own it does not allow for vortex-stretching. Hence, one of the main features thought to be responsible for many of the differences between two-dimensional and three-dimensional turbulence is not included in this model. In Section 8.5 we describe a theory of Callegari and Ting (1978) and Klein and Majda (1991), which does include vortex-stretching, although it is more complicated to formulate the equations. In Section 8.6 we describe a model of Klein, Majda, and Damodaran (1995), which governs the motion of filaments that are nearly parallel. The model is Hamiltonian and combines features of the point vortex model in two dimensions with some of the essential effects leading to vortex-stretching in three dimensions, in particular the linearized LIE. Finally, Section 8.7 briefly summarizes recent work on three-dimensional generalizations of point vortices, generically called "vortons."

8.1   Introduction to Vortex Filaments and the LIE

A simple way to generalize the idea of a point vortex from two to three dimensions is to assume that the vorticity is constrained to lie along a collection of thin tubes, uniformly distributed in each tube. This is the basic structure of a vortex filament.

Definition 8.1.1. *A **vortex filament** is an isolated vortex tube of small[1] cross-sectional area dA, with constant vorticity distribution ω across the*

---

[1]This can be quantified nondimensionally by the statement that the ratio of the vortex core radius, denoted σ, to the radius of curvature of the filament, denoted R, is small, i.e., $\sigma/R \ll 1$.

*tube. The **circulation** of the filament is given by*

$$\Gamma = \int \omega dA.$$

The centerline of the tube is defined by a space-time curve, which we denote $\mathbf{X}(s,t)$*, where s is a parameter along the tube, typically chosen to be the arc length.*

This definition is motivated by a large body of evidence in the computational and experimental literature, highlighting the importance of highly twisted, thin vortex tubes in a wide range of turbulent flows — for a start, see Siggia (1981), Kida (1993). For more general discussions and an introduction to some of the literature, see the book of Chorin (1994) and the article of Majda (1986). Shown in Figure 8.1 are convenient orthogonal coordinates associated with a general filament. At a given point along the filament centerline, one can choose a right handed orthonormal basis $(\mathbf{t}, \mathbf{n}, \mathbf{b})$, where $\mathbf{t}$ is the unit tangent vector, $\mathbf{n}$ is the normal vector, and $\mathbf{b}$ is the binormal. The plane spanned by $\mathbf{t}$ and $\mathbf{n}$ is called the osculating plane, the one spanned by $\mathbf{n}$ and $\mathbf{b}$ is the normal plane, and the one spanned by $\mathbf{b}$ and $\mathbf{t}$ is the rectifying plane. The basic equations from differential geometry that relate the coordinates are the ***Serret–Frenet equations***

$$\mathbf{X}' = \mathbf{t} \tag{8.1.1}$$

$$\mathbf{t}' = \kappa\mathbf{n} \tag{8.1.2}$$

$$\mathbf{n}' = \tau\mathbf{b} - \kappa\mathbf{t} \tag{8.1.3}$$

$$\mathbf{b}' = -\tau\mathbf{n}, \tag{8.1.4}$$

where $' \equiv \partial/\partial s$. Here, $\kappa(s,t)$ is the **curvature** and $\tau(s,t)$ is the **torsion** associated with the filament.

Remarks.

1. It is sometimes convenient to explicitly write the formulas for the triad $(\mathbf{X}', \mathbf{X}'', \mathbf{X}''')$ in terms of the orthonormal vectors $(\mathbf{t}, \mathbf{n}, \mathbf{b})$. The first two are (8.1.1), (8.1.2), while the third, after some algebra, is

$$\mathbf{X}''' = -\kappa^2\mathbf{t} + \kappa'\mathbf{n} + \tau\kappa\mathbf{b}.$$

In terms of $(\mathbf{X}', \mathbf{X}'', \mathbf{X}''')$, the orthonormal vectors are

$$\mathbf{t} = \mathbf{X}'$$

$$\mathbf{n} = \frac{1}{\kappa}\mathbf{X}''$$

$$\mathbf{b} = \frac{1}{\tau\kappa}\left[\kappa^2\mathbf{X}' - \frac{\kappa'}{\kappa}\mathbf{X}'' + \mathbf{X}'''\right].$$

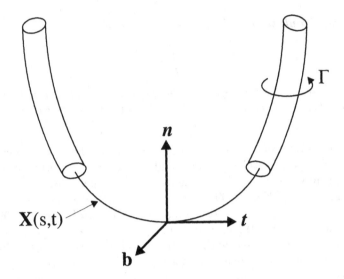

FIGURE 8.1. Vortex filament geometry.

2. It is also sometimes useful to write the curvature and torsion scalars in terms of $(\mathbf{X}', \mathbf{X}'', \mathbf{X}''')$, hence

$$\kappa^2(s,t) = \mathbf{X}'' \cdot \mathbf{X}'' \equiv K(s,t) \tag{8.1.5}$$
$$\tau(s,t) = \mathbf{X}' \cdot (\mathbf{X}'' \times \mathbf{X}''')/K. \qquad \diamond$$

These equations are purely geometric. In order to introduce dynamics, one must use the fact that a filament evolves under the local velocity field

$$\dot{\mathbf{X}} = \mathbf{u}(\mathbf{X}, t),$$

where $\dot{} \equiv \partial/\partial t$. Even in isolation, a vortex filament, unless it is perfectly straight, induces motion on itself and hence contributes to the velocity field $\mathbf{u}$. If more than one filament is present, each induces motion on all the others. In this case, the centerlines are given by a collection of curves, $\{\mathbf{X}_j(s,t)\}_{j=1}^{N}$, and the governing equation for the motion of filament $\mathbf{X}_j$ is

$$\frac{d\mathbf{X}_j}{dt} = \mathbf{u}(\mathbf{X}_1, \mathbf{X}_2, \dots, \mathbf{X}_N, t), \tag{8.1.6}$$

where $\mathbf{u}$ is the total fluid velocity field due to all of the filaments, as well as any additional effects one would like to include, such as, for example, background shear or rotational effects. To obtain the correct local velocity field $\mathbf{u}$, one must make use of the Biot–Savart law, which we now discuss.

In the most extreme case, the cross-sectional area of the tube is zero, which gives a vorticity field due to a filament $\mathbf{X}(s, t)$

$$\omega(\mathbf{x}) = \Gamma \int \delta(\mathbf{x} - \mathbf{X}(s')) \frac{\partial \mathbf{X}}{\partial s'} ds'.$$

This singular case is discussed rigorously in Cottet and Soler (1988). The velocity field is obtained in the usual way from the Biot–Savart law, which we can now write as

$$\mathbf{u}(\mathbf{x}) = -\frac{\Gamma}{4\pi} \int_X \frac{[\mathbf{x} - \mathbf{X}(s')] \times (\partial \mathbf{X}/\partial s')}{\|\mathbf{x} - \mathbf{X}\|^3} ds'. \qquad (8.1.7)$$

One can appreciate some of the complications involved in applying this formula by noting that the closer one gets to the filament, i.e., as $\mathbf{x}$ approaches $\mathbf{X}$, the velocity diverges like $1/\|\mathbf{x} - \mathbf{X}\|$, as shown, for example in Batchelor (1967). In two dimensions this divergence also occurs but causes no problems, since the self-induced velocity of a point vortex is zero (due to symmetry). In three dimensions the same is true if the filament is perfectly straight. The problem arises if one wants to compute the self-induced velocity for a filament that is not perfectly straight, in which case we evaluate $\mathbf{u}(\mathbf{X}(s))$ for s' close to s by using a Taylor expansion method, as done in the work of Arms and Hama (1965),

$$\mathbf{u}(\mathbf{X}(s)) = \frac{\Gamma}{4\pi} \left[\frac{\partial \mathbf{X}}{\partial s} \times \frac{\partial^2 \mathbf{X}}{\partial s^2} \int \frac{ds'}{|s - s'|} + O(1) \right], \qquad (8.1.8)$$

giving a logarithmic divergence. One can look up the details of this calculation in Chapter 2.8 of Pozrikidis (1997). In order to avoid this problem, it is necessary to take into account the vortex core in some way, and there are many ways to do this, as elaborated nicely in Leonard (1985). For example, Rosenhead (1930) desingularized the velocity expression (8.1.7) by introducing a parameter μ into the denominator

$$\mathbf{u}(\mathbf{x}) = -\frac{\Gamma}{4\pi} \int_X \frac{[\mathbf{x} - \mathbf{X}(s')] \times (\partial \mathbf{X}/\partial s')}{(\|\mathbf{x} - \mathbf{X}\|^2 + \mu^2)^{3/2}} ds'. \qquad (8.1.9)$$

Then one is left with how to scale μ with the other parameters in the problem, such as the core radius or the Reynolds number. We turn now to two illustrative examples before we discuss more generally how to treat the dynamics of filaments.

Example 8.1 (Parallel filaments) For the special case in which each filament is exactly parallel (say along the z-axis), and the vorticity is represented by a delta function for each filament,

$$\omega(\mathbf{x}, t) = \sum_{j=1}^{N} \Gamma_j \delta(\mathbf{x} - \mathbf{X}_j(t)),$$

the velocity field is simply

$$\mathbf{u}(\mathbf{x}, t) = \sum_{j=1}^{N} \Gamma_j \frac{(\mathbf{x} - \mathbf{X}_j)^\perp}{|\mathbf{x} - \mathbf{X}_j|^2},$$

where $\mathbf{x}^\perp \equiv (-x, y)$. The equations reduce, in this case, to the familiar planar point vortex equations of Chapter 2, which we write here as

$$\frac{d\mathbf{X}_j}{dt} = \sum_{k \neq j}^{N} \Gamma_k \frac{(\mathbf{X}_j - \mathbf{X}_k)^\perp}{|\mathbf{X}_j - \mathbf{X}_k|^2},$$

$$\mathbf{X}_j(0) = \mathbf{X}_j^{(0)}.$$

When viewed in this way — as special symmetric three-dimensional fila-ments — one can naturally ask about their stability with respect to various kinds of perturbations. Kelvin (1880) considered long wavelength pertur-bations of a single rectilinear vortex with a constant vorticity core, and perturbations in the form of helical disturbances proportional to $\exp(i\Omega t + im\theta + ikx)$, where m is the azimuthal wave number in cylindrical coordi-nates and k is the axial wave number. For an arbitrary core distribution his dispersion relation for the first mode $m = \pm 1$ is

$$\Omega = \pm \frac{\Gamma k^2}{4\pi} \left[\log(\frac{2}{k\sigma}) - \gamma + \int_0^\sigma \chi^2(r) \frac{dr}{r} + \int_\sigma^\infty (\chi^2(r) - 1) \frac{dr}{r} \right],$$

where $\gamma = 0.5772\ldots$ is Euler's constant, and σ is the core radius. Here,

$$\chi(r) = \frac{2\pi}{\Gamma} \int_0^r r' \omega_0(r') dr'$$

represents the fraction of circulation within radius r for a given vorticity distribution ω_0. More recently, Moore and Saffman (1975) and Tsai and Widnall (1976) identified instabilities when rectilinear vortices are placed in external strain fields, or when subject to short wavelength perturbations. Crow (1970) studied the instability associated with sinusoidally perturbed pairs of filaments under their mutual inductance, now called the "Crow instability." Nice discussions of these classical works can be found in the review paper of Widnall (1975) and that of Leonard (1985).   ◊

Example 8.2   (Hill's spherical vortex) An interesting and much refer-enced three-dimensional solution to the Euler equations is the axisymmet-ric *Hill's spherical vortex*. For general axisymmetric flows if one uses cylindrical coordinates (x, r, ϕ) and one assumes that the field is indepen-dent of ϕ, then the equations for the velocity components $u_x \equiv \dot{x}$, $u_r \equiv \dot{r}$ in terms of the Stokes's streamfunction Ψ (see Milne-Thompson (1968)) are

$$u_x = \frac{1}{r} \cdot \frac{\partial \Psi}{\partial r}$$

$$u_r = -\frac{1}{r} \cdot \frac{\partial \Psi}{\partial x}.$$

The vorticity vector points in the azimuthal direction,

$$\boldsymbol{\omega} = \omega \hat{\mathbf{e}}_\phi,$$

with

$$\omega = \frac{1}{r} \left(\frac{\partial^2 \Psi}{\partial x^2} + \frac{\partial^2 \Psi}{\partial r^2} - \frac{1}{r} \cdot \frac{\partial \Psi}{\partial r} \right).$$

Then the Stokes's streamfunction for the Hill's spherical vortex is made up of two parts,

$$\Psi_{in} = \frac{\alpha}{10} r^2 (a^2 - x^2 - r^2)$$

$$\Psi_{out} = -\frac{\alpha}{15} a^2 r^2 \left(1 - \frac{a^3}{(x^2 + r^2)^{3/2}} \right),$$

where Ψ_{in} is the streamfunction inside the sphere of radius a, and Ψ_{out} is the streamfunction outside the sphere. From the exterior streamfunction one can see that the vortex translates steadily along the x-axis with velocity

$$U = \frac{2}{15} \alpha a^2,$$

always retaining its spherical shape. The stability of this special configuration is considered in Pozrikidis (1986). ◊

Example 8.3 (Circular vortex rings) A classical configuration is that of a thin circular vortex ring of constant vorticity inside a thin tube of radius σ, with ring radius R, under the assumption $\sigma/R \ll 1$. We take the vorticity distribution inside the tube to correspond to solid body rotation, hence

$$\omega_0(\mathbf{x}) = \Gamma/\pi\sigma^2, \qquad \mathbf{x} < \sigma$$
$$\omega_0(\mathbf{x}) = 0, \qquad\qquad \mathbf{x} > \sigma.$$

As described in Lamb (1932), the ring moves steadily along the axis perpendicular to the ring's plane, with constant translation velocity U_r given by the formula

$$U_r = \frac{\Gamma}{4\pi R} \left[\log \left(\frac{8R}{\sigma} \right) - \frac{1}{4} + O \left(\frac{\sigma}{R} \right) \right]. \tag{8.1.10}$$

In considering the limit $R/\sigma \gg 1$, it is not too drastic to assume that

$$\log \left(\frac{8R}{\sigma} \right) \sim const,$$

hence the ring's velocity becomes proportional to its curvature, i.e.,

$$U_r \sim \frac{C_1}{R}.$$

For more general vorticity distributions within the ring, the translation velocity formula becomes

$$U_r = \frac{\Gamma}{4\pi R} \left[\log\left(\frac{8R}{\sigma}\right) - \frac{1}{2} + \int_0^\sigma \chi^2(r)\frac{dr}{r} + \int_\sigma^\infty (\chi^2(r) - 1)\frac{dr}{r} \right].$$

If the vorticity is confined to the surface of the ring, one has a so called "hollow vortex," and the $-1/4$ term in (8.1.10) is replaced by $-1/2$, as shown in Hicks (1884). If one uses the desingularized velocity expression (8.1.9) with $\mu^2 = \alpha\sigma^2$, one obtains a translation velocity formula

$$U_r = \frac{\Gamma}{4\pi R} \left[\log\left(\frac{8R}{\sigma\sqrt{\alpha}}\right) - 1 \right].$$

These results can be found in Lamb (1932), Batchelor (1967), as well as more complete discussion in Fraenkel (1970, 1972), Fraenkel and Berger (1974), and Norbury (1972, 1973). In particular, the works of Fraenkel and Norbury complement each other and establish the existence of a one-parameter family of rings, parametrized by the cross-sectional area of the ring. Explicit asymptotic constructions are obtained both in the limit of small area, as well as the upper (finite) limit, in which one obtains the Hill's spherical vortex as an exact solution.

Saffman (1970) calculated the viscous corrections to the translation velocity of a vortex ring, obtaining

$$U_r = \frac{\Gamma}{4\pi R} \left[\log\left(\frac{8R}{\sqrt{4\nu t}}\right) - \gamma + O\left((\frac{\nu t}{R^2})^{1/2} \log\left(\frac{\nu t}{R^2}\right)\right) \right],$$

where ν is the kinematic viscosity. Hence, the ring slows down like $-\log \nu t$ due to viscous vorticity diffusion. See also Maxworthy (1977) and Kambe and Takao (1971) for experimental results, including ring interactions. Rings of elliptic shape were studied in Arms and Hama (1965) and Dhanak and deBernardinis (1981), while inviscid ring interactions were studied in Anderson and Greengard (1989). For a general review see Shariff and Leonard (1992).   ◊

Example 8.4   (Coaxial circular ring interaction) Konstantinov (1994) recently derived and numerically investigated a system of equations for the interaction of coaxial circular rings without swirl. In cylindrical coordinates, if we denote the axial distance by Z, the radial distance by r, and the azimuthal angle by θ, the velocity components in terms of the Stokes's streamfunction are

$$u_z = \frac{1}{r}\frac{\partial \psi}{\partial r}, \qquad u_r = -\frac{1}{r}\frac{\partial \psi}{\partial z},$$

and the vorticity equation for the inviscid swirl-free flow is

$$\frac{D(\omega/r)}{Dt} = 0.$$

The streamfunction equation is given by

$$\frac{\partial^2 \psi}{\partial z^2} + \frac{\partial^2 \psi}{\partial r^2} - \frac{1}{r}\frac{\partial \psi}{\partial r} + r^2 f(\psi) = 0,$$

where $\omega/r = f(\psi)$ for an arbitrary function $f(\psi)$. In terms of the ring coordinates $(Z_i(t), R_i(t))$, $i = 1, \ldots, N$, Konstantinov (1994) writes the evolution equation for N rings, based on the original formulation of Dyson (1893):

$$\dot{Z}_i = \frac{\Gamma_i}{4\pi R_i}\left[\log\left(\frac{8R_i}{a_i}\right) - \frac{1}{4}\right] + \frac{1}{\Gamma_i R_i}\cdot\frac{\partial U}{\partial R_i}$$

$$\dot{R}_i = -\frac{1}{\Gamma_i R_i}\cdot\frac{\partial U}{\partial Z_i}.$$

The circulation in each ring is assumed constant,

$$\Gamma_i = \int \omega\, dr\, dz,$$

and $a_i << R_i$ is the radius of the circular core, which is assumed to remain circular, with $a_i^2 R_i = const$. The potential U is

$$U = \frac{1}{2\pi}\sum_{\substack{i\neq j}}^{N}\Gamma_i\Gamma_j I_{ij},$$

where

$$I_{ij} = \int_0^\pi \frac{R_i R_j \cos\theta'}{\sqrt{(Z_i - z_j)^2 + R_i^2 - 2R_i R_j \cos\theta' + R_j^2}}\, d\theta'$$

$$\equiv \sqrt{R_i R_j}C(k_{ij})$$

$$C(k_{ij}) = \left(\frac{2}{k_{ij}} - k_{ij}\right)K(k_{ij}) - \frac{2}{k_{ij}}E(k_{ij})$$

$$k_{ij}^2 = \frac{4R_i R_j}{(Z_i - Z_j)^2 + (R_i + R_j)^2},$$

with $K(k)$ and $E(k)$ being the complete elliptic integrals of the first and second kinds. One can write the system in canonical Hamiltonian form by introducing the variables

$$p_i = \Gamma_i R_i^2, \quad q_i = Z_i.$$

Then, the Hamiltonian is

$$
\mathcal{H} = \sum_{i=1}^{N} \frac{\Gamma_i^{3/2} p_i^{1/2}}{2\pi} \left[\log \left(\frac{8\Gamma_i^{-3/4} p_i^{3/4}}{A_i} \right) - \frac{7}{4} \right]
$$

$$
+ \frac{1}{\pi} \sum_{i \neq j}^{N} (\Gamma_i \Gamma_j)^{3/4} (p_i p_j)^{1/4} C(k_{ij})
$$

$$
k_{ij}^2 = \frac{4(\Gamma_i \Gamma_j p_i p_j)^{1/2}}{[\Gamma_i \Gamma_j (q_i - q_j)^2 + (\Gamma_j^{1/2} p_i^{1/2} + \Gamma_i^{1/2} p_j^{1/2})^{1/2}]}
$$

$$
A_i = a_i(0)\sqrt{R_i(0)},
$$

and the equations of motion are

$$
\dot{p}_i = -\frac{\partial \mathcal{H}}{\partial q_i}, \quad \dot{q}_i = \frac{\partial \mathcal{H}}{\partial p_i}.
$$

It is straightforward to verify that, in addition to the Hamiltonian, the system conserves the total momentum $\sum_{i=1}^{N} p_i$. The center of vorticity of a system of N interacting coaxial rings is

$$
R_c = \left(\frac{\sum_{i=1}^{N} \Gamma_i R_i^2}{\sum_{i=1}^{N} \Gamma_i} \right)^{1/2}, \quad Z_c = \frac{\sum_{i=1}^{N} \Gamma_i R_i^2 Z_i}{\sum_{i=1}^{N} \Gamma_i R_i^2}.
$$

The paper of Konstantinov (1994) examines the above system numerically, while the more recent work of Marsden and Shashikanth (2000) uses it as a starting point for the analysis of geometric phases and reduction methods for leapfrogging configurations, thereby generalizing the planar leapfrogging configurations described in Example 2.4.   ◊

Aside from Rosenhead's (1930) method, one of the simplest ways to remove the logarithmic divergence associated with (8.1.8) is to use a so-called cut-off procedure (Leonard (1985)), i.e., one "cuts off" the line integral in (8.1.8) for $|s - s'| < \delta\sigma$, where δ is some cut-off parameter.[2] This procedure, first used by Hama (1962, 1963), Arms and Hama (1965), and Crow (1970), physically reflects the fact that portions of the filament that are sufficiently far away do not contribute much to its local motion. With this approximation the singular integral in (8.1.8) is replaced by $\log(L/\sigma)$ where L is the cut-off parameter to be specified. The velocity field (8.1.8) along the filament then leads to

$$
\frac{\partial \mathbf{X}}{\partial t} = \log \left(\frac{L}{\sigma} \right) \frac{\Gamma}{4\pi} \left(\frac{\partial \mathbf{X}}{\partial s} \times \frac{\partial^2 \mathbf{X}}{\partial s^2} \right). \tag{8.1.11}
$$

---

[2]A simple physical way to understand the cut-off idea is the following. Since the velocity of a vortex ring is known exactly, one can use it to desingularize the Biot–Savart integral. One simply removes a small arc interval responsible for the logarithmic singularity, and replaces it with the known contribution of the osculating vortex ring.

This *localized induction approximation* (LIA) is beautifully derived in Arms and Hama (1965). From it, the self-induced motion of a filament (after time rescaling) is given simply by the formula

$$\frac{\partial \mathbf{X}}{\partial t} = \kappa \mathbf{b}, \qquad (8.1.12)$$

since $\partial \mathbf{X}/\partial s \equiv \mathbf{t}$, $\partial^2 \mathbf{X}/\partial s^2 \equiv \kappa \mathbf{n}$, and $\mathbf{b} = \mathbf{t} \times \mathbf{n}$.

Remarks.

1. The approximations made in arriving at (8.1.12) are:

   (a) (Thin filament approximation) The filament is thin, hence the core remains constant in time and the core radius is much smaller than the radius of curvature of the filament. In addition, perturbation wavelengths are much larger than the core radius.

   (b) (Cut-off approximation) Long distance effects on the self-induced motion of the filament are ignored.

   It is important to keep in mind the inherent limitations of these approximations. For example, when short wavelength perturbations need to be considered, one can no longer use the thin filament approach. When filaments collapse or become knotted, one can no longer use the cut-off approximation. Unfortunately, it is precisely these kinds of effects that one would like to model in order to understand fully turbulent three-dimensional flows. See, for example, Chorin (1982) or Siggia (1985) for a discussion of some of the relevant issues.

2. One can understand (8.1.12) heuristically by noting that a general curved filament can be approximated locally by a vortex ring tangent to the filament with curvature $\kappa = 1/R$, hence using the ring velocity formula

$$v \sim \frac{C_1}{R},$$

   along with the fact that a circular ring moves perpendicular to its axis, one obtains (8.1.12).

3. It is common to rescale the time variable in (8.1.11) and write it more simply as

$$\frac{\partial \mathbf{X}}{\partial t} = \mathbf{X}' \times \mathbf{X}''.$$

By differentiating this equation with respect to s, we can also write the LIE in terms of the tangent vector

$$\frac{\partial \mathbf{t}}{\partial t} = \mathbf{t} \times \frac{\partial^2 \mathbf{t}}{\partial s^2},$$

which sometimes is a more useful representation, particularly for numerical purposes (Buttke (1988)).   ◊

One of the deficiencies of the LIE is that it does not allow for self-stretching of an isolated filament. We summarize here the main theorem in Arms and Hama (1965), which addresses this issue.

Theorem 8.1.1 (Arms and Hama (1965)). *Consider the arc length of a portion of filament* $s_1 \le s \le s_2$, *defined as*

$$l(t) = \int_{s_1}^{s_2} \|\mathbf{X}'\| \; ds.$$

Define the projected area of a closed filament curve $\mathcal{C}$ *as*

$$\mathbf{P}(t) = \frac{1}{2} \int_{\mathcal{C}} (\mathbf{X} \times \mathbf{X}') \; ds.$$

Then

$$\frac{dl}{dt} = 0$$
$$\frac{d\mathbf{P}}{dt} = 0.$$

Remarks.

1. Notice the linear momentum of the fluid, given by

$$\mathbf{L} = \frac{1}{2} \int_{\mathbb{R}^3} \mathbf{X} \times \omega dV.$$

For a vortex filament we have $\omega = \omega \mathbf{t}$, $\Gamma = \omega \mathcal{A}$ and $dV = \mathcal{A} ds = (\Gamma/\omega)\,ds$, hence

$$\mathbf{L} = \frac{1}{2}\Gamma \int_{\mathcal{C}} \mathbf{X} \times \mathbf{X}' ds = \mathbf{P}(t),$$

hence the statement is just the conservation of linear momentum. See Ricca (1992) for a simple proof and further discussion.

2. $\mathbf{P}(t)$ is a vector that is perpendicular to the plane of maximum projected area of the curve $\mathcal{C}$, with corresponding magnitude representing the projected area. Hence, the theorem states that the projected area of the closed vortex filament remains constant in magnitude and direction, as does the arc length of the curve. See Ricca (1992) for further discussion.

3. Arms and Hama (1965) define another conserved quantity, $\mathbf{w}(t)$, called the **wobble**, as

$$\mathbf{w}(t) = \frac{2\pi}{l^2} \int_{\mathcal{C}} \mathbf{X} \times \mathbf{X}' \; ds,$$

which by the theorem is constant. One can conclude from this that a noncircular vortex ring never becomes perfectly circular, undergoing uniform translation, since the wobble due to perturbations never completely dies out. A linearized perturbation analysis around a circular ring that reflects this was carried out in Kambe and Takao (1971), where it was shown that the ring is neutrally stable. In addition, they conclude, based on the eigenfunctions of the linear problem, that a deformed vortex ring makes a twisted rotation around its mean circle, with, in general, a slight change in the direction of translation. See also Widnall and Sullivan (1973) for more on the linear stability of vortex rings. ◊

The paper of Arms and Hama (1965) contains a proof of the theorem as well as more detailed discussion, including the fate of perturbed vortex rings. Despite the fact that the LIA has deficiencies with regard to understanding three-dimensional turbulent flows, it is and continues to be quite useful due to its simplicity and beautiful analytical structure, which we describe in the following sections. For an interesting discussion of the history of the LIA, see Ricca (1991, 1996).

We end this section by describing a family of torus knot solutions to (8.1.12) as derived in Ricca (1993) and discussed in Keener (1990).

Example 8.5 (Torus knots) When written in terms of its Cartesian components $X = (x, y, z)$, (8.1.12) becomes

$$\dot{x} = y'z'' - z'y''$$
$$\dot{y} = z'x'' = x'z''$$
$$\dot{z} = x'y'' = y'x''.$$

In cylindrical polar coordinates $(r(s,t), \alpha(s,t), z(s,t))$ with $x = r\cos\alpha$, $y = r\sin\alpha$, the equations are

$$\dot{r} = r\alpha'z'' - 2r'\alpha'z' - r\alpha''z' \tag{8.1.13}$$

$$r\dot{\alpha} = -r'z'' + r''z' - r(\alpha')^2 z' \tag{8.1.14}$$

$$\dot{z} = 2(r')^2\alpha' + rr'\alpha'' - rr''\alpha' + r^2(\alpha')^3. \tag{8.1.15}$$

We can now look for periodic solutions that are near the unperturbed circle torus state given by

$$r = r_0$$
$$\alpha = s/r_0$$
$$z = t/r_0,$$

by inserting the perturbed quantities

$$r = r_0 + \epsilon r_1$$
$$\alpha = s/r_0 + \epsilon\alpha_1$$
$$z = t/r_0 + \epsilon z_1$$

$(0 < \epsilon << 1)$ into (8.1.13), (8.1.14), (8.1.15), and keeping only first order terms in ϵ. This yields the linearized system

$$\dot{r}_1 = z_1''$$
$$\dot{\alpha}_1 = -\frac{1}{r_0^2}z_1'$$
$$\dot{z}_1 = -r_1'' - \frac{r_1}{r_0^2}.$$

The first and last of these can be combined to give a single equation for r_1,

$$\ddot{r}_1 = \dot{z}_1'' = -r_1'''' - \frac{1}{r_0^2}r_1''.$$

We now look for periodic traveling wave solutions in the new variable $\xi = s - at$, which yields

$$r_1'' + (a^2 + \frac{1}{r_0^2})r_1 = A\xi + B,$$

where $' \equiv d/d\xi$ and $r_1(s) = r_1(s+L)$, hence $A = 0, B = 0$. Here L denotes the length of the perturbed curve. Then

$$r_1(\xi) = k\sin\left(\frac{2\pi q}{L}\xi + \beta\right)$$

where q represents the number of times the knot wraps around the torus in the meridian plane. The wavespeed is then given by

$$a^2 = -\frac{1}{r_0^2} + \left(\frac{2\pi q}{L}\right)^2,$$

and the remaining coordinates are

$$z_1 = \frac{akL}{2\pi q} \cos\left(\frac{2\pi q}{L}\xi + \beta\right)$$

$$\alpha_1 = \frac{kL}{2\pi q r_0^2} \cos\left(\frac{2\pi q}{L}\xi + \beta\right).$$

In order to form a torus-knot solution, we require that $\alpha(s+L) = \alpha(s)+2p\pi$ where p is the number of wraps around the torus in the longitudinal plane and p and q are coprime. Since α_1 is periodic with period L, we have that

$$\frac{L}{r_0} = 2p\pi,$$

which then gives

$$a^2 = -\frac{1}{r_0^2} + \left(\frac{q}{pr_0}\right)^2 = \frac{1}{r_0^2}\left[\left(\frac{q}{p}\right)^2 - 1\right].$$

The full torus-knot solutions can then be written

$$r = r_0 + \epsilon k \sin\left[\left(\frac{q}{p}\right)\frac{\xi}{r_0} + \beta\right]$$

$$\alpha = \frac{s}{r_0} + \epsilon\left(\frac{p}{q}\right)\frac{k}{r_0}\cos\left[\left(\frac{q}{p}\right)\frac{\xi}{r_0} + \beta\right]$$

$$z = \frac{t}{r_0} + \epsilon k\sqrt{1 - \left(\frac{p}{q}\right)^2}\cos\left[\left(\frac{q}{p}\right)\frac{\xi}{r_0} + \beta\right]$$

with $s/r_0 \in [0, 2p\pi]$. From these solutions it is straightforward to calculate the curvature and torsion associated with the family of torus knots,

$$\kappa(s,t) \sim \frac{1}{r_0}\left[1 + \epsilon\frac{k}{r_0}\left[\left(\frac{q}{p}\right)^2 - 1\right]\sin\gamma\right]$$

$$\tau(s,t) \sim \frac{\epsilon\frac{k}{r_0^2}\sqrt{(q/p)^2 - 1}\left[(q/p)^2 - 1\right]\sin\gamma\cos\alpha}{\left[\left(1 + \epsilon\frac{k}{r_0}\left[(q/p)^2 - 1\right]\sin\gamma\right)\cos\alpha + \epsilon(k/r_0)(q/p)\cos\gamma\sin\alpha\right]},$$

where

$$\alpha = \frac{s}{r_0} + \epsilon\left(\frac{p}{q}\right)\frac{k}{r_0}\cos\gamma$$

$$\gamma = \left[\left(\frac{q}{p}\right)\frac{s - at}{r_0} + \beta\right].$$

The full family of exact torus-knot solutions in terms of elliptic functions was derived originally in Kida (1981b). The linear stability of the small amplitude torus knot solutions is considered in Ricca (1995) where it is shown that if $q/p > 1$, the knots are stable, while if $q/p \leq 1$, they are unstable. Thus, it is interesting to point out that by interchanging p and q one has a pair of knots that are topologically equivalent, but dynamically distinct. The nonlinear time evolution of these knot solutions has recently been studied by Ricca et al (1999). Further properties of these solutions and simpler representations can be found in Ricca (1993) and Keener (1990). Stability results for the general class of uniformly translating solutions can be found in Kida (1982).   ◊

8.2   DaRios–Betchov Intrinsic Equations

In this section we explore the consequences of using the LIE

$$\dot{\mathbf{X}} = \mathbf{X}' \times \mathbf{X}'' \tag{8.2.1}$$

to describe the evolution of a general filament. We focus on a classic paper of Betchov (1965) in which the general evolution equations for the curvature $K(s,t)$ and torsion $\tau(s,t)$ of a filament evolving under (8.2.1) are derived. These equations were originally derived in DaRios (1906), and so we refer to the system as the DaRios–Betchov equations. The coupled partial differential equations are given by

$$\dot{K} = -2(K'\tau + K\tau') \tag{8.2.2}$$

$$\dot{\tau} + 2\tau\tau' = \frac{1}{2}\left(K' + \frac{K'''}{K} + \frac{K'^3}{K^3} - 2\frac{K'K''}{K^2} \right), \tag{8.2.3}$$

where $\dot{} \equiv \partial/\partial t$ and $' \equiv \partial/\partial s$. K and τ are defined through the relations (8.1.5).

The derivation is an exercise in vector algebra. One starts with the identities

$$\mathbf{X}' \cdot \mathbf{X}' = 1$$
$$\mathbf{X}' \cdot \mathbf{X}'' = 0.$$

The first normalizes the tangent vector to unity, while the second asserts that the tangent vector and the normal vector are perpendicular. By successive differentiation of these with respect to s, one obtains

$$\mathbf{X}' \cdot \mathbf{X}''' = -K$$
$$\mathbf{X}' \cdot \mathbf{X}^{iv} = -\frac{3}{2}K'$$
$$\mathbf{X}' \cdot \mathbf{X}^{v} = -K'' - K''' - M + 3M',$$

where

$$M \equiv \mathbf{X}''' \cdot \mathbf{X}'''.$$

The first intrinsic equation (8.2.2) is obtained by time differentiating K and using (8.2.1), hence

$$\dot{K} = 2\dot{\mathbf{X}}'' \cdot \mathbf{X}'' = 2(\mathbf{X}' \times \mathbf{X}^{iv}) \cdot \mathbf{X}'' = -2\Delta', \qquad (8.2.4)$$

where

$$\Delta \equiv |\mathbf{X}', \mathbf{X}'', \mathbf{X}'''| = \mathbf{X}' \cdot (\mathbf{X}'' \times \mathbf{X}''') = \tau K.$$

It is then straightforward to show that (8.2.4) is equivalent to (8.2.2), since

$$\dot{K} = -2\Delta' = -2(\tau K)' = -2(K'\tau + \tau'K).$$

To derive the second equation (8.2.3), we take the time derivative of Δ,

$$\dot{\Delta} = \frac{1}{2}K''' + \frac{9}{2}KK' - 2M'. \qquad (8.2.5)$$

Using the fact that

$$\dot{\Delta} = \dot{\tau}K + \dot{K}\tau,$$

we have

$$K\dot{\tau} = -\dot{K}\tau + \frac{1}{2}K''' + \frac{9}{2}KK' - 2M'$$

$$= 2\tau(K'\tau + K\tau') + \frac{1}{2}K''' + \frac{9}{2}KK' - 2M'$$

upon application of the first intrinsic equation. Then, owing to the fact that

$$M = K^2 + \frac{1}{4} \cdot \frac{(K')^2}{K} + \frac{\Delta^2}{K},$$

we can differentiate and obtain

$$M' = 2KK' + \frac{1}{2} \cdot \frac{K'K''}{K} - \frac{1}{4} \cdot \frac{(K')^3}{K^2} + \frac{2\Delta\Delta'}{K} - \frac{\Delta^2}{K^2}K'$$

$$= 2KK' + \frac{1}{2} \cdot \frac{K'K''}{K} - \frac{1}{4} \cdot \frac{(K')^3}{K^2} + 2K\tau\tau' + K'\tau^2.$$

Hence,

$$K\dot{\tau} = 2K'\tau^2 + 2K\tau\tau' + \frac{1}{2}K''' + \frac{9}{2}KK' - 4KK' - \frac{K'K''}{K} + \frac{1}{2}\frac{(K')^3}{K^2}$$

$$- 4K\tau\tau' - 2K'\tau^2,$$

which leads to the second intrinsic equation (8.2.3).

We now consider some special solutions of the equations (8.2.2), (8.2.3).

Example 8.6   (Helicoidal filament) Notice first that if K' and τ' are everywhere zero at some instant of time, they will remain zero in the future. Hence, a helical filament defined by

$$K = const$$
$$\tau = const$$

is a special solution to (8.2.2), (8.2.3). As discussed further in Betchov (1965), the helicoid will move with translation velocity V_T along its axis, in a screwlike fashion with rotational velocity V_R, where

$$\frac{V_R}{V_T} = \frac{\tau}{K^{1/2}}.$$

As shown in Mezić, Leonard, and Wiggins (1998),

$$V_T = \frac{\Gamma}{\pi}\kappa\gamma\sin\beta, \qquad V_R = \frac{\Gamma}{\pi}\kappa\gamma\cos\beta,$$

where $\tan\beta = a/k$, with a being the radius of the cylinder circumscribed by the helix, and $2\pi k$ is the pitch. Γ is the circulation, and γ is a dimensionless parameter that depends on the core radius. We also have that

$$\kappa = \frac{a}{a^2 + k^2}.$$

To analyze the basic stability of such a state, one need only linearize (8.2.2), (8.2.3) around a constant state $(K, \tau) = (K_0, \tau_0)$. If we denote the perturbation coordinates as (K, τ) (this should not cause confusion), one obtains the linear system

$$\dot{K} + 2\tau_0 K' = -2K_0\tau',$$
$$\dot{\tau} + 2\tau_0\tau' = \frac{1}{2}\left(K' + \frac{K'''}{K_0}\right).$$

The terms on the left immediately tell us that perturbations propagate along the filament with velocity $2\tau_0$. If we then shift to a new arc length parameter

$$\xi = s - 2\tau_0 t$$

by standard methods, we obtain the new equations

$$\dot{K} = -2K_0\tau'$$
$$\dot{\tau} = \frac{1}{2}\left(K' + \frac{K'''}{K_0}\right)$$

(Prime now denotes differentiation with respect to ξ) which upon successive differentiation and manipulation yields the linear system

$$\ddot{K} + K^{iv} + K_0 K'' = 0 \qquad (8.2.6)$$
$$\ddot{\tau} + \tau^{iv} + K_0\tau'' = 0. \qquad (8.2.7)$$

Thus, if the perturbation to the helicoid has wavelength λ and frequency Ω, then the dispersion relation becomes

$$\Omega^2 = \left(\frac{2\pi}{\lambda}\right)^2 \left[\left(\frac{2\pi}{\lambda}\right)^2 - K_0\right].$$

Since $K_0 \equiv 1/R^2$, we know that the perturbation propagates if $\lambda < 2\pi R$, and grows exponentially if $\lambda > 2\pi R$. For a circular ring the wavelength cannot be larger than the circumference of the ring, hence a vortex ring is neutrally stable according to this theory. (See also the classical work of Levy and Forsdyke (1928), and Widnall (1972).) The particle motion around a helix in a strain field is studied in Mezić, Leonard, and Wiggins (1998), and uses the streamfuction formulation obtained in Hardin (1982).

$\diamond$

Example 8.7 (Planar filaments) If $\Delta = 0$, the filament is planar and the only solutions of (8.2.2), (8.2.3) satisfy

$$\dot{K} = 0$$

$$\frac{K'''}{K} + \frac{(K')^3}{K^3} - 2\frac{K'K''}{K^2} + K' = 0.$$

This second equation can be integrated twice to yield

$$s - s_0 = \pm \int_{s_0}^{s} \frac{1}{(cK + bK^2 - K^3)^{1/2}} dK,$$

where b and c are constants of integration. If we let $c = 0$, then

$$K = b\left[\cosh \frac{b^{1/2}}{2}(s - s_0)\right]^{-2}.$$

The particular (arbitrary) choice $s_0 = 0$, $b = 4$ yields a single looped filament described parametrically as

$$x_1 = s - 2\tanh s$$
$$x_2 = \frac{2}{\cosh s}$$

and shown in the original paper of Betchov (1965). This single-looped filament rotates about the x_1-axis with a velocity proportional to $K^{1/2}$. $\diamond$

Remarks.

1. Betchov's equations (8.2.2), (8.2.3) can be put in conservation form

$$\frac{\partial K}{\partial t} + \frac{\partial}{\partial s}(KT) = 0$$

$$\frac{\partial T}{\partial t} + \frac{\partial}{\partial s}\left[\frac{1}{2}T^2 - K + \frac{(K')^2}{2K^2} - \frac{K''}{K}\right] = 0,$$

where $T \equiv 2\tau$. This was originally observed in Hasimoto (1972) in a slightly different form.

2. We mention the work of Ricca (1994), who considers the role of torsion on the dynamics of a helical thin filament. In particular, when the curvature is held constant and one varies the pitch of the helix, there are two limiting cases. Infinite pitch corresponds to a rectilinear filament, while zero pitch corresponds to a cylindrical vortex sheet. The dynamics associated with these two geometries, as well as the intermediate cases, are considered in Ricca (1994).

3. The recent papers of Boersma and Wood (1999) and Kuibin and Okulov (1998) consider the self-induced motion of a helical filament. They derive a formula for the binormal velocity U_h as a function of the pitch p, that is,

$$U_h = -\frac{1}{4} \log 2 + 2p^2 - 2p(p^2 + 1)^{1/2}$$
$$-\frac{1}{2} \log(p^2 + 1) + (p^2 + 1)^{3/2} W(p),$$

where

$$W(p) = \int_0^\infty \left[\frac{2(1 - \cos t)}{(p^2 t^2 + 2(1 - \cos t))^{3/2}} - \frac{1}{(p^2 + 1)^{3/2}} \cdot \frac{H(1 - t)}{t} \right] dt,$$

H being the Heaviside function.

4. Hardin (1982) derives an explicit expression for the velocity field and streamfunction associated with a helical filament, written in the form of an infinite series representation. $\Diamond$

8.3 Hasimoto's Transformation

Our goal in this section is to relate the evolution equation governing the filament function $\mathbf{X}(s, t)$ to the nonlinear Schrödinger equation, as achieved first in the paper of Hasimoto (1972). A more general derivation was given recently by Klein and Majda (1991 I,II) and Majda (1998), in which one starts from the perturbed binormal evolution equation

$$\frac{\partial \mathbf{X}}{\partial t} = \kappa \mathbf{b} + \epsilon \mathbf{u}, \tag{8.3.1}$$

where $\mathbf{u}$ represents some background velocity field, and ϵ is considered small. We will follow this approach here as it leads to several interesting insights.

Start by combining the last two Serret–Frenet equations (8.1.3), (8.1.4) into complex form

$$(\mathbf{n} + i\mathbf{b})' = -i\tau(\mathbf{n} + i\mathbf{b}) - \kappa\mathbf{t}. \tag{8.3.2}$$

This motivates the introduction of the complex function $\mathbf{N}(s, t)$,

$$\mathbf{N}(s, t) \equiv (\mathbf{n} + i\mathbf{b}) \exp(i\Phi),$$

where $\Phi(s, t)$ is defined in terms of a complex *filament function* ψ.

Definition 8.3.1. *The **filament function** $\psi(s, t)$ is defined in terms of the curvature $\kappa(s, t)$ and torsion $\tau(s, t)$ as*

$$\psi(s, t) \equiv \kappa(s, t) \exp(i\Phi) \tag{8.3.3}$$

$$\Phi = \int_0^s \tau(s, t)ds. \tag{8.3.4}$$

Then

$$\kappa = |\psi|$$
$$\tau = \Phi'.$$

Note from the definition of $\mathbf{N}(s, t)$, we can replace the orthogonal vectors $(\mathbf{n}, \mathbf{b})$ with the new vectors $(\mathbf{N}, \mathbf{N}^*)$, which gives a new basis set $(\mathbf{t}, \mathbf{N}, \mathbf{N}^*)$, with corresponding orthogonality relations

$$\mathbf{t} \cdot \mathbf{t} = 1, \quad \mathbf{N} \cdot \mathbf{N} = 0, \quad \mathbf{N} \cdot \mathbf{N}^* = 2. \tag{8.3.5}$$

(8.3.2) can be written as

$$\mathbf{N}' + i\tau\mathbf{N} = -\psi\mathbf{t}, \tag{8.3.6}$$

while the second of the Serret–Frenet equations (8.1.2) is written

$$\mathbf{t}' = \frac{1}{2}(\psi^*\mathbf{N} + \psi\mathbf{N}^*). \tag{8.3.7}$$

(8.3.6) and (8.3.7) are the variational equations along the filament $\mathbf{X}(s, t)$, expressed in terms of the basis set $(\mathbf{t}, \mathbf{N}, \mathbf{N}^*)$.

We can now compute the temporal variations of $\mathbf{t}$ and $\mathbf{N}$ along the filament, expressing them in terms of $(\mathbf{t}, \mathbf{N}, \mathbf{N}^*)$. The temporal variation of $\mathbf{t}$ is obtained by taking the s derivative of the perturbed binormal law (8.3.1),

$$\dot{\mathbf{t}} = \kappa'\mathbf{b} - \kappa\tau\mathbf{n} + \epsilon\mathbf{u}'$$
$$= \frac{1}{2}i(\psi'\mathbf{N}^* + \psi^{*\prime}\mathbf{N}) + \epsilon\mathbf{u}'. \tag{8.3.8}$$

We write the temporal variation of $\mathbf{N}$ as

$$\dot{\mathbf{N}} = \alpha \mathbf{N} + \beta \mathbf{N}^* + \gamma \mathbf{t},$$

which, from (8.3.5) and (8.3.8), yields

$$\alpha + \alpha^* = \frac{1}{2}(\mathbf{N} \cdot \mathbf{N}^*)\dot{} = 0$$

$$\beta = \frac{1}{4}(\mathbf{N} \cdot \mathbf{N})\dot{} = 0$$

$$\gamma = \mathbf{t} \cdot \dot{\mathbf{N}} = -\dot{\mathbf{t}} \cdot \mathbf{N} = -i\psi' - \epsilon \mathbf{N} \cdot \mathbf{u}'.$$

The evolution equation for $\mathbf{N}$ is then

$$\dot{\mathbf{N}} = i[R\mathbf{N} - (\psi' - i\epsilon \mathbf{N} \cdot \mathbf{u}')\mathbf{t}] \tag{8.3.9}$$

where $R(s,t)$ is a real scalar function yet to be determined.

We can now obtain two representations of $\dot{\mathbf{N}}'$ by differentiating (8.3.6) with respect to t and (8.3.9) with respect to s. Since the two representations must be equivalent, by equating the coefficients of $\mathbf{t}$ and $\mathbf{N}$ we obtain the compatibility conditions

$$\dot{\psi} + \epsilon \psi \mathbf{u}' \cdot \mathbf{t} = i[R\psi + \psi'' - i\epsilon(\mathbf{N} \cdot \mathbf{u}')'] \tag{8.3.10}$$

$$R' = \frac{1}{2}(\psi'\psi^* + \psi\psi^{*\prime}) + \frac{1}{2}i\epsilon(\psi\mathbf{N}^* - \psi^*\mathbf{N}) \cdot \mathbf{u}'. \tag{8.3.11}$$

From (8.3.11) we recover Hasimoto's result that, for $\epsilon = 0$,

$$R = \frac{1}{2}|\psi|^2. \tag{8.3.12}$$

When inserted into (8.3.10) (with $\epsilon = 0$), this yields the cubic Schrödinger equation for the filament function

$$\frac{1}{i}\dot{\psi} = \psi'' + \frac{1}{2}|\psi|^2\psi. \tag{8.3.13}$$

More generally ($\epsilon \neq 0$), we can integrate (8.3.11) and insert the result into (8.3.10) to obtain the perturbed Schrödinger equation first derived in Klein and Majda (1991 I,II),

$$\frac{1}{i}\dot{\psi} = \psi'' + \frac{1}{2}|\psi|^2\psi - \epsilon\left(i\left[(\mathbf{N} \cdot \mathbf{u}')' - \psi\mathbf{u}' \cdot \mathbf{t}\right] + \psi \int_{s_0}^{s} \Im(\psi\mathbf{N}^*) \cdot \mathbf{u}' ds\right). \tag{8.3.14}$$

Remarks.

1. Note that we have not used the fact that ϵ is small in (8.3.14), hence it is an *exact* equation.

2. Upon integrating (8.3.11), one can introduce an arbitrary integration constant $A(t)$. However, it can be eliminated without loss through a phase shift in $\Phi(s, t)$, i.e., by defining

$$\hat{\Phi}(s, t) = \Phi(s, t) - \int_0^t A(t)dt. \qquad \diamond$$

Example 8.8 (Solitary wave on a filament) Based on our extensive knowledge of properties and solutions of the nonlinear Schrödinger equation (see Lamb (1980), or Newell (1985)), we can construct solutions that correspond to traveling loops along the filament. In the simplest case, we look for solutions to (8.3.13) that propagate steadily down the filament, with speed c, where we require the filament to be straight at infinity; hence

$$\kappa = 0 \quad \text{as} \quad s \to \pm\infty.$$

We then introduce the traveling variable

$$\xi = s - ct,$$

and assume that the filament function depends only on that variable, i.e., $\psi(\xi)$. Using

$$\psi(\xi) = \kappa(\xi) \exp\left[i \int_0^s \tau(\xi)ds\right]$$

in (8.3.13), the real and imaginary parts yield

$$-c\kappa[\tau(\xi) - \tau(-ct)] = \kappa'' - \kappa\tau^2 + \frac{1}{2}\kappa^2\kappa,$$

$$c\kappa' = 2\kappa'\tau + \kappa\tau'.$$

The second of these can be integrated to give

$$(c - 2\tau)\kappa^2 = 0.$$

Assuming κ is not identically zero, we obtain

$$\tau = \tau_0 = \frac{1}{2}c = const,$$

i.e., the torsion is constant along the filament and the velocity of the solitary wave is twice the torsion. The first equation can be integrated twice to yield

$$\kappa = \pm 2\tau_0 \text{sech}(\pm\tau_0\xi).$$

By substituting these values of κ and τ into the Serret–Frenet equations, the shape of the filament is then obtained. As shown in Hasimoto, this gives parametric expressions for the variables $(\mathbf{X}, \mathbf{t}, \mathbf{n}, \mathbf{b})$. We show in Figure 8.2 the filament shape associated with this single traveling loop, as obtained in Hasimoto's (1972) original paper and reprinted from Ricca (1996). $\diamond$

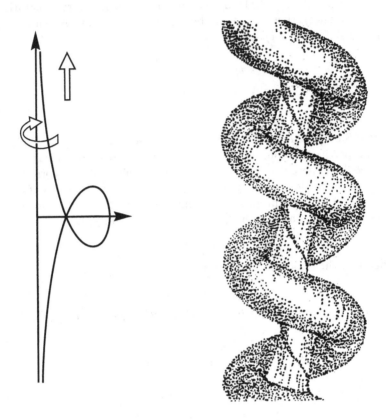

FIGURE 8.2. Single soliton loop on a filament.

Remarks.

1. Very interesting discussions of such solitary wave loops traveling on thin filaments can be found in Aref and Flinchem (1984) as well as Hopfinger and Browand (1982). In particular, Aref and Flinchem (1984) are concerned with solitary waves propagating on a filament with the additional effects of background shear, hence the perturbed binormal law (8.3.1) is relevant for their work. Hopfinger and Browand (1982) are the first to observe such solitary waves in the laboratory, where they do careful comparisons with the analytical model. See also Leibovich and Ma (1983).

2. Knowing the integrable properties of the nonlinear Schrödinger equation, one might look for N-soliton solutions traveling on the filament and interacting. These are described in Aref and Flinchem (1984), and a general prescription for obtaining them is described in Levi,

Sym, and Wojciechowski (1983). See also the work of Fukumoto and Miyazaki (1986, 1991) and Miyazaki and Fukumoto (1988).

3. Lamb (1977) has generalized the work of Hasimoto (1972) and is able to relate other well-known integrable equations, such as the Sine–Gordon equation, the Hirota equation, and the modified KdV equation, to the evolution of helical space curves.

4. Hasimoto (1972) gives a simple relationship between the nonlinear Schrödinger equation and Betchov's intrinsic equations. The two are related by the following transformations. If we let

$$\psi = \log W$$

$$W = \sqrt{K} \exp(\frac{1}{2} i \Phi)$$

$$\Phi = 2 \int_0^s \tau \, ds$$

in the nonlinear Schrödinger equation (8.3.13), one obtains

$$\frac{1}{2i}\left(\frac{\dot{K}}{K} + i\dot{\Phi}\right) = \frac{1}{2}\left(\frac{K''}{K} - \left(\frac{K'}{K}\right)^2 + i\Phi''\right) + \frac{1}{4}\left(\frac{K'}{K} + i\Phi'\right)^2 + \frac{1}{2}K.$$

Then, by comparing the real and imaginary parts and differentiating the real part with respect to s, one obtains Betchov's equations. ◊

We end this section by showing a calculation in Majda (1998) that linearizes the LIE about a straight line filament parallel to the z-axis. We write (8.1.11)

$$\dot{\mathbf{X}} = C_0 \mathbf{t} \times \mathbf{X}'' \tag{8.3.15}$$

where C_0 is a constant that sets the time scale. Consider small perturbations of the form

$$\mathbf{X} = \mathbf{X}_0 + \epsilon \mathbf{X}_1,$$

$$= (0, 0, s) + \epsilon(x(s,t), y(s,t), 0),$$

with $\epsilon << 1$. Inserting this into (8.3.15) and dropping higher powers in ϵ gives the linearized self-induction equation along a straight line filament,

$$\dot{\mathbf{X}}_1 = J\left[C_0 \mathbf{X}_1''\right], \tag{8.3.16}$$

where $\mathbf{X}_1 = (x(s,t), y(s,t))$, and J is the skew symmetric matrix

$$J = \begin{pmatrix} 0 & -1 \\ 1 & 0 \end{pmatrix}.$$

With the complex function $\psi(s,t) = x(s,t) + iy(s,t)$, we can write (8.3.16) as a linear Schrödinger equation

$$\frac{1}{i}\dot\psi = C_0\psi_{ss},$$

showing the dispersive character of small amplitude waves propagating along a straight filament.

Remarks.

1. The linearized self-induction equation around a straight line filament was first derived and used in Hama (1963), and around a circular ring by Arms and Hama (1965) and Kambe and Takao (1971).

2. Kida (1981) solves the LIE for solutions that evolve dynamically without changing form, i.e., relative equilibrium configurations. His general solution can be written in terms of Jacobian elliptic functions and contains as limiting cases all the previously discovered classical solutions, such as the circular vortex ring (Lamb (1932)), the helicoidal filament (Betchov (1965)), the planar sinusoidal filament (Kelvin (1880)), Euler's elastic (Hasimoto (1971)), and the solitary wave filaments of Hasimoto (1972). Particularly interesting are filaments that wind around a torus with two independent frequencies. This family includes ones that are closed (rationally related frequencies), as well as ones that densely fill the torus surface (irrationally related frequencies). A linearized stability analysis of all of these structures is carried out in Kida (1982).   ◇

8.4 LIA Invariants

Since the system of equations obtained via the LIA is not identical to the full Euler equations, one might well ask what are the conserved quantities associated with the LIE or the DaRios–Betchov system. As discussed in Section 1.1.2, the conserved quantities for the Euler equations are the total kinetic energy E and total vorticity W_1,

$$E = \frac{1}{2}\int_{\mathbb{R}^3}\|\mathbf{u}\|^2 d\mathbf{x} \quad = E(0)$$

$$W_1 = \int_{\mathbb{R}^3}\omega(\mathbf{x})d\mathbf{x} \quad = W_1(0).$$

In addition, one also has conservation of linear momentum L, angular momentum I, and helicity J, where

$$L(t) = \frac{1}{2} \int_{\mathbb{R}^3} (\mathbf{x} \times \boldsymbol{\omega}) \, d\mathbf{x} \quad = L(0)$$

$$I(t) = -\frac{1}{2} \int_{\mathbb{R}^3} \|\mathbf{x}\|^2 \boldsymbol{\omega} \, d\mathbf{x} \quad = I(0),$$

$$J(t) = \int_{\mathbb{R}^3} \mathbf{u} \cdot \boldsymbol{\omega} \, d\mathbf{x} \quad\quad = J(0).$$

We know from our previous discussions that the total length l of a vortex filament is conserved under the LIA, as is the linear momentum $L(t)$ for a closed filament loop. It is straightforward to show that the total vorticity W_1 is conserved under LIA as well. It is shown in Fukumoto (1987) that the angular momentum $I(t)$ and the helicity $J(t)$ for closed filament loops are also conserved under the LIE. In fact, the helicity invariant $J(0)$ must have value zero under LIA owing to the fact that one ignores the far-field effect of the filament at any given point along it, hence global information about the degree of knottedness of a filament under LIA is lost. It is interesting to point out that despite the fact that the total enstrophy of the three-dimensional Euler equations

$$W_2 = \int_{\mathbb{R}^3} \|\boldsymbol{\omega}\|^2 \, d\mathbf{x}$$

is not conserved, it is conserved under the LIE, as discussed and shown in Ricca (1992). He also shows that the LIE conserves the total kinetic energy of the flow. Hence, all of the known conserved quantities for the Euler equations are preserved under the LIA.

There are additional conserved quantities associated with the LIE that are not present in the Euler equations. For example, from the conservation form of (8.2.3) we obtain that the total torsion [3] associated with a filament is constant,

$$T_0 \equiv \int_X \tau ds = const$$

To obtain additional, more subtle invariants, one can make use of the integrability of the cubic Schrödinger equation (8.3.13) which is known to lead to an infinite number of polynomial conservation laws (Zakharov and Shabat (1972)), as discussed, for example, in Newell (1985) or Lamb (1980). One can write down a recursion formula (see Moffatt and Ricca (1991)) for

---

[3]The total torsion of a filament measures the global twisting of the Seret–Frenet orthonormal frame along the filament centerline.

the conserved quantities as follows:

$$f_{n+1} = q\frac{\partial}{\partial s}\left(\frac{1}{q}f_n\right) + \sum_{j+k=n} f_j f_k,$$

where $f_n(s,t) \in \mathbb{C}$, $n = 1, 2, \ldots$, and

$$q = \frac{1}{2}i\psi$$
$$f_1 = \frac{1}{4}|\psi|^2 = \frac{1}{4}\kappa^2.$$

The integral invariants are then written in terms of f_n,

$$(2i)^n I_n \equiv \int_X f_n(s,t)ds = const$$

As obtained by Ricca (1992), the first five of these are:

$$iI_1 = \frac{1}{8}\int_X |\psi|^2 ds = \frac{1}{8}\int_X \kappa^2 ds$$

$$iI_2 = -\frac{1}{32}\int_X (\psi^*\psi' - \psi\psi^{*\prime})ds = -\frac{1}{16}\int_X \kappa^2\tau ds$$

$$iI_3 = \frac{1}{32}\int_X \left(\frac{\kappa^4}{4} - (\kappa')^2 - \kappa^2\tau^2\right) ds$$

$$iI_4 = -\frac{1}{64}\int_X \left(\frac{3}{4}\kappa^4 - (\kappa')^2 - \kappa^2\tau^2 + 2\kappa\kappa''\right)\tau ds$$

$$iI_5 = \frac{1}{128}\int_X \left[\kappa^2\left(\frac{\kappa^4}{8} - (\kappa')^2\right) - \frac{3}{2}\kappa(\kappa')^2 + (\kappa'')^2 + \kappa^2\left(\tau^4 - \frac{3}{2}\kappa\tau + (\tau')^2\right)\right.$$
$$\left. - 2\tau^2(2\kappa\kappa'' + (\kappa')^2)\right] ds.$$

The first shows that the total squared curvature is conserved. Fukumoto (1987) uses this to prove the following theorem regarding the degree of knottedness of a closed filament.

Theorem 8.4.1 (Fukumoto (1987)). *If initially we have*

$$\left(\int_X \kappa^2 ds\right)^{1/2} l^{1/2} \leq 4\pi,$$

then the closed filament is unknotted and always remains unknotted.

More génerally, Moffatt and Ricca (1991) obtain the following bounds on the total curvature.

Theorem 8.4.2 (Moffatt and Ricca (1991)). *If one defines the total curvature of a filament as*

$$\int_X \kappa ds,$$

then

1. *For an unknotted filament, the total curvature is bounded above and below by*

$$2\pi \leq \int_X \kappa ds \leq \left(l \int_X \kappa^2 ds \right)^{1/2} = const$$

2. *For a knotted filament, the total curvature is bounded above and below by*

$$4\pi \leq \int_X \kappa ds \leq \left(l \int_X \kappa^2 ds \right)^{1/2} = const$$

The physical meaning of the higher invariants I_2 and I_3 are interpreted by Ricca (1992) as the "pseudohelicity" and the Lagrangian associated with the LIE. We refer the interested reader to these papers for more thorough discussion.

8.5   Vortex-Stretching Models

In this section we describe recent work whose main goal is to improve the LIE (8.1.12) by including higher order effects in the velocity field associated with a vortex filament. The motivation for these higher order theories is based on the main deficiency of the LIE, that it does not incorporate any self-stretching of an isolated filament. This stands in contrast to observations of three-dimensional turbulent flows that conclude vortex-stretching plays a prominent role in the generation of small scales (Siggia (1985), Chorin (1982)). Our description in this section is based on work of Callegari and Ting (1978) (see also the monograph of Ting and Klein (1991)), and more recently Klein and Majda (1991 I,II), which seeks to improve upon the LIE by including higher order effects. We follow closely the exposition given in Majda (1998).

We start by making the simple observation regarding self-stretching of a filament $\mathbf{X}(s,t)$, which evolves under the generic flow

$$\dot{\mathbf{X}} = \beta(s,t)\mathbf{n} + \gamma(s,t)\mathbf{b} + \alpha(s,t)\mathbf{t}.$$

The LIA corresponds to the choice $\beta = 0, \gamma = \kappa, \alpha = 0$. By differentiating this equation with respect to s, along with use of the Serret–Frenet equations, we obtain

$$\frac{\partial l}{\partial t} = (-\kappa\beta + \alpha')l \tag{8.5.1}$$

where

$$l \equiv (\mathbf{X}' \cdot \mathbf{X}')^{1/2}.$$

Hence, if $\alpha = const$ and $\beta = 0$, i.e., the tangential component of velocity is constant (with respect to s) and the normal component is zero, then there is no vortex-stretching. It is interesting to point out that if the normal and tangential components are balanced so that

$$\alpha' = \kappa\beta,$$

then there is also no stretching, although this is probably a pathological situation. Hence, higher order theories that attempt to remedy the deffi- ciencies of the LIA must avoid these velocity fields. This result also shows, however, that if the LIE is used in conjunction with a suitable background flow, vortex-stretching does occur, a fact discussed and used in Aref and Flinchem (1984), for example.

To compute the velocity field near a vortex filament, one must expand the Biot–Savart integral in the neighborhood of the filament and keep enough terms to improve upon the leading order LIA. This was done originally in Bliss (1970), as reviewed in Widnall (1975). It was done systematically in Callegari and Ting (1978) by using a matched asymptotic expansion procedure, where they derive the following expression for the local velocity field near a filament

$$\mathbf{u} = \frac{\Gamma}{2\pi\sigma}\mathbf{q} + \frac{\Gamma}{4\pi R} \log\left(\frac{L}{\sigma}\right)\mathbf{b} + \mathbf{Q}. \qquad (8.5.2)$$

The first term is simply the rotational velocity around the filament axis, while the second is the binormal component already discussed in the LIA. The last term, $\mathbf{Q}$, denotes the finite term contributions due to induction effects from distant parts of the filament that were ignored in using the cut- off procedure. The self-induced filament motion (suitably scaled), derived in Callegari and Ting (1978), then has the form

$$\dot{\mathbf{X}} = \kappa\mathbf{b}\left[\log(1/\sigma) + \tilde{C}(t)\right] + \mathbf{Q}(s,t) + O(1).$$

Remarks.

1. Callegari and Ting's (1978) analysis relies on asymptotics based on two length scales in the flow. The smallest scale, σ, is the vortex core thickness, and the large scale, R, is the radius of curvature of the filament. It is assumed that $\sigma << R$.

2. The contribution $\mathbf{Q}$ is a nonlocal, nonlinear functional of the filament curve, while the function $\tilde{C}(t)$ is obtained by asymptotically matching

the induced velocity near the filament to an inner solution valid inside the viscous vortex core. Viscous effects enter through a balancing of the core size σ and the Reynolds number R_e, where it is assumed $\sigma = R_e^{-1/2}$.

3. The LIE (8.1.12) takes this form upon dropping $\mathbf{Q}$ and rescaling time to absorb the coefficient $\left[\log(1/\sigma) + \tilde{C}(t)\right]$.

4. If we choose $\alpha' = 0$ in the stretching formula (8.5.1), then the local rate of stretching can be written

$$\frac{\dot{l}}{l} = -\kappa \left(\dot{\mathbf{X}} \cdot \mathbf{n}\right),$$

which shows that the self-stretching is proportional to $\mathbf{Q} \cdot \mathbf{n}$. Therefore any higher order theories based on expansions of the Biot–Savart law will lead to nonlocal self-stretching terms. $\Diamond$

A recent theory that includes self-stretching effects was proposed by Klein and Majda (1991 I, II). Their idea involves the introduction of a third length scale in the problem, which allows the filament to have perturbation wavelengths of scale ϵ that are small relative to the radius of curvature of the filament, which is normalized to unity, hence $\epsilon \ll 1$. This length scale is in turn large relative to the core thickness, hence $\sigma \ll \epsilon$. The perturbations are small amplitude and short wavelength perturbations of a straight line filament aligned with the z-axis and take the form

$$\mathbf{X}(s,t) = s\hat{z} + \epsilon^2 \mathbf{X}_2(\frac{s}{\epsilon}, \frac{t}{\epsilon^2}) + O(\epsilon^2)$$

$$\mathbf{X}_2 = \left(x^{(2)}(\frac{s}{\epsilon}, \frac{t}{\epsilon^2}), y^{(2)}(\frac{s}{\epsilon}, \frac{t}{\epsilon^2}), 0 \right).$$

Their analysis requires the following balance of core thickness, Reynolds number, and perturbation wavelength:

1. $\epsilon^2 [\log(2\epsilon/\sigma) + c] = 1$.

2. $\sigma = R_e^{-1/2}$.

3. $\sigma \ll 1$.

The second condition is that introduced in Callegari and Ting (1978). Under these assumptions the following perturbed binormal law emerges:

$$\dot{\mathbf{X}} = \kappa \mathbf{b} + \epsilon^2 I[\mathbf{X}_2] \times \hat{z}, \tag{8.5.3}$$

where $I[w]$ is a *linear* nonlocal operator defined as

$$I[w] \equiv \int_{-\infty}^{x} \frac{1}{|h^3|} \Big[w(\sigma + h) - w(\sigma) - h w_\sigma(\sigma + h)$$
$$+ \frac{1}{2} h^2 H(1 - |h|) w_{\sigma\sigma}(\sigma) \Big] \, dh,$$

where H is the Heaviside function. Furthermore, if one subjects this nonlocal perturbed binormal evolution equation to the Hasimoto transformation via the filament function ψ, one obtains the following perturbed, nonlocal, nonlinear Schrödinger equation

$$\frac{1}{i} \psi_{\hat{t}} = \psi_{\hat{s}\hat{s}} + \epsilon^2 \left(\frac{1}{2} |\psi|^2 \psi - I[\psi] \right),$$

where $\hat{s} = s/\epsilon^2$, $\hat{t} = t/\epsilon^2$. What is interesting here is that the cubic term is of the same order of magnitude as the nonlocal term. Furthermore, it is shown in Klein and Majda (1991 I,II) that the local curve-stretching associated with this model is

$$\frac{\dot{l}}{l} = \frac{\epsilon^2}{4} i \int_{-\infty}^{\infty} \frac{1}{|h|} \left[\psi^*(\sigma + h) \psi(\sigma) - \psi(\sigma + h) \psi^*(\sigma) \right] dh.$$

The price one pays for incorporating self-stretching of a filament is to replace the relatively simple LIE (8.1.12) with the nonlocal, integro-differential equation (8.5.3). Properties of this equation have been studied in Klein and Majda (1991 I,II), Klein and Majda (1993), and Klein, Majda, and Damodaran (1995). We also refer the reader to the paper of Zhou (1997), in which careful numerical comparisons are made between the various models for filament evolution. In general, it is an interesting question to understand the differences between the dynamics of various structures under the full Euler equations via the Biot–Savart law versus their evolution under the various approximations of this exact law, the most radical of these being the LIA. These issues are not yet fully understood but discussions and some results can be found in Ricca et al (1999) for torus-knot evolution and also in Zhou (1997).

8.6   Nearly Parallel Filaments

In this section, we describe a set of equations that model the interaction of a collection of N filaments nearly but not perfectly parallel to the z-axis. Hence, consider the centerline associated with each filament,

$$\mathbf{X}_j(s,t) = (x_j(s,t), y_j(s,t)), \qquad 1 \leq j \leq N,$$

where s parametrizes the z direction. We know that the dynamics associated with the jth filament is determined by the total velocity field, according to (8.1.6). The model, introduced by Klein, Majda, and Damodaran (1995), is based on the following idea of decomposing the velocity field $\mathbf{u} = \mathbf{u}_1 + \mathbf{u}_2$:

1. Since each of the filaments is nearly parallel, we model the self-interaction of each one by using the *linearized* self-induction equation; hence we take the velocity field $\mathbf{u}_1$ to be

$$\mathbf{u}_1 = J\Gamma_j \frac{\partial^2 \mathbf{X}_j}{\partial s^2}, \qquad 1 \le j \le N \qquad (8.6.1)$$

$$J = \begin{pmatrix} 0 & 1 \\ -1 & 0 \end{pmatrix}. \qquad (8.6.2)$$

2. The mutual interaction of the filaments is modeled by the point vortex equations for exactly parallel filaments, hence we take $\mathbf{u}_2$ to be

$$\mathbf{u}_2 = J \left[\sum_{k \neq j}^{N} 2\Gamma_k \frac{(\mathbf{X}_j - \mathbf{X}_k)}{|\mathbf{X}_j - \mathbf{X}_k|^2} \right], \qquad 1 \le j \le N.$$

Then, the equations of motion for N interacting nearly parallel filaments is given by the coupled partial differential equations

$$\frac{\partial \mathbf{X}_j}{\partial t} = \mathbf{u}_1 + \mathbf{u}_2 \qquad (8.6.3)$$

$$= J \left[\Gamma_j \frac{\partial^2 \mathbf{X}_j}{\partial s^2} \right] + J \left[\sum_{k \neq j}^{N} 2\Gamma_k \frac{(\mathbf{X}_j - \mathbf{X}_k)}{|\mathbf{X}_j - \mathbf{X}_k|^2} \right]. \qquad (8.6.4)$$

The details of the derivation, including the precise asymptotic assumptions, are described in Klein, Majda, and Damodaran (1995).

The system (8.6.4) inherits many of the features of the full Euler equations, including its Hamiltonian structure. In particular, (8.6.4) can be written

$$\Gamma_j \frac{\partial \mathbf{X}_j}{\partial t} = J \frac{\delta \mathcal{H}}{\partial \mathbf{X}_j}, \qquad 1 \le j \le N,$$

where the Hamiltonian is given by

$$\mathcal{H} = -\sum_{j=1}^{N} \frac{\alpha_j \Gamma_j^2}{2} \int |\frac{\partial \mathbf{X}_j}{\partial s}|^2 ds + 2 \sum_{j<k}^{N} \int \Gamma_j \Gamma_k \log |\mathbf{X}_j(s) - \mathbf{X}_k(s)| ds.$$

In addition to the Hamiltonian, one has the following conserved quantities:

$$M = \int \sum_{j=1}^{N} \Gamma_j \mathbf{X}_j(s) ds \qquad (8.6.5)$$

$$A = \int \sum_{j=1}^{N} \Gamma_j |\mathbf{X}_j(s)|^2 ds, \qquad (8.6.6)$$

where M can be interpreted as the mean center of vorticity of the system, and A the mean angular momentum. An additional conserved quantity is

$$W = \int \sum_{j=1}^{N} \Gamma_j (J\mathbf{X}_j(s)) \cdot \frac{\mathbf{X}_j(s)}{\partial s} ds. \qquad (8.6.7)$$

Finally, if we define the average distance functional

$$I(t) = \frac{1}{2} \int \sum_{j,k} \Gamma_j \Gamma_k (\Gamma_k - \Gamma_j) \int \left(\frac{\partial x_j}{\partial s} \frac{\partial y_k}{\partial s} - \frac{\partial x_k}{\partial s} \frac{\partial y_j}{\partial s} \right) ds, \quad (8.6.8)$$

and in particular, if all the filaments are identical, i.e., $\Gamma_k = \Gamma_j$, then

$$\frac{dI}{dt} = 0.$$

Example 8.9   (Interacting filament pairs) We now specialize to the case of two interacting, nearly parallel filaments. Without loss, we can assume one has circulation $\Gamma_1 = 1$ and the other $\Gamma_2 = \Gamma$, where $-1 \le \Gamma \le 1$ ($\Gamma \ne 0$). Writing the coordinates in complex form,

$$\psi_j = x_j(s,t) + iy_j(s,t),$$

one arrives at the coupled nonlinear Schrödinger system

$$\frac{1}{i} \frac{\partial \psi_1}{\partial t} = \frac{\partial^2 \psi_1}{\partial s^2} + 2\Gamma \frac{\psi_1 - \psi_2}{|\psi_1 - \psi_2|^2} \qquad (8.6.9)$$

$$\frac{1}{i} \frac{\partial \psi_2}{\partial t} = \Gamma \frac{\partial^2 \psi_2}{\partial s^2} - 2 \frac{\psi_1 - \psi_2}{|\psi_1 - \psi_2|^2}. \qquad (8.6.10)$$

Using the variables

$$\psi_1 = \frac{(\phi + \psi)}{2} \qquad \psi_2 = \frac{(\phi - \psi)}{2},$$

one arrives at

$$\frac{1}{i} \phi_t = \frac{(1 + \Gamma)}{2} \phi_{ss} + \frac{(1 - \Gamma)}{2} \left[\psi_{ss} - 4 \frac{\psi}{|\psi|^2} \right] \qquad (8.6.11)$$

$$\frac{1}{i} \psi_t = \frac{(1 - \Gamma)}{2} \phi_{ss} + \frac{(1 + \Gamma)}{2} \left[\psi_{ss} + 4 \frac{\psi}{|\psi|^2} \right]. \qquad (8.6.12)$$

First, consider the special case of planar point vortex interactions, where ψ_1 and ψ_2 are independent of the spatial variable. In this case, the coupled system reduces to

$$\frac{1}{i}\phi_t = \frac{2(\Gamma - 1)\psi}{|\psi|^2}$$

$$\frac{1}{i}\psi_t = \frac{2(\Gamma + 1)\psi}{|\psi|^2}.$$

Suppose we place the vortices a distance d apart initially, that is,

$$\psi_1(0) = (\frac{d}{2}, 0)$$

$$\psi_2(0) = (-\frac{d}{2}, 0),$$

then for $\Gamma \neq -1$, we have the exact solutions

$$\phi(t) = d\left[1 - \exp\left(\frac{2i(1+\Gamma)t}{d^2}\right)\right]\left(\frac{1-\Gamma}{1+\Gamma}\right),$$

$$\psi(t) = d\exp\left(\frac{2i(1+\Gamma)t}{d^2}\right).$$

This is the generic configuration of two single point vortices moving in concentric circles about their centers of vorticity. The case where $\Gamma = -1$ is solved by

$$\phi(t) = -\frac{4i}{d^2}t$$

$$\psi(t) = (d, 0),$$

which corresponds to uniform translation of the vortex pair.

More generally, allowing for spatial dependence, one can show that the system (8.6.9), (8.6.10) has the following conserved quantities:

$$\mathcal{H} = -\frac{1}{2}\int|\frac{\partial\psi_1}{\partial s}|^2 ds - \frac{\Gamma^2}{2}\int|\frac{\partial\psi_2}{\partial s}|^2 ds$$

$$+ 2\int\Gamma\log|\psi_1 - \psi_2|ds \qquad (8.6.13)$$

$$M = \int\psi_1 ds + \int\Gamma\psi_2 ds \qquad (8.6.14)$$

$$A = \int|\psi_1|^2 ds + \int\Gamma|\psi_2|^2 ds \qquad (8.6.15)$$

$$W = i\int\psi_1\frac{\partial\psi_1^*}{\partial s} + \Gamma\psi_2\frac{\partial\psi_2^*}{\partial s}ds. \qquad (8.6.16)$$

If $\Gamma = 1$ so that the pairs are corotating, we have

$$I = \int |\psi_1 - \psi_2|^2 ds.$$

In this case, the equations decouple as

$$\frac{1}{i}\phi_\tau = \phi_{ss}$$
$$\frac{1}{i}\psi_\tau = \psi_{ss} + 4\frac{\psi}{|\psi|^2}.$$

Letting

$$\psi = B\exp[i(ks + \omega t)]$$

gives the dispersion relation

$$\omega = \frac{4}{B^2} - k^2$$

and the family of exact solutions

$$\psi_1 = \phi(s,t) + \frac{B}{2}\exp[i(ks + \omega t)]$$
$$\psi_2 = \phi(s,t) - \frac{B}{2}\exp[i(ks + \omega t)],$$

where $\phi(s,t)$ is an arbitrary solution to the linear Schrödinger equation.

$\Diamond$

We can now linearize the filament equations (8.6.11), (8.6.12) around these exact solutions and prove the following theorem of Klein, Majda, and Damodaran (1995).

Theorem 8.6.1 (Klein, Majda, Damodaran (1995)). *For any negative circulation ratio* $-1 \le \Gamma \le 0$, *exactly parallel filament configurations are unstable to long wavelength perturbations. For any positive circulation ratio* $0 < \Gamma \le 1$, *the filaments are neutrally stable.*

The same paper goes on to numerically study the system in the nonlinear regime, providing evidence for self-similar, finite-time collapse if the filament equations have negative circulation ratio, with no collapse for positive ratios. See also Klein and Majda (1993) and a recent statistical equilibrium mechanics formulation for nearly parallel filaments in Lions and Majda (2000).

8.7   The Vorton Model

There is a growing body of work that goes under the name of "vortex particle methods" in which the motion of a two-dimensional point vortex is generalized to a three-dimensional *vorton* (Novikov (1983)). The vorticity field is represented as

$$\omega(\mathbf{x}, t) = \sum_{i=1}^{N} \alpha_i(t) \delta(\mathbf{x} - \mathbf{x}_i(t)). \tag{8.7.1}$$

The evolution equations for the vorton positions $\mathbf{x}_i(t)$ and the strengths $\alpha_i(t)$ are given by (see Winckelmans and Leonard (1993))

$$\dot{\mathbf{x}}_i = \mathbf{u}_i(\mathbf{x}_i(t), t) \tag{8.7.2}$$

$$\dot{\alpha}_i = (\alpha_i \cdot \nabla) \, \mathbf{u}_i(\mathbf{x}_i, t). \tag{8.7.3}$$

Here, $\mathbf{u}_i$ represents the velocity field without the ith element. Alternative formulations of (8.7.3) include the so-called transpose particle equation,

$$\dot{\alpha}_i = \left(\alpha_i \cdot \nabla^{\perp} \right) \mathbf{u}_i(\mathbf{x}_i, t), \tag{8.7.4}$$

and the mixed particle equation

$$\dot{\alpha}_i = \frac{1}{2} \left(\alpha_i \cdot (\nabla + \nabla^{\perp}) \right) \mathbf{u}_i(\mathbf{x}_i, t). \tag{8.7.5}$$

There are several unusual features of these three-dimensional models that we mention here without proof. First, the vorticity decomposition (8.7.1) is not divergence-free and as a result, neither is the corresponding streamfunction. This causes some problems, since then the vorticity field is generally not the curl of the velocity field, as it should be. Thus, since $\omega \neq \nabla \times \mathbf{u}$, the three formulations (8.7.3), (8.7.4), and (8.7.5) are not equivalent. In particular, the classical scheme (8.7.3) and the mixed scheme (8.7.5) do not lead to the conservation of total vorticity, whereas the transpose scheme (8.7.4) does. In addition, if one defines the delta function as a singular limit of a suitably regularized radially symmetric sequence of functions, as pointed out in Greengard and Thomann (1988), the transpose scheme is a weak solution of the vorticity evolution equations, whereas the other two are not. See Saffman and Meiron (1986) and Winckelmans and Leonard (1988) for further discussion of this issue. Hence, the transpose vorton method (8.7.4) is the most promising from both a theoretical and a numerical point of view. Some of the original work on these ideas can be found in Novikov (1983) and Aksman, Novikov, and Orszag (1985). A comprehensive discussion of these methods, along with appropriate references, can be found in the review paper of Winckelmans and Leonard (1993) and the book of Cottet and Koumoutsakos (2000).

8.8 Exercises

1. Derive the curvature and torsion formulas (8.1.5).

2. Derive formula (8.1.8) and use it to prove that the velocity field of a curved filament diverges logarithmically near the filament. [Hint: See Arms and Hama (1965).]

3. Prove Theorem 8.1.1.

4. Prove that a helicoid will move with translation velocity V_T and rotational velocity V_R in the ratio

$$\frac{V_R}{V_T} = \frac{\tau}{K^{1/2}}$$

   under LIE.

5. Prove that a single-looped filament rotates about its axis at a velocity proportional to κ.

6. Go though the details of the derivation of the cubic Schrödinger equation (8.3.13) for the filament function ψ.

7. Derive the nonlocal perturbed Schrödinger equation (8.3.14) governing the filament function ψ.

8. Prove that the total vorticity W_1, angular momentum I, and helicity J for a filament evolving under the LIE are conserved.

9. Prove that the helicity for a closed filament evolving under the LIE is zero.

10. Prove that the total enstrophy W_2 and the total kinetic energy are both conserved under the LIE.

11. Derive the five integral invariants $I_1, \dots, I_5$ obtained by the recursion formula based on the cubic Schrödinger equation, and show that I_3 corresponds to the Lagrangian associated with the nonlinear Schrödinger equation.

12. Derive the general stretching formula (8.5.1).

13. Using model (8.6.4) for the evolution of nearly parallel filaments, prove that M, A, and W are conserved, as defined by (8.6.5), (8.6.6), (8.6.7). Also derive the formula (8.6.8) for $I(t)$.

14. Derive the coupled Schrödinger system (8.6.9), (8.6.10) for interacting filament pairs and verify the conserved quantities (8.6.13), (8.6.14), (8.6.15), (8.6.16).

15. (a) Derive the vorton evolution equations (8.7.2), (8.7.3).

    (b) Derive the alternative formulations (8.7.4), (8.7.5).

References

[1] R.H. Abraham, J.E. Marsden (1978). *Foundations of Mechanics*, Benjamin/Cummings, Menlo Park, CA.

[2] R. H. Abraham, J.E. Marsden, T.Ratiu (1988). *Manifolds, Tensor Analysis, and Applications*, 2nd ed., Springer–Verlag, New York.

[3] M. Adams, T. Ratiu (1988). The three–point vortex problem: commutative and noncommutative integrability, in "Hamiltonian Dynamical Systems," *Cont. Math.* **81**, 245–257, American Mathematical Society, Providence.

[4] Y. Aharonov, J. Anandan (1987). Phase change during a cyclic quantum evolution, *Phys. Rev. Lett.* **58**(16), 1593–1596.

[5] L.A. Ahlfors (1979). *Complex Analysis*, McGraw-Hill, New York.

[6] M.J. Aksman, E.A. Novikov, S.A. Orszag (1985). Vorton method in three dimensional hydrodynamics, *Phys. Rev. Lett.* **54**, 2410–2413.

[7] M.S. Alber, J.E. Marsden (1992). On geometric phases for soliton equations, *Comm. Math. Phys.* **149**, 217–240.

[8] J. Anandan, L. Stodolsky (1987). Some geometric considerations of Berry's phase, *Phys. Rev.* **D** **35**(8), 2597–2600.

[9] J.L. Anderson (1992). Multiple time scale methods for adiabatic systems, *Am. J. Phys.* **60**(10), 923–927.

[10] C. Anderson, C. Greengard (1989). The vortex ring merger problem at infinite Reynolds number, *Comm. Pure App. Math.* **42**(8), 1123–1139.

[11] H. Aref (1979). Motion of three vortices, *Phys. Fluids* **22**(3), 393–400.

[12] H. Aref (1982). Point vortex motions with a center of symmetry, *Phys. Fluids* **25**, 2183–2187.

[13] H. Aref (1983). Integrable, chaotic, and turbulent vortex motion in two–dimensional flows, *Ann. Rev. Fluid Mech.* **25**, 345–389.

[14] H. Aref (1985). Chaos in the dynamics of a few vortices – fundamentals and applications, in *Theoretical and Applied Mechanics IUTAM 1985* F.I. Niordson, N. Olhoff, eds., 43–68, Elsevier Publishers, Amsterdam.

[15] H. Aref (1989). Three vortex motion with zero total circulation–addendum, *J. App. Math. Phys. (ZAMP)*, **40**, 495–500.

[16] H. Aref (1992). Trilinear coordinates in fluid mechanics, in *Studies in Turbulence*, T.B. Gatski, S. Sarkar, C.G. Speziale, eds., 568–581, Springer–Verlag, New York.

[17] H. Aref (1995). On the equilibrium and stability of a row of point vortices *J. Fluid Mech.* **290**, 167–181.

[18] H. Aref, M. Bröns (1998). On stagnation points and streamline topology in vortex flows, *J. Fluid Mech.* **370**, 1–27.

[19] H. Aref, E.P. Flinchem (1984). Dynamics of a vortex filament in a shear flow, *J. Fluid Mech.* **148**, 477–497.

[20] H. Aref, S.W. Jones, S. Mofina, I. Zawadzki (1989). Vortices, kinematics and chaos, *Physica D* **37**, 423–440.

[21] H. Aref, J.B. Kadtke, I. Zawadzki, L.J. Campbell (1988). Point vortex dynamics – recent results and open problems, *Fluid Dyn. Res.* **3**, 63–74.

[22] H. Aref, N. Pomphrey (1980). Integrable and chaotic motions of four vortices, *Phys. Lett.* **78A**(4), 297–300.

[23] H. Aref, N. Pomphrey (1982). Integrable and chaotic motions of four vortices I. The case of identical vortices, *Proc. Roy. Soc. Lond.* **A 380**, 359–387.

[24] H. Aref, N. Rott, H. Thomann (1992). Gröbli's solution of the three vortex problem, *Ann. Rev. Fluid Mech.* **24**, 1–20.

[25] H. Aref, M.A. Stremler (1996). On the motion of three point vortices in a periodic strip, *J. Fluid Mech.* **314**, 1–25.

[26] H. Aref, M.A. Stremler (1999). Four–vortex motion with zero total circulation and impulse, *Phys. Fluids* **11**(12), 3704–3715.

[27] H. Aref, D.L. Vainchtein (1998). Point vortices exhibit asymmetric equilibria, *Nature* **392**, 23 April, 769–770.

[28] H. Aref, I. Zawadzki, Linking of vortex rings (1991). *Nature* **354**, 7 Nov., 50–53.

[29] R.J. Arms, F.R. Hama (1965). Localized induction concept on a curved vortex and motion of an elliptic vortex ring, *Phys. Fluids* **8**, 553–559.

[30] V.I. Arnold (1961). Small denominators I. Mapping the circle onto itself, *Izv. Akad. Nauk SSSR Ser. Mat.* **25**, 21–86.

[31] V.I. Arnold (1962). On the behavior of an adiabatic invariant under slow periodic variation of the Hamiltonian, *Soviet Math. Dok.* **3**(1), 136–139.

[32] V.I. Arnold (1963a). Proof of a theorem of A.N. Kolmogorov on the invariance of quasi-periodic motions under small perturbations of the Hamiltonian, *Russ. Math. Surveys* **18**(5), 9–36.

[33] V.I. Arnold (1963b). Small denominators and problems of stability of motion in classical and celestial mechanics, *Russ. Math. Surveys* **18**(6), 85–192.

[34] V.I. Arnold (1964). Instability of dynamical systems with several degrees of freedom, *Soviet Math. Dok.* **5**(3), 581–585.

[35] V.I. Arnold (1965). Conditions for nonlinear stability of stationary plane curvilinear flows of an ideal fluid, *Dokl. Nat. Nauk.* **162**, 773–777.

[36] V.I. Arnold (1966a). Sur un principe variationel pour les écoulements stationaires des liquides parfaits et ses applications aux problemes des stabilité non–lineares, *J. Mécanique* **5**, 29–43.

[37] V.I. Arnold (1966b). Sur la geometrie differentielle des groupes de Lie de dimension infinie et ses applications á l'hydrodynamique des fluids parfait, *Ann. Inst. Fourier (Grenoble)* **16**, 319–361.

[38] V.I. Arnold (1966c). On an a priori estimate in the theory of hydrodynamic stability, *American Mathematical Society Transl.* **79**, 267–269.

[39] V.I. Arnold (1969). Hamilton character of the Euler equations of solid body and ideal fluid, *Usp. Mat. Nauk* **24**(3), 225–226.

[40] V.I. Arnold (1974). The asymptotic Hopf invariant and its applications, (Russian), *Sel. Math. Sov.* **5**, 327 (English).

[41] V.I. Arnold (1978). *Mathematical Methods of Classical Mechanics*, Springer–Verlag, New York.

[42] V.I. Arnold (1988). ed. *Encyclopedia of Mathematical Science, Vol. 3, Dynamical Systems III*, Springer–Verlag, New York.

[43] V.I. Arnold (1988). *Geometrical Methods in the Theory of Ordinary Differential Equations*, 2nd ed., Springer–Verlag, New York.

[44] V.I. Arnold, A. Avez (1989). *Ergodic Problems in Classical Mechanics*, Advanced Book Classics, Addison–Wesley, Redwood City.

[45] V.I. Arnold, B.A. Khesin (1992). Topological methods in hydrodynamics, *Ann. Rev. Fluid Mech.* **24**, 145–166.

[46] V.I. Arnold, B.A. Khesin (1998). *Topological Methods in Hydrodynamics*, **AMS 125**, Springer–Verlag, New York.

[47] A. Babiano, G. Boffetta, G. Provenzale, A. Vulpiani (1994). Chaotic advection in point vortex models and two-dimensional turbulence, *Phys. Fluids* **6**, 2465–2474.

[48] A.A. Bagrets, D.A. Bagrets (1997). Nonintegrability of two problems in vortex dynamics, *Chaos* **7**(3), 368–375.

[49] A.C. Bakai, Y.P. Stepanovski (1981). *Adiabatic invariants*, Naukova Dumka, Kiev.

[50] C. Bardos (1972). Existence et unicité de la solution de l'équation d'Euler en dimension deux, *J. Math. Anal. App.* **40**, 769–790.

[51] J. Barrow–Green (1997). *Poincaré and the Three Body Problem*, History of Math. **11**, American Mathematical Society, Providence.

[52] G.K. Batchelor (1967). *An Introduction to Fluid Mechanics*, Cambridge University Press, Cambridge.

[53] A. Battacharjee, T. Sen (1988). Geometric angles in cyclic evolutions of a classical system, *Phys. Rev. A* **38**, 4389–4394.

[54] L. Bauer, G.K. Morikawa (1976). Stability of rectilinear geostrophic vortices in stationary equilibrium, *Phys. Fluids* **19**, 929–942.

[55] J.T. Beale, T. Kato, A. Majda (1984). Remarks on the breakdown of smooth solutions for the three dimensional Euler equations, *Comm. Math. Phys.* **94**, 61–66.

[56] J.T. Beale, A. Majda (1982). Vortex methods. I. Convergence in three dimensions, Vortex methods. II. Higher order accuracy in two or three dimensions, *Math. Comp.* **39**, 1–52.

[57] D. Beigie, A. Leonard, S. Wiggins (1994). Invariant manifold templates for chaotic advection,*Chaos, Solitons, and Fractals* **4**(6), 749–868.

[58] C.M. Bender, S.A. Orszag (1978). *Advanced Mathematical Methods for Scientists and Engineers*, McGraw–Hill, New York.

[59] G. Benfatto, P. Picco, M. Pulvirenti (1987). On the invariant measures for the two–dimensional Euler flows, *J. Stat. Phys.* **46**, 729–742.

[60] T.B. Benjamin (1984). Impulse, flow force and variational principles, *IMA J. App. Math.* **32**, 3–68.

[61] T.B. Benjamin, P.J. Olver (1982). Hamiltonian structure, symmetries and conservation laws for water waves, *J. Fluid Mech.* **125**, 137–185.

[62] G. Bennettin, L. Galgani, A. Giorgilli (1985). A proof of Nekhoroshev's theorem for the stability times in nearly integrable Hamiltonian systems, *Celest. Mech.* **37**, 1–25.

[63] R. Benzi, M. Collela, M. Briscolini, P. Santangelo (1992). A simple point vortex model for two-dimensional decaying turbulence, *Phys. Fluids* **A4**, 1036–1039.

[64] V. Berdishevsky (1995). Statistical mechanics of point vortices, *Phys. Rev. E* **3**, 51, no. 5, part A, 4432–4452.

[65] V. Berdishevsky (1998). Statistical mechanics of vortex lines, *Phys. Rev. E* **3**, 57, no. 3, part A, 2885–2905.

[66] M.V. Berry (1984). Quantal phase factors accompanying adiabatic changes, *Proc. Roy. Soc. Lond.* **A 392**, 45–57.

[67] M.V. Berry (1985). Classical adiabatic angles and quantal adiabatic phase, *J. Phys. A: Math. Gen.* **18**, 15–27.

[68] M.V. Berry (1987a). Quantum phase corrections from adiabatic iterations, *Proc. Roy. Soc. Lond.* **A 414** 31–46.

[69] M.V. Berry (1987b). The adiabatic phase and Pancharatnan's phase for polarized light, *J. Mod. Opt.* **34**(11), 1401–1407.

[70] M.V. Berry (1988). The geometric phase, *Scientific American*, December, 46–52.

[71] M.V. Berry (1990). Anticipations of the geometric phase, *Physics Today* **43**(12), 34–40.

[72] M.V. Berry (1997). Aharonov–Bohm geometric phases for rotated rotators, *J. Phys. A: Math. Gen.* **30**(23), 8355–8362.

[73] M.V. Berry, J.H. Hannay (1988). Classical nonadiabatic angles, *J. Phys. A: Math. Gen.* **21**, L325–331.

[74] M.V. Berry, S. Klein (1996). Geometric phases from stacks of crystal plates, *J. Mod. Opt.* **43**(1), 165–180.

[75] M.V. Berry, M.A. Morgan (1996). Geometrical angle for rotated rotators, and the Hannay angle of the world, *Nonlinearity* **9**, 787–799.

[76] M.V. Berry, J.M. Robbins (1997). Indistinguishability for quantum particles: spin, statistics and the geometric phase, *Proc. Roy. Soc. Lond.* **A 453**, 1771–1790.

[77] G. Bertin, C.C. Lin (1996). *Spiral Structures in Galaxies: A Density Wave Theory*, MIT Press, Cambridge.

[78] A.L. Bertozzi (1988). Heteroclinic orbits and chaotic dynamics in planar fluid flows, *SIAM J. Math. Anal.* **19**(6), 1271–1294.

[79] A. Bertozzi, P. Constantin (1993). Global regularity for vortex patches, *Comm. Math. Phys.* **152**, 19–28.

[80] R. Betchov (1965). On the curvature and torsion of an isolated vortex filament, *J. Fluid Mech.* **22**, part 3, 471–479.

[81] G.D. Birkhoff (1927). *Dynamical Systems*, American Mathematical Society, Providence.

[82] D.B. Bliss (1970). The dynamics of curved, rotational vortex lines, M.S. thesis, MIT.

[83] G.W. Bluman, S. Kumei (1989). *Symmetries and Differential Equations*, **AMS 81**, Springer–Verlag, New York.

[84] S. Boatto, R.T. Pierrehumbert (1999). Dynamics of a passive tracer in a velocity field of four identical point vortices, *J. Fluid Mech.* **394**, 137–174.

[85] G.J. Boer (1983). Homogeneous and isotropic turbulence on the sphere, *J. Atmos. Sci.* **40**, 154–163.

[86] G.J. Boer, T.G. Shepherd (1983). Large–scale two–dimensional turbulence in the atmosphere, *J. Atmos. Sci.* **40**, 164–184.

[87] J. Boersma, D.H. Wood (1999). On the self–induced motion of a helical vortex, *J. Fluid Mech.* **384**, 263–280.

[88] N.N. Bogoliubov, Y.A. Mitropolsky (1961). *Asymptotic Methods in the Theory of Non–linear Oscillations*, Gordon and Breach, New York.

[89] V.A. Bogomolov (1977). Dynamics of vorticity at a sphere, *Fluid Dynamics* **6**, 863–870.

[90] V.A. Bogomolov (1979). Two dimensional fluid dynamics on a sphere, *Izv. Atmos. Oc. Phys.* **15**(1), 18–22.

[91] V.A. Bogomolov (1985). On the motion of a vortex on a rotating sphere, *Izv. Atmos. Oc. Phys.* **21**(4), 298–302.

[92] A.V. Borisov, V.G. Lebedev (1998). Dynamics of three vortices on a plane and a sphere – II, *Regular and Chaotic Dynamics* **T.3**, 99–114.

[93] A.V. Borisov, A.E. Pavlov (1998). Dynamics and statics of vortices on a plane and a sphere – I, *Regular and Chaotic Dynamics* **T.3**, 1, 28–38.

[94] H. Born, V. Fock (1928). Beweis des Adiabatensatzes, *Z. Physik* **51**, 165–180.

[95] F.J. Bourland, R. Haberman (1988). The modified phase shift for strongly nonlinear, slowly varying, and weakly damped oscillators, *SIAM J. App. Math.* **48**, 737–748.

[96] F.J. Bourland, R. Haberman (1989). The slowly varying phase shift for perturbed, single and multi–phased, strongly nonlinear dispersive waves, *Physica D* **35**, 1–2, 127–147.

[97] F.J. Bourland, R. Haberman (1990). Separatrix crossing: time invariant potentials with dissipation, *SIAM J. App. Math.* **50**, 1716–1744.

[98] F.J. Bourland, R. Haberman (1994). Connection across a separatrix with dissipation, *Stud. App. Math.* **91**(2), 95–124.

[99] F.J. Bourland, R. Haberman, W. Kath (1991). Averaging methods for the phase shift of arbitrarily perturbed, strongly nonlinear oscillators with an application to capture, *SIAM J. App. Math.* **51**(4), 1150–1167.

[100] P.L. Boyland, H. Aref, M.A. Stremler (2000). Topological fluid mechanics of stirring, *J. Fluid Mech.* **403**, 277–304.

[101] P.L. Boyland, M.A. Stremler, H. Aref (1999). Topological fluid mechanics of point vortex motions, TAM Report # **917**, UILU–ENG–99–6019.

[102] H. Brands, P.H. Chavanis, R. Pasmanter, J. Sommeria (1999). Maximum entropy versus minimum enstrophy vortices, *Phys. Fluids* **11**(11), 3465–3477.

[103] H. Brands, J. Stulemeyer, R. Pasmanter, T. Schep (1997). A mean field prediction of the asymptotic state of decaying 2D turbulence, *Phys. Fluids* **9**(10), 2815–2817.

[104] F. Bretherton (1970). A note on Hamilton's principle for perfect fluids, *J. Fluid Mech.* **40**, part 1, 19–31.

[105] D. Brouwer, G.M. Clemence (1961). *Methods of Celestial Mechanics*, Academic Press, New York.

[106] J.M. Burgers (1948). A mathematical model illustrating the theory of turbulence, in *Advances in Applied Mechanics*, **1**, Academic Press, New York.

[107] K. Burns, V.J. Donnay (1997). Embedded surfaces with ergodic geodesic flow, *Int. Journal of Bif. and Chaos* **7**(7), 1509–1527.

[108] T.F. Buttke (1988). Numerical study of superfluid turbulence in the self–induction approximation, *J. Comp. Phys.* **76**, 301–326.

[109] R. Caflisch (1988). Mathematical analysis of vortex dynamics, in *Proc. Workshop on Math. Aspects of Vortex Dynamics*, 1–24, Society for Industrial and Applied Math, Philadelphia.

[110] E. Caglioti, P.L. Lions, C. Marchioro, M. Pulvirenti (1992). A special class of stationary flows for two-dimensional Euler equations: a statistical mechanics description, *Comm. Math. Phys.* **143**, 501–525.

[111] E. Caglioti, P.L. Lions, C. Marchioro, M. Pulvirenti (1995). A special class of stationary flows for two–dimensional Euler equations: a statistical mechanics description: Part II., *Comm. Math. Phys.* **174**, 229–260.

[112] A.J. Callegari, L. Ting (1978). Motion of a curved vortex filament with decaying vortical core and axial velocity, *SIAM J. App. Math.* **35**(1), 148–175.

[113] L.J. Campbell, J.B. Kadtke (1987). Stationary configurations of point vortices and other logarithmic objects in two-dimensions, *Phys. Rev. Lett.* **58**, 670–673.

[114] L.J. Campbell, K. O Neil (1991). Statistics of 2–D point vortices and high energy vortex states, *J. Stat. Phys.* **65**(3-4), 495–529.

[115] L.J. Campbell, R.M. Ziff (1978). A catalog of 2–D vortex patterns, *Los Alamos Sci. Report*, **No. LA–7384–MS**.

[116] L.J. Campbell, R.M. Ziff (1979). Vortex patterns and energies in a rotating superfluid, *Phys. Rev. B* **20**, 1886–1902.

[117] B. Cantwell (1981). Organized motion in turbulent flow, *Ann. Rev. Fluid Mech.* **13**, 457–515.

[118] G.F. Carnevale (1982). Statistical features of the evolution of two–dimensional turbulence, *J. Fluid Mech.* **122**, 143–153.

[119] G.F. Carnevale, J.C. McWilliams, Y. Pomeau, J.B. Weiss, W.R. Young (1991). Evolution of vortex statistics in two-dimensional turbulence, *Phys. Rev. Lett.* **66**, 2735–2737.

[120] J. Cary, D. Escande, J. Tennyson (1986). Adiabatic invariant change due to separatrix crossing, *Phys. Rev A* **34**, 4256–4275.

[121] J.R. Cary, R.T. Skodje (1989). Phase change between separatrix crossings, *Physica D* **36**(3), 287–316.

[122] M.S.A.C. Castilla, V. Moauro, P. Negrini, W.M. Oliva (1993). The four positive vortices problem – Regions of chaotic behavior and non–integrability, *Ann. de l'Institut Henri Poincaré– Phys. Theor.* **59**(1), 99–115.

[123] A. Celletti, C. Falcolini (1988). A remark on the KAM theorem applied to a four–vortex system, *J. Stat. Phys.* **52**, 1-2, 471–477.

[124] D. Chandler (1987). *Introduction to Modern Statistical Physics*, Oxford University Press, Oxford.

[125] D.M.F. Chapman (1978). Ideal vortex motion in two dimensions: Symmetries and conservation laws, *J. Math. Phys.* **19**(9), 1988–1992.

[126] P.H. Chavanis (1996). Contribution á la mécanique statistique des tourbillons bidimensionnels. Analogie avec la relaxation violente des systémes stellaires , Ph.D. thesis, Ecole Normale Supérieure de Lyon.

[127] P.H. Chavanis (1998a). From Jupiter's great red spot to the structure of galaxies: statistical mechanics of two-dimensional vortices and stellar systems, in *Nonlinear Dynamics and Chaos in Astrophysics: A Festschrift in Honor of George Contopoulas*, Annals of the New York Academy of Sciences, **867**, 120–140.

[128] P.H. Chavanis (1998b). Systematic drift experienced by a point vortex in two–dimensional turbulence, *Phys. Rev. E*, **58**(2), R1199–R1202.

[129] P. H. Chavanis (2000). Quasilinear theory of the 2D Euler equation, *Phys. Rev. Lett.* **84**(24), 5512–5515.

[130] P.H. Chavanis, C. Sire (2000a). The spatial correlations in the velocities arising from a random distribution of point vortices, preprint http://xxx.lanl.gov/abs/cond–mat/0004410

[131] P.H. Chavanis, C. Sire (2000b). The statistics of velocity flucuations arising from a random distribution of point vortices: the speed of flucuations and the diffusion coefficient, http://xxx.lanl.gov/abs/cond–mat/9911032 , to appear, Phys. Rev. E, **62**(1).

[132] P.H. Chavanis, J. Sommeria (1996). Classification of self–organized vortices in two dimensional turbulence: the case of a bounded domain, *J. Fluid Mech.* **314**, 267–297.

[133] P.H. Chavanis, J. Sommeria (1998). Classification of robust isolated vortices in two–dimensional hydrodynamics, *J. Fluid Mech.* **356**, 259–296.

[134] P.H. Chavanis, J. Sommeria, R. Robert (1996). Statistical mechanics of two dimensional vortices and collisionless stellar systems, *Astrophys. J.* **471**, 385–399.

[135] J.Y. Chemin (1991). Sur le mouvement des particules d'un fluide parfait incompressible bidimensionnel, *Invent. Math.* **103**, 599–629.

[136] J. Y. Chemin (1998). *Perfect Incompressible Fluids*, Oxford Lecture Series in Math and Applications, **14**, Clarendon Press, Oxford.

[137] T.M. Cherry (1924). On Poincaré's theorem of the non–existence of uniform integrals of dynamical equations, *Proc. Cambridge Phil. Soc.* **22**, 287.

[138] N.G. Chetaev (1989). *Theoretical Mechanics*, Mir Publishers, Springer–Verlag, New York.

[139] B. Chirikov (1979). A universal instability of many dimensional oscillator systems, *Phys. Rep.* **52**, 265–379.

[140] J.Y.K. Cho, L.M. Polvani (1996). The emergence of jets and vortices in freely evolving, shallow water turbulence on a sphere, *Phys. Fluids* **8**(6), 1531–1552.

[141] A.J. Chorin (1982). The evolution of a turbulent vortex, *Comm. Math. Phys.* **83**, 517–535.

[142] A.J. Chorin (1991a). Statistical mechanics and vortex motion, AMS Lectures in Applied Mathematics 28, 85–101.

[143] A.J. Chorin (1991b). Equilibrium statistics of a vortex filament with applications, *Comm. Math. Phys.* **141**, 619–631.

[144] A.J. Chorin (1994). *Vorticity and Turbulence*, **AMS 103**, Springer–Verlag, New York.

[145] A.J. Chorin (1996a). Partition functions and equilibrium measures in two–dimensional and quasi–three dimensional turbulence, *Phys. Fluids* **8**, 2656–2660.

[146] A.J. Chorin (1996b). Vortex methods, in *Computational Fluid Dynamics, Les Houches Session LVIX 1993*, 65–106, Elsevier Publishers, Amsterdam.

[147] A.J. Chorin, J. Marsden (1979). *A Mathematical Introduction to Fluid Mechanics*, Springer–Verlag, New York.

[148] S.N. Chow, J.K. Hale, J. Mallet–Paret (1980). An example of bifurcation to homoclinic orbits, *J. Diff. Eqns.* **37**, 351–373.

[149] J.F. Christiansen, N.J. Zabusky (1973). Instability, coalescence and fission of finite area vortex structures, *J. Fluid Mech.* **61**, 219–243.

[150] R.R. Clements (1973). An inviscid model of two–dimensional vortex shedding, *J. Fluid Mech.* **57**, part 2, 321–336.

[151] G.H.A. Cole (1967). *An Introduction to the Statistical Theory of Classical Simple Dense Fluids*, Pergamon Press, New York.

[152] A.T. Conlisk, Y.G. Guezennec, G.S. Elliot (1989). Chaotic motion of an array of vortices above a flat wall, *Phys. Fluids* **A1**(4), 704–717.

[153] A.T. Conlisk, D. Rockwell (1981). Modelling of vortex–corner interactions using point vortices, *Phys. Fluids* **24**, 2133– 2142.

[154] P. Constantin, P.D. Lax, A. Majda (1985). A simple one–dimensional model for the three–dimensional vorticity equation, *Comm. Pure App. Math.* **38**, 715–724.

368   References

[155] P. Constantin, E.S. Titi (1988). On the evolution of nearly circular vortex patches, *Comm. Math. Phys.* **119**, 177–198.

[156] L. Cortelezzi (1996). Nonlinear feedback control of the wake past a plate with a suction point on the downstream wall, *J. Fluid Mech.* **327**, 303–324.

[157] L. Cortelezzi, A. Leonard (1993). Point vortex model for the unsteady separated flow past a semi–infinite plate with transverse motion, *Fluid Dyn. Res.* **11**, 263–295.

[158] L. Cortelezzi, A. Leonard, J.C. Doyle (1994). An example of active circulation control of the unsteady separated flow past a semi–infinite plate, *J. Fluid Mech.* **260**, 127–154.

[159] G.H. Cottet, P.D. Koumoutsakos (2000). *Vortex Methods: Theory and Practice*, Cambridge University Press, Cambridge.

[160] G.H. Cottet, J. Soler (1988). Three–dimensional Navier–Stokes equation for singular filament initial data, *J. Diff. Eqns.* **74**, 234–253.

[161] Y. Couder, C. Basdevant (1986). Experimental and numerical study of vortex couples in two–dimensional flows, *J. Fluid Mech.* **173**, 225–251.

[162] R. Courant, D. Hilbert (1953). *Methods of Mathematical Physics, Vol. I,II*, Wiley, New York.

[163] W. Craig (1996). KAM theory in infinite dimensions, in *Dynamical Systems and Probabilistic Methods in PDE*, P. Deift, C.D. Levermore, C.E. Wayne, eds., Lecture in Applied Math Vol. 31, American Mathematical Society, Providence.

[164] S.C. Crow (1970). Stability theory for a pair of trailing vortices, *AIAA J.* **8**, 1731.

[165] L.S. DaRios (1906). Sul moto 'un liquido indefinito con un filetto vorticoso di forma qualunque, *Rendiconti del Circolo Matematico Palermo* **22**, 117–135.

[166] M.D. Dahleh (1992). Exterior flow of the Kida ellipse, *Phys. Fluids* **A4** (9), 1979–1985.

[167] U. Dallman (1988). Three–dimensional vortex structures and vorticity topology, *Fluid Dyn. Res.* **3**, 183–189.

[168] P. Deift (1996). Integrable Hamiltonian systems , in *Dynamical Systems and Probabilistic Methods in PDE*, P. Deift, C.D. Levermore, C.E. Wayne, eds., Lecture in Applied Math Vol. 31, American Mathematical Society, Providence.

[169] V. Del Prete, O.H. Hald (1978). Convergence of vortex methods for Euler's equations, *Math. Comp.* **32**,791–809.

[170] J.W. Dettman (1965). *Applied Complex Variables*, Dover, New York.

[171] M.R. Dhanak, B. de Bernardinis (1981). The evolution of an elliptic vortex ring, *J. Fluid Mech.* **109**, 189–216.

[172] M.R. Dhanak, M.P. Marshall (1993). Motion of an elliptical vortex under applied periodic strain, *Phys. Fluids* **A5**, 1224–1230.

[173] M.T. DiBattista, L.M. Polvani (1998). Barotropic vortex pairs on a rotating sphere, *J. Fluid Mech.* **358**, 107–133.

[174] M. DiBattista, A.J. Majda, B. Turkington (1998). Prototype geophysical vortex structures via large scale statistical theory, *Geo. Astro. Fluid Dyn.* **89**(3-4), 235–283.

[175] R. DiPerna, A. Majda (1987). Concentrations and regularizations for 2D incompressible flow, *Comm. Pure App. Math.* **40**, 301–345.

[176] C.R. Doering, J.D. Gibbon (1995). *Applied Analysis of the Navier–Stokes Equations*, Cambridge University Press, Cambridge.

[177] D.G. Dritschel (1985). The stability and energetics of co–rotating uniform vortices, *J. Fluid Mech.* **157**, 95–134.

[178] D.G. Dritschel (1988a). Nonlinear stability bounds for inviscid two–dimensional parallel or circular flows with monotonic vorticity, and the analogous three–dimensional quasi–geostrophic flows, *J. Fluid Mech.* **191**, 575–581.

[179] D.G. Dritschel (1988b). Contour dynamics/surgery on the sphere, *J. Comp. Phys.* **79**, 477–483.

[180] D.G. Dritschel (1992). 2D turbulence: new results for $Re \to \infty$, in *Topological Aspects of the Dynamics of Fluids and Plasmas*, NATO Adv. Sci. Inst. Ser. E, App. Sci, 218, Kluwer, Amsterdam.

[181] D.G. Dritschel (1993). Vortex properties of two–dimensional turbulence, *Phys. Fluids* **A(4)**, 984–997.

[182] D.G. Dritschel (1995). A general theory for two dimensional vortex interactions, *J. Fluid Mech.* **293**, 269–303.

[183] D.G. Dritschel, L.M. Polvani (1992). The roll–up of vorticity strips on the surface of a sphere, *J. Fluid Mech.* **234**, 47–69.

[184] D.G. Dritschel, N. Zabusky (1996). On the nature of vortex interactions and models in unforced nearly inviscid two-dimensional turbulence, *Phys. Fluids* **8(5)**, 1252–1256.

[185] H.S. Dumas (1993). A Nekhoroshev like theory of classical particle channeling in perfect crystals, in *Dynamics Reported: Expositions in Dynamical Systems Vol. 2*, C.K.R.T. Jones, U. Kirchgraber, H.O. Walther, eds., Springer–Verlag, New York.

[186] A.M. Dyhne (1960). Quantum passages in adiabatic approach, *JETP* **38**(2), 570–578.

[187] F.W. Dyson (1893). The potential of an anchor–ring, Pt. II , *Phil. Trans. R. Soc. Lond.* **A 184**, 1041.

[188] D. Ebin, J.E. Marsden (1970). Groups of diffeomorphisms on the motion of an incompressible fluid, *Ann. of Math.* **92**, 102–163.

[189] B. Eckhardt (1988). Integrable four vortex motion, *Phys. Fluids* **31**(10), 2796–2801.

[190] B. Eckhardt, H. Aref (1988). Integrable and chaotic motion of four vortices. II. Collision dynamics of vortex pairs, *Phil. Trans. R. Soc. Lond.* **A 326**, 655–696.

[191] S.F. Edwards, J.B. Taylor (1974). Negative temperature states of two–dimensional plasmas and vortex fluids, *Proc. Roy. Soc. Lond.* **A 336**, 257–271.

[192] A.R. Elcrat, C. Hu, K.G. Miller (1997). Equilibrium configurations of point vortices for channel flows past interior obstacles, *Eur. J. Mech. B/Fluids* **16**(2), 277–292.

[193] A.R. Elcrat, K.G. Miller (1995). Rearrangements in steady multiple vortex flows, *Comm. in PDE* **20**, 1481–1490.

[194] R. Ellis (1985). *Entropy, Large Deviations, and Statistical Mechanics*, Springer–Verlag, New York.

[195] Y. Elskens, D. Escande (1991). Slowly pulsating separatrices sweep homoclinic tangles where islands must be small: an extension of classical adiabatic theory, *Nonlinearity* **4**, 615.

[196] G.J. Erikson, C.K. Smith (eds.) (1988). *Maximum Entropy and Bayesian Methods in Science and Engineering*, Kluwar, Dordrecht.

[197] R.M. Everson, K.R. Sreenivasan (1992). Accumulation rates of spiral–like structures in fluid flows, *Proc. Roy. Soc. Lond.* **A 437**, 391–401.

[198] G. Eyink (1994). The renormalization group method in statistical hydrodynamics, *Phys. Fluids* **6**(9), 3063–3078.

[199] G. Eyink (1996). Exact results on stationary turbulence in 2D: consequences of vorticity conservation, *Physica D* **91**, no. 1–2, 97–142.

[200] G. Eyink, H. Spohn (1993). Negative–temperature states and large–scale long lived vortices in two–dimensional turbulence, *J. Stat. Phys.* **70**, no. 3–4, 833–886.

[201] L.D. Faddeev, L.A. Takhtajan (1987). *Hamiltonian Methods in the Theory of Solitons*, Springer–Verlag, New York.

[202] M.V. Fedorjuk (1976). Adiabatic invariant of a system of linear oscillators and scattering theory, *Diff. Eqns.* **12**(6), 1012–1018.

[203] N.M. Ferrers (1866). *An Elementary Treatise on Tri–Linear Coordinates. The Method of Reciprocal Polars and the Theory of Projection*, 2nd ed., Macmillan, New York.

[204] R.P. Feynman (1998). *Statistical Mechanics*, Addison–Wesley, MA.

[205] K.S. Fine, A.C. Cass, W.G. Flynn, C.F. Driscoll (1995). Relaxation of 2D turbulence to vortex crystals, *Phys. Rev. Lett.* **75**(18), 3277–3280.

[206] M. Flucher (1999). *Variational Problems with Concentration, Progress in Nonlinear Differential Equations and their Applications*, **36**, Birkhäuser, Basel.

[207] M. Flucher, B. Gustafsson (1997). Vortex motion in two-dimensional hydrodynamics, **TRITA–MAT–1997–MA–02**, Royal Institute of Technology, preprint.

[208] A.T. Fomenko, T.Z. Nguyen (1991). Topological classification of integrable non–degenerate Hamiltonians on isoenergy three–dimensional spheres, in *Advances in Soviet Mathematics*, **6**, ed. A.T. Fomenko, 267–296, American Mathematical Society, Providence.

[209] A.T. Fomenko (1991). Topological classification of all integrable Hamiltonian differential equations of general type with two degrees of freedom, in *The Geometry of Hamiltonian Systems*, ed. T. Ratiu, MSRI Publications, **22**, 131–340, Springer–Verlag, New York.

[210] L.E. Fraenkel (1970). On steady vortex rings of small cross–section in an ideal fluid, *Proc. Roy. Soc. Lond.* **A 316**, 29–62.

[211] L.E. Fraenkel (1972). Examples of steady vortex rings of small cross section in an ideal fluid, *J. Fluid Mech.* **51**, part 1, 119–135.

[212] L.E. Fraenkel, M.S. Berger (1974). A global theory of steady vortex rings in an ideal fluid, *Acta Math.* **132**, 13–51.

[213] J.S. Frederiksen, B.L. Sawford (1980). Statistical dynamics of two–dimensional inviscid flow on a sphere, *J. Atmos. Sci.* **37**, 717–732.

[214] M.H. Freedman, Z–X He (1991a). Links of tori and the energy of incompressible flows, *Topology* **30**(2), 283–287.

[215] M.H. Freedman, Z–X He (1991b). Divergence free fields: energy and asymptotic crossing number, *Ann. of Math.* **134**, 189–229.

[216] S. Friedlander (1975). Interaction of vortices in a fluid on the surface of a rotating sphere, *Tellus* **XXVII**, 15–24.

[217] J. Frohlich, D. Ruelle (1982). Statistical mechanics in an inviscid 2–D fluid, *Comm. Math. Phys.* **87**, 1–36.

[218] Y. Fukumoto (1987). On integral invariants for vortex motion under the Localized Induction Approximation, *J. Phys. Soc. Jap.* **56**(12), 4207–4209.

[219] Y. Fukumoto, M. Miyazaki (1986). N–solitons on a curved vortex filament, *J. Phys. Soc. Japan* **55**, 4152–4155.

[220] Y. Fukumoto, M. Miyazaki (1991). Three dimensional distortions of a vortex filament with axial velocity, *J. Fluid Mech.* **222**, 369–416.

[221] C.S. Gardner (1959). Adiabatic invariants of periodic classical systems, *Phys. Rev.* **115**, 791–794.

[222] I.M. Gel'fand (1964). *Generalized Functions*, Academic Press, New York.

[223] I.M. Gel'fand, S.V. Fomin (1963). *Calculus of Variations*, Prentice–Hall, Englewood Cliffs, N.J.

[224] R.W. Ghrist, P.J. Holmes, M.C. Sullivan (1997). *Knots and Links in Three Dimensional Flows*, Springer Lecture Notes in Mathematics, **1654**, Springer–Verlag, New York.

[225] A. Gilbert (1988). Spiral structures and spectra in two-dimensional turbulence, *J. Fluid Mech.* **193**, 475–497.

[226] A.E. Gill (1982). *Atmosphere–Ocean Dynamics*, International Geophysics Series **30**, Academic Press, New York.

[227] H. Goldstein (1980). *Classical Mechanics*, 2nd Ed., Addison–Wesley, MA.

[228] S. Golin (1988). Existence of the Hannay angle for single–frequency systems, *J. Phys. A: Math. Gen.* **21**, 4535–4547.

[229] S. Golin (1989). Can one measure Hannay angles?, *J. Phys. A: Math. Gen.* 4573–4580.

[230] S. Golin, A. Knauf, S. Marmi (1989). The Hannay angles: geometry, adiabaticity, and an example, *Comm. Math. Phys.* **123**, 95–122.

[231] S. Golin, A. Knauf, S. Marmi (1990). Hannay angles and classical perturbation theory, in J.M. Luck, P. Moussa, M. Waldschmidt eds., *Number Theory and Physics*, Lecture Notes in Physics, Springer-Verlag, Berlin.

[232] S. Golin, S. Marmi (1989). Symmetries, Hannay angles, and the precession of orbits, *Europhys. Lett.* **8**, 399–404.

[233] S. Golin, S. Marmi (1990). A class of systems with measurable Hannay angles, *Nonlinearity* **3**, 507–518.

[234] J. Goodman, T.Y. Hou, J. Lowengrub (1990). Convergence of the point vortex method for 2–D Euler equations, *Comm. Pure App. Math.* **43**, 415–430.

[235] E. Gozzi, W.D. Thacker (1987a). Classical adiabatic holonomy in a Grassmanian system, *Phys. Rev. D* **35**(8), 2388–2397.

[236] E. Gozzi, W.D. Thacker (1987b). Classical adiabatic holonomy and its canonical structure, *Phys. Rev. D* **35**(8), 2398–2406.

[237] R. Grauer, T. Sideris (1991). Numerical computation of three dimensional incompressible ideal fluids with swirl, *Phys. Rev. Lett.* **67**, 3511–3514.

[238] R. Grauer, T. Sideris (1995). Finite time singularities in ideal flows with swirl, *Physica D* **88**(2), 116–132.

[239] S.I. Green (1995). Introduction to Vorticity, in *Fluid Vortices*, ed. S.I. Green, Kluwar, Dordrecht.

[240] J.M. Greene (1982). Noncanonical Hamiltonian mechanics, in *Mathematical Methods in Hydrodynamics and Integrability in Dynamical Systems*, M. Tabor, Y.M. Treve, eds., 91–98, AIP Conf. Proc. **88**.

[241] C. Greengard (1985). The core spreading method approximates the wrong equation, *J. Comp. Phys.* **61**, 345–348.

[242] C. Greengard, E. Thomann (1988). Singular vortex systems and weak solutions of the Euler equations, *Phys. Fluids* **31**(10), 2810–2813.

[243] A.G. Greenhill (1878). Plane vortex motion, *Q.J. Math.* **15**, 10.

[244] D.T. Greenwood (1997). *Classical Dynamics*, Dover, New York.

[245] I.S. Gromeka (1952). On vortex motions of liquid on a sphere, *Collected Papers* Moscow, AN USSR, 296.

[246] V.M. Gryanik (1983). Dynamics of singular geostrophic vortices in a two–level model of the atmosphere (or ocean), *Bull. Izv. Acad. Sci. USSR, Atmospheric and Oceanic Physics* **19**(3), 171–179.

[247] J. Guckenheimer, P. Holmes (1983). *Nonlinear Oscillations, Dynamical Systems and Bifurcations of Vector Fields*, **AMS 43**, Springer–Verlag, New York.

[248] D. Gurarie (1995). Symmetries and conservation laws of two–dimensional hydrodynamics, *Physica D* **86**, 621–636.

[249] B. Gustafsson (1979). On the motion of a vortex in two–dimensional flow of an ideal fluid in simply and multiply connected domains, **TRITA–MAT–1979–7**, Royal Institute of Technology, preprint.

[250] B. Gutkin, U. Smilansky, E. Gutkin (1999). Hyperbolic billiards on surfaces of constant curvature, *Comm. Math. Phys.* **208**, 65–90.

[251] R. Haberman (1988). The modulated phase shift for weakly dissipated nonlinear oscillatory waves of the KdV type, *Stud. App. Math.* **78**(1), 73–90.

[252] R. Haberman (1991). Phase shift modulations for stable, oscillatory, traveling, strongly nonlinear waves, *Stud. App. Math* **84**(1), 57–69.

[253] R. Haberman (1992). Phase shift modulations for perturbed, strongly nonlinear oscillatory dispersive waves, 73–82, in *Nonlinear Dispersive Wave Systems*, World Scientific, Singapore.

[254] R. Haberman, F.J. Bourland (1988). Variation of wave action: modulation of the phase shift for strongly nonlinear dispersive waves with weak dissipation, *Physica D* **32**(1), 72–82.

[255] O.H. Hald (1979). Convergence of vortex methods for Euler's equations , II, *SIAM J. Num. Anal.* **16**, 726–755.

[256] O. H. Hald (1991). Convergence of Vortex Methods, in *Vortex Methods and Vortex Motion*, K.E. Gustafson, J.A. Sethian, eds., Society for Industrial and Applied Math., Philadelphia.

[257] J. Hale (1980). *Ordinary Differential Equations*, R. Krieger Publishing Co., Huntington, NY.

[258] G. Haller, I. Mezić (1998). Reduction of three-dimensional, volume preserving flows with symmetry, *Nonlinearity* **11**, 319–339.

[259] D. Hally (1980). Stability of streets of vortices on surfaces of revolution with a reflection symmetry, *J. Math. Phys.* **21**, 211–217.

[260] P. Halmos (1958). *Lectures on Ergodic Theory*, Chelsea, NY.

[261] F.R. Hama (1962). Progressive deformation of a curved vortex filament by its own induction,*Phys. Fluids* **5**, 1156–1162.

[262] F.R. Hama (1963). Progressive deformation of a perturbed line vortex filament, *Phys. Fluids* **6**(4), 526–534.

[263] J.H. Hannay (1985). Angle variable holonomy in adiabatic excursion of an integrable Hamiltonian, *J. Phys. A: Math. Gen.* **18**, 221–230.

[264] J.C. Hardin (1982). The velocity field induced by a helical vortex filament, *Phys. Fluids* **25**, 1949–1952.

[265] J.K. Harvey, F.J. Perry (1971). Flowfield produced by trailing vortices in the vicinity of the ground, *AIAA Journal* **9**, 1659–1660.

[266] H. Hasimoto (1971). Motion of a vortex filament and its relation to Elastica, *J. Phys. Soc. Japan* **31**(1), 293–294.

[267] H. Hasimoto (1972). A soliton on a vortex filament, *J. Fluid Mech.* **51**, 477–485.

[268] H. Hasimoto (1988). Elementary aspects of vortex motion, *Fluid Dyn. Res.* **3**, 1–12.

[269] H. Hasimoto, K. Ishii, Y. Kimura, M. Sakiyama (1984). Chaotic and coherent behaviors of vortex filaments in bounded domains, in *Turbulence and Chaotic Phenomena in Fluids*, T. Tatsumi, ed., 231–237, Elsevier Publishers, Amsterdam.

[270] T.H. Havelock (1931). The stability of motion of rectilinear vortices in ring formation, *Phil. Mag.* **7**(11), 617–633.

[271] H. von Helmholtz (1858). On the integrals of the hydrodynamical equations which express vortex motion, *Phil Mag.* **4**(33), 485–512.

[272] J. Henrard (1993). The Adiabatic Invariant in Classical Mechanics, in *Dynamics Reported: Expositions in Dynamical Systems Vol. 2*, C.K.R.T. Jones, U. Kirchgraber, H.O. Walther, eds., Springer-Verlag, New York.

[273] J.R. Herring (1977). On the statistical theory of two–dimensional topography turbulence, *J. Atmos. Sci.* **34**, 1731–1750.

[274] W.M. Hicks (1884). On the steady motion and the small vibrations of a hollow vortex, *Phil. Trans. A* **175**, 183.

[275] N. Hogg, H. Stommel (1985a). The heton, an elementary interaction between discrete baroclinic geostrophic vortices, and its applications concerning eddy heat–flow, *Proc. Roy. Soc. A* **397**, 1–20.

[276] N. Hogg, H. Stommel (1985b). Hetonic explosions: The breakup and spread of warm pools as explained by baroclinic point vortices, *J. Atmos. Sci.* **42**(14), 1465–1476.

[277] G. Holloway (1986). Eddies, waves, circulation, and mixing: Statistical geofluid mechanics, *Ann. Rev. Fluid Mech.* **18**, 91–147.

[278] D.D. Holm (1997). *Hamiltonian Geophysical Fluid Dynamics*, Los Alamos Report.

[279] D.D. Holm, J.E. Marsden, T.S. Ratiu (1986). The Hamiltonian structure of continuum mechanics in material, spatial and convective representations , Séminaire de Mathématiques Supérieures, Les Presses de L'Univ. de Montréal 100, 11–122.

[280] D.D. Holm, J.E. Marsden, T.S. Ratiu (1998a). Euler–Poincaré models of ideal fluids with nonlinear dispersion, *Phys. Rev. Lett.* **349**, 4173–4177.

[281] D.D. Holm, J.E. Marsden, T.S. Ratiu (1998b). Euler–Poincaré equations and semidirect products with applications to continuum theories, *Adv. in Math.* **137**, 1–81,

[282] D.D. Holm, J.E. Marsden, T.S. Ratiu, A. Weinstein (1985). Nonlinear stability of fluid and plasma equilibria, *Phys. Rep.* **123**, 1–116.

[283] P.J. Holmes (1980). Averaging and chaotic motions in forced oscillations, *SIAM J. App. Math.* **38**, 68–80, and **40**, 167–168.

[284] P.J. Holmes (1990a). Poincaré, celestial mechanics, dynamical systems theory and chaos, *Phys. Rep.* **193**(3), 137–163.

[285] P.J. Holmes (1990b). Nonlinear dynamics, chaos, and mechanics, *Applied Mech. Rev.* **43**, no. 5, part 2, S23–S39.

[286] P.J. Holmes, J.E. Marsden (1982a). Melnikov's method and Arnold diffusion for perturbations of integrable Hamiltonian systems, *J. Math. Phys.* **23**(4), 669–675.

[287] P.J. Holmes, J.E. Marsden (1982b). Horseshoes in perturbations of Hamiltonian systems with two degrees of freedom, *Comm. Math. Phys.* **82**, 523–544.

[288] E. Hopf (1952). Statistical hydromechanics and functional calculus, *J. Rat. Mech. Anal.* **1**, 87–123.

[289] E. Hopfinger, F. Browand (1982). Vortex solitary waves in a rotating turbulent flow, *Nature* **295**, 393–395.

[290] T.Y. Hou (1991). A survey on convergence analysis for point vortex methods, in *Vortex Dynamics and Vortex Methods*, C. Anderson, C. Greengard, eds., Lectures in Applied Mathematics **28**, 327–340, American Mathematical Society, Providence.

[291] T.Y. Hou, J. Lowengrub (1990). Convergence of the point vortex method for 3–D Euler equations, *Comm. Pure App. Math.* **43**, 965–981.

[292] L.N. Howard (1957/58). Divergence formulas involving vorticity, *Arch. Rat. Mech. Anal.* **1**, 113–123.

[293] K. Ide, S. Wiggins (1995). The dynamics of elliptically shaped regions of uniform vorticity in time–periodic, linear external velocity fields, *Fluid Dyn. Res.* **15**, 205–235.

[294] D. Iftimie, T.C. Sideris, P. Gamblin (1999). On the evolution of compactly supported planar vorticity, *Comm. PDE* **24**, 9,10, 1709–1730.

[295] D.C. Ives (1976). A modern look at conformal mapping, including multiply connected regions, *AIAA Journal* **14**, 1006–1011.

[296] J.D. Jackson (1963). *Classical Electrodynamics*, Wiley, New York.

[297] M. Jamaloodeen (2000). Hamiltonian methods for some geophysical vortex dynamics models , Ph.D. thesis, Department of Mathematics, University of Southern California.

[298] E.T. Jaynes (1957). Information theory and statistical mechanics, *Phys. Rev.* **106**, 620–630.

[299] E.T. Jaynes (1979). Where do we stand on maximum entropy, in R.D. Levine, M. Tribus, eds., *The Maximum Entropy Formalism*, MIT Press, Cambridge.

[300] J.H. Jeans (1933). *The Mathematical Theory of Electricity and Magnetism*, 5th edition, Cambridge Press, Cambridge.

[301] J. Jiménez (ed.) (1991). *The Global Geometry of Turbulence,* Plenum Press, New York.

[302] R. Jost (1964). Poisson brackets (An unpedagogical lecture), *Rev. Mod. Phys.*, 572–579.

[303] G. Joyce, D. Montgomery (1973). Negative temperature states for the two–dimensional guiding center plasma, *J. Plasma Phys.* **10**, 107–121.

[304] B. Juttner, A. Thess, J. Sommeria (1995). On the symmetry of self–organized structures in two–dimensional turbulence, *Phys. Fluids* **7**(9), 2108–2110.

[305] J.B. Kadtke, L.J. Campbell (1987). A method for finding stationary states of point vortices, Phys. Rev. A **36**, 4360–4370.

[306] J.B. Kadtke, E.A. Novikov (1993). Chaotic capture of vortices by a moving body. I. The single point vortex case, *Chaos* **3**(4), 543–553.

[307] T. Kambe, T. Takao (1971). Motion of distorted vortex rings, *J. Phys. Soc. Japan* **31**(2), 591–599.

[308] T.R. Kane, M.P. Sher (1969). A dynamical explanation of the falling cat phenomenon, *Int. J. Solids and Structures* **5**, 663–670.

[309] T.J. Kaper, S. Wiggins (1992). On the structure of separatrix swept regions in singularly perturbed Hamiltonian systems, *Diff. and Int. Eqns.* **5**(6), 1363–1381.

[310] M. Karweit (1975). Motion of a vortex pair approaching an opening in a boundary, *Phys. Fluids* **18**, 1604–1606.

[311] T. Kato (1950). On the adiabatic theorem of quantum mechanics, *Phys. Soc. Japan* **5**, 435–439.

[312] T. Kato (1967). On classical solutions of the two dimensional non-stationary Euler equations, *Arch. Rat. Mech. Anal.* **25**, 188–200.

[313] T. Kato (1972). Non–stationary flows of viscous and ideal fluids in $\mathbb{R}^3$, *J. Funct. Anal.* **9**, 296–305.

[314] A. Katok, B. Hasselblatt (1995). *Introduction to the Modern Theory of Dynamical Systems*, Cambridge Press, Cambridge.

[315] A. Kawakami, M. Funakoshi (1999). Chaotic motion of fluid particles around a rotating elliptic vortex in a linear shear flow, *Fluid Dyn. Res.* **25**, 167–193.

[316] J.P. Keener (1990). Knotted vortex filaments in an ideal fluid, *J. Fluid Mech.* **211**, 629–651.

[317] J.B. Keller (1953). The scope of the image method, Comm. Pure and App. Math, **VI**, 505–512.

[318] O.D. Kellogg (1954). *Foundations of Potential Theory*, Dover, New York.

[319] Lord Kelvin (1867). On vortex atoms, *Proc. R. Soc. Edinburgh* **6**, 94–105.

[320] Lord Kelvin (1868). On vortex motion, *Trans. Royal Soc. Edinburgh* **25**, 217–260.

[321] Lord Kelvin (1878). Floating magnets, *Nature* **18**, 13–14.

[322] Lord Kelvin (1880). On the vibrations of a columnar vortex, *Phil. Mag.* **5**, 155.

[323] Lord Kelvin (1910). *Mathematical and Physical Papers*, Cambridge Press, Cambridge.

[324] R. Kerr (1993). Evidence for a singularity of the three dimensional incompressible Euler equations, *Phys. Fluids* A **5**(7), 1725–1746.

[325] J. Kevorkian (1982). Adiabatic invariance and passage through resonance for nearly periodic Hamiltonian systems, *Stud. App. Math.* **66**, 95–119.

[326] J. Kevorkian (1987). Perturbation techniques for oscillatory systems with slowly varying coefficients, *SIAM Rev.* **29**, 391–461.

[327] J. Kevorkian (1990). *Partial Differential Equations: Analytical Solution Techniques*, Wadsworth and Brooks/Cole Mathematics Series.

[328] J. Kevorkian, J.D. Cole (1996). *Multiple Scale and Singular Perturbation Methods*, **AMS 114**, Springer–Verlag, New York.

[329] K.M. Khanin (1982). Quasi–periodic motions of vortex systems, *Physica D* **4**, 261–269.

[330] K.M. Khazin (1976). Regular polygons of point vortices, *Sov. Phys. Dokl.* **21**, 567–569.

[331] A. Khein, D.F. Nelson (1993). Hannay angle study of the Foucault pendulum in action–angle variables, *Am. J. Physics* **61**(2), 170–174.

[332] B.A. Khesin, Yu. V. Chekanov (1989). Invariants of the Euler equations for ideal and barotropic hydrodynamics and superconductivity in D dimensions, *Physica D* **40**, 119–136.

[333] S. Kida (1975). Statistics of the system of line vortices, *J. Phys. Soc. Japan* **39**, 1395–3404.

[334] S. Kida (1981a). Motion of an elliptical vortex in a uniform shear flow, *J. Phys. Soc. Japan* **50**, 3517–3520.

[335] S. Kida (1981b). A vortex filament moving without changes of form, *J. Fluid Mech.* **112**, 397–409.

[336] S. Kida (1982). Stability of a steady vortex filament, *J. Phys. Soc. Japan* **51**, 1655–1662.

[337] S. Kida (1993). Tube Like Structures in Turbulence, in Lecture Notes in Numerical Applied Analysis, Springer–Verlag, New York.

[338] S. Kida, M. Takaoka (1994). Vortex reconnection, *Ann. Rev. Fluid Mech.* **26**, 169–189.

[339] R. Kidambi (1999). Integrable vortex motion on a sphere, Ph.D. thesis, Department of Aerospace Engineering, University of Southern California.

[340] R. Kidambi, P.K. Newton (1998). Motion of three point vortices on a sphere, *Physica D* **116**, 143–175.

[341] R. Kidambi, P.K. Newton (1999). Collision of three vortices on a sphere, *Il Nuovo Cimento* **22** **C**(6), 779–791.

[342] R. Kidambi, P.K. Newton (2000a). Streamline topologies for integrable vortex motion on a sphere, *Physica D* **140**, 95–125.

[343] R. Kidambi, P.K. Newton (2000b). Vortex motion on a sphere with solid boundaries, *Phys. Fluids* **12**(3), 581–588.

[344] M. Kiessling (1993). Statistical mechanics of classical particles with logarithmic interactions, *Comm. Pure App. Math.* **46**, 27–56.

[345] Y. Kimura (1988). Chaos and collapse of a system of point vortices, *Fluid Dyn. Res.* **3**, 98–104.

[346] Y. Kimura (1999). Vortex motion on surfaces with constant curvature, *Proc. Roy. Soc. Lond.* **A** **455**, 245–259.

[347] Y. Kimura, R.H Kraichnan (1993). Statistics of an advected passive scalar, *Phys. Fluids* **A5**(9), 2264–2277.

[348] Y. Kimura, H.Okamoto (1987). Vortex motion on a sphere, *J. Phys. Soc. Japan* **56**(12), 4203–4206.

[349] G.R. Kirchhoff (1876). Vorlesungen über Mathematische Physik, **I**, Teubner, Leipzig.

[350] F. Kirwan (1988). The topology of reduced phase spaces of the motion of vortices on a sphere, *Physica D* **30**, 99–123.

[351] R. Klein, O.M. Knio (1995). Asymptotic vorticity structure and numerical simulation of slender vortex filaments, *J. Fluid Mech.* **284**, 275–321.

[352] R. Klein, O.M. Knio, L. Ting (1996). Representations of core dynamics in slender vortex filament simulations, *Phys. Fluids* **8**(9), 2415–2425.

[353] R. Klein, A.J. Majda (1991a). Self–stretching of a perturbed vortex filament. I. The asymptotic equation for deviations from a straight line, *Physica D* **49**, 323–352.

[354] R. Klein, A.J. Majda (1991b). Self–stretching of a perturbed vortex filament. II. Structure of solutions, *Physica D* **53**, 267–294.

[355] R. Klein, A.J. Majda (1993). An asymptotic theory for the nonlinear instability of anti–parallel pairs of vortex filaments, *Phys. Fluids* **A5**, 369–387.

[356] R. Klein, A. Majda, K. Damodaran (1995). Simplified equations for the interaction of nearly parallel vortex filaments, *J. Fluid Mech.* **288**, 201–248.

[357] K.V. Klyatskin, G.M. Reznik (1989). Point vortices on a rotating sphere, *Oceanology* **29**(1), 12–16.

[358] O. Knio, A. Ghoniem (1990). Three–dimensional vortex methods, *J. Comp. Phys.* **86**, 75–106.

[359] H. Kober (1952). *Dictionary of Conformal Representations*, Dover, New York.

[360] N.E. Kochin, I.A. Kibel, N.V. Rose (1965). *Theoretical Hydromechanics*, Interscience Publishers, New York.

[361] J. Koiller, S. Carvalho (1985). Chaos and non–integrability of point vortex motions, Laboratorio de Computacao Cientifica, preprint 010–85.

[362] J. Koiller, S.P. Carvalho (1989). Non–integrability of the 4–vortex system: analytical proof, *Comm. Math. Phys.* **120**(4), 643–652.

[363] J. Koiller, S. Carvalho, R. da Silva Rodriques, L.C. Concalves de Olivera (1985). On Aref's vortex motions with a symmetry center, *Physica D* **16**, 27–61.

[364] J. Koiller, K. Ehlers, R. Montgomery (1996). Problems and progress in microswimming, *J. Nonlinear Sci.* **6**, 507–541.

[365] A.N. Kolmogorov (1954). On the preservation of quasi–periodic motions under a small variation of Hamilton's function, *Dokl. Akad. Nauk. SSSR* **98**, 525.

[366] M. Konstantinov (1994). Chaotic phenomena in the interaction of vortex rings, *Phys. Fluids* **6**(5), 1752–1767.

[367] V.V. Kozlov (1983). Integrability and non–integrability in Hamiltonian mechanics, *Russ. Math. Surveys*, **38**(1), 1–76.

[368] R.H. Kraichnan (1975). Statistical dynamics of two–dimensional flow, *J. Fluid Mech.* **67**, 155–175.

[369] R. H. Kraichnan (1993). Stochastic modeling of turbulence, in *Turbulence in Fluid Flows*, IMA **55**, Springer–Verlag, New York.

[370] R. H. Kraichnan, Y.S. Chen (1989). Is there a statistical mechanics of turbulence, *Physica D* **37**, no. 1–3, 160–172.

[371] R.H. Kraichnan, D. Montgomery (1980). Two dimensional turbulence, *Rep. Prog. Phys.* **43**(5), 547–619.

[372] R. Krasny (1986). A study of singularity formation in a vortex sheet by the point vortex approximation, *J. Fluid Mech.* **167**, 65–93.

[373] M. Kruskal (1962). Asymptotic theory of Hamiltonian and other systems with all solutions nearly periodic, *J. Math. Phys.* **3**, 806–828.

[374] M. Kugler (1989). Motion in noninertial systems: Theory and demonstrations, *Am. J. Phys.* **57**, 249–251.

[375] P.A. Kuibin, V.L. Okulov (1998). Self–induced motion and asymptotic expansion of the velocity field in the vicinity of a helical vortex filament, *Phys. Fluids* **10**, 607–614.

[376] S. Kuksin (1993). *Nearly Integrable Infinite Dimensional Hamiltonian Systems*, Springer Lecture Notes in Math 1552, Springer–Verlag, New York.

[377] R. Kulsrud (1957). Adiabatic invariant of the harmonic oscillator, *Phys. Rev* **106**, 205–207.

[378] I. Kunin, F. Hussain, Z. Zhou, D. Kovich (1990). Dynamics of point vortices in a special rotating frame, *Int. J. Eng. Sci.* **28**(9), 965–970.

[379] I. Kunin, F. Hussain. X. Zhou (1994). Dynamics of a pair of vortices in a rectangle, *Int. J. Eng. Sci.* **32**(11), 1835–1844.

[380] I. Kunin, F. Hussain, X. Zhou, S.J. Prishepionok (1992). Centroidal frames in dynamical systems I. Point vortices, *Proc. Roy. Soc. Lond.* **A 439**, 441–463.

[381] L. Kuznetsov, G.M. Zavlavsky (1998). Regular and chaotic advection in the flowfield of a three–vortex system, *Phys. Rev. E* **58**(6), 7330–7349.

[382] L. Kuznetsov, G.M. Zavlavsky (2000). Passive particle transport in three–vortex flow, *Phys. Rev. E* **61**(4), 3777–3792.

[383] O.A. Ladyzhenskaya (1969). *The Mathematical Theory of Viscous Incompressible Flow*, 2nd ed., Gordon and Breach, New York.

[384] M. Lagally (1921). Über ein verfahren zur transformation ebener wirbelprobleme, *Math. Z.* **10**, 231–239.

[385] J. Lagrange (1788). *Mécanique Analytique*, Veuve Desaint, Paris.

[386] H. Lamb (1932). *Hydrodynamics*, Dover, New York.

[387] G.L. Lamb (1976). Solitons and the motion of helical curves , *Phys. Rev. Lett.* **37**, 235–237.

[388] G.L. Lamb (1977). Solitons on moving space curves, *J. Math. Phys.* **18**, 1654–1661.

[389] G.L. Lamb (1980). *Elements of Soliton Theory*, Wiley, New York.

[390] L.D. Landau, E.M. Lifschitz (1959). *Fluid Mechanics*, Pergamon Press, Oxford.

[391] L.D. Landau, E.M. Lifschitz (1976). *Mechanics*, Pergamon Press, Oxford.

[392] L. Landau, E. Lifshitz (1980). *Statistical Physics*, Pergamon Press, Oxford.

[393] P.S. Laplace (1799). *Traité de Mécanique Céleste*, Duprat, Paris.

[394] J. Larmor (1889). On the images of vortices in a spherical vessel, *Quart. J. Math.* **23**, 338.

384   References

[395] J.P. Laval, P.H. Chavanis, B. Dubrulle, C. Sire (2000). Scaling laws and vortex profiles in 2D decaying turbulence, preprint E–print http://xxx.lanl.gov/abs/cond–mat/0005468.

[396] G. Leal (1992). *Laminar Flow and Convective Transport Processes*, Butterworth–Heinemann, MA.

[397] J.L. Lebowitz, H. Rose, E. Speer (1988). Statistical mechanics of the nonlinear Schrödinger equation, *J. Stat. Phys.* **50**, 657–687.

[398] T.D. Lee (1952). On some statistical properties of hydromechanical and magnetohydromechanical fields, *Q. App. Math.* **10**, 69–74.

[399] S. Leibovich, H.Y. Ma (1983). Soliton propagation on vortex cores and the Hasimoto soliton, *Phys. Fluids* **26**, 3173–3179.

[400] C.E. Leith (1984). Minimum enstrophy vortices, *Phys. Fluids* **27**, 1388–1395.

[401] A. Lenard (1959). Adiabatic invariance to all orders, *Annals of Physics* **6**, 261–276.

[402] A. Leonard (1985). Computing three-dimensional incompressible flows with vortex elements, *Ann. Rev. Fluid Mech.* **17**, 523–559.

[403] N.E. Leonard, J.E. Marsden (1997). Stability and drift of underwater vehicle dynamics: mechanical systems with rigid motion symmetry, *Physica D* **105**, 130–162.

[404] X. Leoncini, L. Kuznetsov, G. Zaslavsky (2000). Motion of three vortices near collapse, *Phys. Fluids* **12**(8), 1911–1927.

[405] M. Lesieur (1983). Introduction a la turbulence bidimensionelle, in Journal de Mécanique théorique et appliquée, Numéro spécial, 5–20.

[406] M. Levi (1981). Adiabatic invariants of linear Hamiltonian systems with periodic coefficients, *J. Diff. Eqns* **42**(1), 47–71.

[407] M. Levi (1993). Geometric phases in the motion of rigid bodies, *Arch. Rat. Mech. Anal.* **122**, 213–219.

[408] D. Levi, A. Sym, S. Wojciechowski (1983). N–solitons on a vortex filament, *Phys. Lett A* **94**, 408–411.

[409] R.D. Levine, M. Tribus (eds.) (1979). *The Maximum Entropy Formalism*, MIT Press, Cambridge.

[410] H. Levy, A.G. Forsdyke (1928). The steady motion and stability of a helical vortex, *Proc. Roy. Soc. Lond.* **A 120**, 670–690.

[411] D. Lewis, J.E. Marsden, R. Montgomery, T.S. Ratiu (1986). The Hamiltonian structure for dynamic free boundary problems, *Physica D* **18**, 391–404.

[412] D. Lewis, T. Ratiu (1996). Rotating n-gon/kn-gon vortex configurations, *J. Nonlinear Sci.* **6**, 385–414.

[413] Y. Li, D. McLaughlin, J. Shatah, S. Wiggins (1996). Persistent homoclinic orbits for a perturbed NLS, *Comm. Pure App. Math.* **49**(11), 1175–1255.

[414] A.J. Lichtenberg, M.A. Lieberman (1982). *Regular and Stochastic Motion*, Springer–Verlag, New York.

[415] W. Lick (1969). Two–variable expansions and singular perturbation problems, *SIAM J. App. Math.* **17**, 815–825.

[416] M.J. Lighthill (1958). *An Introduction to Fourier Analysis and Generalized Functions*, Cambridge Press, Cambridge.

[417] C.C. Lim (1990). Existence of KAM tori in the phase space of lattice vortex systems , *J. App. Math. and Phys. (ZAMP)* **41**, 227–244.

[418] C.C. Lim (1991a). Binary trees, symplectic matrices and the Jacobi coordinates of celestial mechanics, *Arch. Rat. Mech. Anal.* **115**, 153–165.

[419] C.C. Lim (1991b). Graph theory and a special class of symplectic transformations: the generalized Jacobi variables, *J. Math. Phys.* **32**(1), 1–7.

[420] C.C. Lim (1993a). A combinatorial perturbation method and whiskered tori in vortex dynamics, *Physica D* **64**, 163–184.

[421] C.C. Lim (1993b). Nonexistence of Lyapunov functions and the instability of the von Karman vortex streets, *Phys. Fluids* A **5**(9), 2229–2233.

[422] C.C. Lim (1999). Mean field theory and coherent structures for point vortices in the plane, *Phys. Fluids* **11**(5), 1201–1207.

[423] C.C. Lim (1998). Relative equilibria of symmetric n–body problems on a sphere: inverse and direct results, *Comm. Pure App. Math.* **51**(4), 341–371.

[424] C.C. Lim (2000). Exact solutions of an energy–enstrophy theory for the barotropic vorticity equation on a rotating sphere, preprint.

[425] C.C. Lim, A.J. Majda (2000). Point vortex dynamics for coupled surface–interior QG and propagating heton clusters in models for ocean convection, preprint.

[426] C.C. Lim, J. Montaldi, M.R. Roberts (1999). Relative equilibria of point vortices on a sphere, preprint.

[427] C.C. Lin (1941a). On the motion of vortices in 2D I. Existence of the Kirchhoff– Routh function, *Proc. Natl. Acad. Sci.* **27**, 570–575.

[428] C.C. Lin (1941b). On the motion of vortices in 2D II. Some further investigations on the Kirchhoff–Routh function, *Proc. Natl. Acad. Sci.* **27**, 575–577.

[429] C.C. Lin (1976). Theory of spiral structures, in *Theoretical and Applied Mechanics*, W.T. Koiter, ed., 57–69, North–Holland, Amsterdam.

[430] P.L. Lions (1996). *Mathematical Topics in Fluid Mechanics, Vol. 1 Incompressible Models*, Oxford Lecture Series in Mathematics and its Applications, 3, Oxford.

[431] P.L. Lions, A. Majda (2000). Equilibrium statistical theory for nearly parallel vortex filaments , *Comm. Pure App. Math.*, **Vol. LIII**(1), 76–142.

[432] R.G. Littlejohn (1982). Singular Poisson Tensors, in *Mathematical Methods in Hydrodynamics and Integrability in Dynamical Systems*, M. Tabor, Y.M. Treve, eds., AIP Conf. Proc. **88**, 47–66.

[433] R.G. Littlejohn (1988). Phase anholonomy in the classical adiabatic motion of charged particles, *Phys. Rev. A* **38**, 6034–6045.

[434] R.G. Littlejohn, W.G. Flynn (1991). Geometric phases in the asymptotic theory of coupled wave equations, *Phys. Rev. A* **3**(8), 5239–5256.

[435] R.G. Littlejohn, M. Reinsch (1997). Gauge fields in the separation of rotations and internal motions in the n–body problem, *Rev. Mod. Phys.* **69**(1), 213–275.

[436] J.E. Littlewood (1963). Lorentz's pendulum problem, *Annals of Phys.* **21**, 233–242.

[437] P. Lochak (1990). Effective speed of Arnold's diffusion and small denominators, *Phys. Lett. A* **143**(1), 29–42.

[438] P. Lochak, C. Meunier (1988). *Multiphase Averaging for Classical Systems*, Springer–Verlag, New York.

[439] P. Lochak, A. Neishtadt (1992). Estimates of stability time for nearly integrable systems with a quasi–convex Hamiltonian, *Chaos* 4(2), 494–500.

[440] M. Lopes–Filho, H. Nussenzveig Lopes (1998). An extension of C. Marchioro's bound on the growth of a vortex patch to flows with L^p vorticity, *SIAM J. Math. Anal.* 29(3), 596–599.

[441] A.E.H. Love (1893). On the stability of certain vortex motion, *Proc. Roy. Soc. Lond.* 18–42.

[442] A.E.H. Love (1894). On the motion of paired vortices with a common axis, *Proc. Lond. Math. Soc.* 25, 185.

[443] H.H. Luithardt, J.B. Kadtke, G. Pedrizzetti (1994). Chaotic capture of vortices by a moving body. II. Bound pair model, *Chaos* 4 (4), 681–691.

[444] T.S. Lundgren (1982). Strained spiral vortex model for turbulent fine structure, *Phys. Fluids*, 25(12), 2193–2203.

[445] T.S. Lundgren (1993). A small scale turbulence model, *Phys. Fluids* A5(6), 1472–1483.

[446] T.S. Lundgren, Y.B. Pointin (1977a). Non–Gaussian probability distributions for a vortex fluid, *Phys. Fluids* 20(3), 356–363.

[447] T.S. Lundgren, Y.B. Pointin (1977b). Statistical mechanics of two–dimensional vortices, *J. Stat. Phys.* 17, 323–355.

[448] D. Lynden–Bell (1967). Statistical mechanics of violent relaxation in stellar systems, *Mon. Not. R. Astr. Soc.* 136, 101–121.

[449] R.S. MacKay, J.D. Meiss (1987). *Hamiltonian Dynamical Systems*, Adam Hilgar, Bristol.

[450] A.J. Majda (1986). Vorticity and the mathematical theory of incompressible fluid flow, *Comm. Pure App. Math.* 39, S187–S220.

[451] A. Majda (1988). Vortex Dynamics: Numerical Analysis, Scientific Computing, and Mathematical Theory, in ICIAM 87, J. McKenna, R. Temam, eds., Society for Industrial and Applied Math., Philadelphia.

[452] A. Majda (1991). Vorticity, turbulence, and acoustics in fluid flow, *SIAM Review* 33, 349–388.

[453] A.J. Majda (1998). Simplified asymptotic equations for slender vortex filaments , in *Recent Advances in Partial Differential Equations, Venice 1996*, Proceedings of Symposia in Applied Math., **54**, R. Spigler, S. Venakides, eds., 237–280, American Mathematical Society, Providence.

[454] A.J. Majda, A. Bertozzi (2001). *Vorticity and Incompressible Fluid Flow*, Cambridge Texts in Applied Mathematics **27**, Cambridge Press, Cambridge.

[455] A.J. Majda, M. Holen (1997). Dissipation, topography and statistical theories of large scale coherent structures, *Comm. Pure App. Math.* **L**, 1183–1234.

[456] A.J. Majda, P.R. Kramer (1999). Simplified models for turbulent diffusion: Theory, numerical modelling, and physical phenomena, *Phys. Reports* **314**, 237–574.

[457] C. Marchioro (1988). Euler evolution for singular initial data and vortex theory: A global solution, *Comm. Math. Phys.* **116**, 45–55.

[458] C. Marchioro (1994). Bounds on the growth of the support of a vortex patch, *Comm. Math. Phys.* **164**, 507–524.

[459] C. Marchioro, E. Pagani (1986). Evolution of two concentrated vortices in a two–dimensional bounded domain, *Math. Meth. App. Sci.* **8**, 328–344.

[460] C. Marchioro, M. Pulvirenti (1983). Euler evolution for singular initial data and vortex theory, *Comm. Math. Phys.* **91**, 563–572.

[461] C. Marchioro, M. Pulvirenti (1984). *Vortex Methods in Two Dimensional Fluid Dynamics*, Lecture Notes in Physics **203**, Springer–Verlag, New York.

[462] C. Marchioro, M. Pulvirenti (1985). Some considerations on the nonlinear stability of stationary planar Euler flows, *Comm. Math. Phys.* **100**, 343–354.

[463] C. Marchioro, M. Pulvirenti (1994). *Mathematical Theory of Incompressible Nonviscous Fluids*, **AMS 96**, Springer–Verlag, New York.

[464] B. Marcu, E. Meiburg, P.K. Newton (1995). Dynamics of heavy particles in a Burgers vortex, *Phys. Fluids* **7**(2), 400–410.

[465] H. Marmanis (1998). The kinetic theory of point vortices, *Proc. R. Soc. Lond.* **A 454**, 587–606.

[466] J.E. Marsden (1992). *Lectures on Mechanics*, London Mathematical Society Lecture Note Series **174**, Cambridge Press, Cambridge.

[467] J.E. Marsden, M.J. Hoffman (1987). *Basic Complex Analysis*, 2nd Edition, W.H. Freeman and Co., New York.

[468] J.E. Marsden, R. Montgomery (1986). Covariant Poisson brackets for classical fields, *Ann. Phys.* **169**, 29–47.

[469] J.E. Marsden, R. Montgomery, T. Ratiu (1989). Cartan–Hannay–Berry phases and symmetry, *Cont. Math.* AMS, **97**, 279–295.

[470] J.E. Marsden, R. Montgomery, T. Ratiu (1990). Reduction, symmetry and phases in mechanics, AMS Memoirs, **436**.

[471] J.E. Marsden, O.M. O Reilly, F.J. Wicklin, B.W. Zombro (1991) Symmetry, stability, geometric phases and mechanical integrators (Part II)., *Nonlinear Science Today* **1**(1).

[472] J.E. Marsden, S. Pekarsky, S. Shkoller (1999). Stability of relative equilibria of point vortices on a sphere and symplectic integrators, *Il Nuovo Cimento* **22** **C**(6), 793–802.

[473] J.E. Marsden, T.S. Ratiu (1999). *Introduction to Mechanics and Symmetry*, Second Edition, **TAM 17**, Springer–Verlag, New York.

[474] J.E. Marsden, T.S. Ratiu, S. Shkoller (2000). The geometry and analysis of the averaged Euler equations and a new diffeomorphism group, *Geom. Func. Analysis*, vol. 10, 582-599.

[475] J.E. Marsden, J. Scheurle (1995). Pattern evocation and geometric phases in mechanical systems with symmetry, *Dyn. and Stab. of Sys.* **10**, 315–338.

[476] J.E. Marsden, J. Scheurle, J. Wendlandt (1996). Visualization of orbits and pattern evocation for the double spherical pendulum, ICIAM 95: Mathematical Research, Academie Verlag, K. Kirchgässner, O. Mahrenholtz, R. Mennicken, eds., **87**, 213–232.

[477] J.E. Marsden, B.N. Shashikanth (2000). On the Hamiltonian model of leap–frogging vortex rings: geometric phases and discrete reduction, CDS preprint.

[478] J.E. Marsden, S. Shkoller (2000). The anisotropic averaged Euler equations, preprint.

[479] J. Marsden, A. Weinstein (1983). Coadjoint orbits, vortices and Clebsch variables for incompressible fluids, *Physica D* **7**, 305–323.

390   References

[480] A. Masotti (1931). *Atti. Pont. Accad. Sci. Nuovi Lincei* 84, 209–216, 235–245, 464–467, 468–473, 623–631.

[481] T. Maxworthy (1974). Turbulent vortex rings, *J. Fluid Mech.* 64, 227–239.

[482] T. Maxworthy (1977). Some experimental studies of vortex rings, *J. Fluid Mech.* 81, 465–495.

[483] W.D. McComb (1989). *Physical Theories of Turbulence*, Cambridge Press, Cambridge.

[484] B.E. McDonald (1974). Numerical calculation of nonunique solutions of a two–dimensional Sinh–Poisson equation, *J. Comp. Phys.* 16, 360–370.

[485] H.P. McKean, K.L. Vaninsky (1997). Action–angle variables for the cubic Schrödinger equation, *Comm. Pure App. Math* L(6), 489–562.

[486] D. McLaughlin, E. Overman (1995). Whiskered tori for integrable PDE's : Chaotic behavior in near integrable pde's, *Surveys in App. Math.* 1, 83–203, Plenum Press, New York.

[487] D. McLaughlin, J. Shatah (1996). Melnikov analysis for PDE's, Lectures in Applied Math., 31, American Mathematical Society, Providence.

[488] J.C. McWilliams (1983). On the relevance of two–dimensional turbulence to geophysical fluid motions, in Journal de Mécanique théorique et appliquée, Numéro spécial, 83–97.

[489] J.C. McWilliams (1984). The emergence of isolated coherent structures in turbulent flow, *J. Fluid Mech.* 146, 21–43.

[490] E. Meiburg, P.K. Newton (1991). Particle dynamics and mixing in a viscously decaying shear layer, *J. Fluid Mech.* 227, 211-244.

[491] E. Meiburg, P.K. Newton, N. Raju, G. Ruetsch (1995). Unsteady models for the nonlinear evolution of the mixing layer, *Phys. Rev. E* 52(2), 1639–1657.

[492] L. Meirovitch (1970). *Methods of Analytical Dynamics*, McGraw–Hill, New York.

[493] M.V. Melander, J.C. McWilliams, N.J. Zabusky (1986). A model for symmetric vortex merger, *Trans. 3rd Army Conf. on App. Math. and Comp.* ARO Rept. 867.

[494] M.V. Melander, J.C. McWilliams, N.J. Zabusky (1987). Axisymmetrization and vorticity gradient intensification of an isolated two dimensional vortex through filamentation, *J. Fluid Mech.* **178**, 137–159.

[495] M.V. Melander, A.S. Styczek, N.J. Zabusky (1984). Elliptically desingularized vortex model for the two–dimensional Euler equations, *Phys. Rev. Lett.* **53**, 1222–1225.

[496] M.V. Melander, N.J. Zabusky, A.S. Styczek (1986). A moment model for vortex interactions of two–dimensional Euler equations. Part 1. Computational validation of a Hamiltonian elliptical representation, *J. Fluid Mech.* **167**, 95–115.

[497] V.V. Meleshko, M. Yu. Konstantinov, A.A. Gurzhi, T.P. Konovalijuk (1992). Advection of a vortex pair atmosphere in a velocity field of point vortices, *Phys. Fluids* **A4**(12), 2779–2797.

[498] V.K. Melnikov (1963). On the stability of the center for time periodic perturbations, *Trans. Mosc. Math. Soc.* **12**, 1–57.

[499] G.J. Mertz (1978). Stability of body–centered polygonal configurations of ideal vortices, *Phys. Fluids* **21**, 1092–1095.

[500] K.R. Meyer, G.R. Hall (1992). *Introduction to Hamiltonian Dynamical Systems and the N–Body Problem,* **AMS 90**, Springer–Verlag, New York.

[501] R. Meyer (1976). Adiabatic variation – Part V: Nonlinear near–periodic oscillator, *ZAMP* **27**, 181–195.

[502] R. Meyer (1980). Exponential asymptotics , *SIAM Review* **22**, 213–224.

[503] I. Mezić (1994). On the geometrical and statistical properties of dynamical systems: theory and applications, Ph.D. thesis, Caltech.

[504] I. Mezić, A. Leonard, S. Wiggins (1998). Regular and chaotic particle motion near a helical vortex filament, *Physica D* **111**, 179–201.

[505] I. Mezić, I. Min (1997). On the equilibrium distribution of like-signed vortices in two dimensions, in Proceedings of the Workshop on Chaos, Kinetics and Nonlinear Dynamics in Fluids and Plasmas, Marseille France.

[506] I. Mezić, S. Wiggins (1994). On the integrability and perturbation of three–dimensional fluid flows with symmetry, *J. Nonlinear Sci.* **4**, 157–194.

[507] J. Michel, R. Robert (1994). Large deviations for Young measures and statistical mechanics of infinite dimensional dynamical systems with conservation law, *Comm. Math. Phys.* **159**, 195–215.

[508] A. Mikelic, R. Robert (1998). On the equations describing a relaxation towards a statistical equilibrium state in the two–dimensional perfect fluid dynamics, *SIAM J. Math. Anal.* **29**(5), 1238–1255.

[509] J. Miller (1990). Statistical mechanics of Euler equations in two dimensions, *Phys. Rev. Lett.* **65**, 2137–2140.

[510] J. Miller, P. Weichman, M.C. Cross (1992). Statistical mechanics, Euler's equations and Jupiter's red spot, *Phys. Rev. A* **45**, 2328–2359.

[511] K. G. Miller (1996). Stationary corner vortex configurations, *Z angew Math Phys (ZAMP)* **47**, 39–56.

[512] L.M. Milne–Thompson (1940). Hydrodynamic images, *Proc. Camb. Phil. Soc.* **36**, 246–247.

[513] L.M. Milne–Thompson (1968). *Theoretical Hydrodynamics*, MacMillan, New York.

[514] T. Miloh, D.J. Shlein (1977). Passage of a vortex ring through a circular aperature in an infinite plane, *Phys. Fluids* **20**, 1219–1227.

[515] I.A. Min, I. Mezić, A. Leonard (1996). Lévy stable distributions for velocity and velocity difference in systems of vortex elements, *Phys. Fluids* **8**(5), 1169–1180.

[516] T. Miyazaki, Y. Fukumoto (1988). N–solitons on a curved vortex filament with axial flow, *J. Phys. Soc. Japan* **57**, 3365–3376.

[517] H.K. Moffatt (1969). The degree of knottedness of tangled vortex lines, *J. Fluid Mech.* **35**, part 1, 117–129.

[518] H.K. Moffatt (1985). Magnetostatic equilibria and analogous Euler flows of arbitrarily complex topology. Part I. Fundamentals, *J. Fluid Mech.* **159**, 359–378.

[519] H.K. Moffatt (1990a). Structure and stability of solutions of the Euler equations: a Lagrangian approach, *Phil. Trans. Roy. Soc. A* **333**(1), 321–342.

[520] H.K. Moffatt (1990b). The energy spectrum of knots and links, *Nature* **347**, 367–369.

[521] H.K. Moffatt (1993). Spiral structures in turbulent flow, in *New Approaches and Concepts in Turbulence*, T. Dracos, A. Tsinobar, eds.

[522] H.K. Moffatt, R.L. Ricca (1991). Interpretation of invariants of the Betchov–DaRios equations and of the Euler equations, in *The Global Geometry of Turbulence*, J. Jimenez, ed., 257–264, Plenum Press, New York.

[523] H.K. Moffatt, A. Tsinobar, eds. (1990). *Topological Fluid Mechanics*, Cambridge Press, Cambridge.

[524] H.K. Moffatt, A. Tsinobar (1992). Helicity in laminar and turbulent flows, *Ann. Rev. Fluid Mech.* **24**, 281–312.

[525] J. Montaldi, R.M. Roberts (1999). Relative equilibria of molecules, *J. Nonlinear Sci.* **9**, 53–88.

[526] D. Montgomery, G. Joyce (1974). Statistical mechanics of negative temperature states, *Phys. Fluids* **17**, 1139–1145.

[527] D. Montgomery, W.T. Matthaeus, D. Martinez, S. Oughton (1992). Relaxation in two dimensions and the sinh–Poisson equation, *Phys. Fluids* **A4**, 3–6.

[528] R. Montgomery (1988). The connection whose holonomy is the classical adiabatic angle of Hannay and Berry and its generalization to the nonintegrable case, *Comm. Math. Phys.* **120**, 269–294.

[529] R. Montgomery (1991a). How much does the rigid body rotate? A Berry's phase from the 18th century, *Am. J. Phys.* **59** (5), 394–398.

[530] R. Montgomery (1991b). Optimal control of deformable bodies and its relation to gauge theory, in *The Geometry of Hamiltonian Systems*, 403–438, Springer–Verlag, New York.

[531] R. Montgomery (1993). Gauge theory of the falling cat, in *Dynamics and Control of Mechanical Systems*, 193–218, American Mathematical Society, Providence.

[532] R. Montgomery (1996). The geometric phase of the three body problem, *Nonlinearity* **9**, 1341–1360.

[533] D. Moore, P.G. Saffman (1972). *Aircraft Wake Turbulence*, J. Olsen, A. Goldberg, N. Rogers, eds., 339–354, Plenum Press, New York.

[534] D.W. Moore, P.G. Saffman (1972). The motion of a vortex filament with axial flow, *Phil. Trans. Roy. Soc. Lond.* **A 333**, 491–508.

[535] D.W. Moore, P.G. Saffman (1975). The instability of a straight vortex filament in a strain field, *Proc. Roy. Soc. Lond.* **A 346**, 413–425.

[536] R. Moreau (ed.) (1983). *Two Dimensional Turbulence*, Journal de Mécanique Théorique et Appliquée, Special Issue.

[537] G.K. Morikawa, E.V. Swenson (1971). Interacting motion of rectilinear geostrophic vortices, *Phys. Fluids* **14**, 1058–1073.

[538] J.A. Morrison (1966). Comparison of the modified method of averaging and the two variable expansion procedure, *SIAM Rev.* **8**, 66–85.

[539] P.J. Morrison (1982). Poisson brackets for fluids and plasmas, in *Mathematical Methods in Hydrodynamics and Integrability in Dynamical Systems*, M. Tabor, Y.M. Treve, eds., AIP Conf. Proc. **88**, 13–46.

[540] P.J. Morrison (1998). Hamiltonian description of the ideal fluid, *Rev. Mod. Phys.* **70**(2), 467–521.

[541] P.M. Morse, H. Feshbach (1953). *Methods of Theoretical Physics, Parts I,II*, McGraw–Hill, New York.

[542] J. Moser (1962). On invariant curves of area–preserving mappings on an annulus, *Nachr. Akad. Wiss. Goettingen Math. Phys.* **K1**, 1.

[543] J. Moser (1966). A rapidly converging iteration method and nonlinear differential equations, *Annali della Scuolo Nor. Sup. de Pisa* **(3)**, 20, 265–315; 499–535.

[544] J. Moser (1967). Convergent series expansions for quasi–periodic motions, *Math. Ann.* **169**, 136–76.

[545] J. Moser (1973). *Stable and Random Motions in Dynamical Systems*, Princeton U. Press, Princeton.

[546] J. Moser, C.L. Seigel (1971). *Lectures on Celestial Mechanics*, Springer–Verlag, Berlin.

[547] F.R. Moulton (1902). *An Introduction to Celestial Mechanics*, Macmillan, New York.

[548] W. Müller (1930). Bewegung von wirbeln in einer idealen flüssigkeit unter dem einfluss von ebenen wänden, *Z. Angew Math. Mech.* **10**, 227.

[549] T. Needham (1997). *Visual Complex Analysis*, Clarendon Press, Oxford.

[550] Z. Nehari (1952). *Conformal Mapping*, Dover, New York.

[551] A.I. Neishtadt (1975). Passage through a separatrix in a resonance problem with a slowly varying parameter, *PMM* 39, 4, 594–605.

[552] A.I. Neishtadt (1981). On the degree of precision of the adiabatic invariant conservation, *Prikl. Mat. Mekh.* 45(1), 80–87.

[553] A.I. Neishtadt, D.K. Chaikovskii, A.A. Chernikov (1991). Adiabatic chaos and particle diffusion, *Sov. Phys. JETP* 22(3), 423–430.

[554] N.N. Nekhoroshev (1971). The behavior of Hamiltonian systems that are close to integrable ones, *Func. Anal. and its App.* 5,4.

[555] N.N. Nekhoroshev (1972). Action–angle variables and their generalizations, *Trans. Moscow Math. Soc.* 26, 180–98.

[556] N.N. Nekhoroshev (1977). An exponential estimate of the time stability of nearly integrable Hamiltonian systems, *Russ. Math. Surveys* 32(6), 1–65.

[557] J. Neu (1984). The dynamics of a columnar vortex in an imposed strain, *Phys. Fluids* 27(10), 2397–2402.

[558] Z. Neufeld, T. Tél (1997). The vortex dynamics analogue of the restricted three–body problem: advection in the field of three identical point vortices, *J. Phys. A: Math. Gen.* 30, 2263–80.

[559] Z. Neufeld, T. Tél (1998). Advection in chaotically time–dependent open flows, *Phys. Rev. E* 57(3), 2832–2842.

[560] A.C. Newell (1985). *Solitons in Mathematics and Physics*, CBMS–NSF Regional Conference Series in Applied Math, 48, Society for Industrial and Applied Math., Philadelphia.

[561] I. Newton (1687). *Philosophiae Naturalis Principia Mathematica*, Royal Society of London.

[562] P.K. Newton (1994). Hannay–Berry phase and the restricted three-vortex problem, *Physica D* 79, 416–423.

[563] P.K. Newton (2001). Constrained N-vortex problems and particle interactions on a hoop, preprint.

[564] P.K. Newton, E. Meiburg (1991). Particle dynamics in a viscously decaying cat s eye: The effect of finite Schmidt numbers, *Phys. Fluids* A3(5), 1068–1072.

[565] P.K. Newton, I. Mezić, (1999). Streamline patterns and statistical mechanics of vortex motion on a sphere, Bull. APS.

[566] P.K. Newton, I. Mezić (2000). Kinetic point vortex theory on a sphere, preprint.

[567] P.K. Newton, B. Shashikanth (1996). Vortex problems, rotating spiral structures, and the Hannay–Berry phase, Second International Workshop on Vortex Flows and Related Numerical Methods, Montréal Canada, (http://www.emath.fr/Maths/Proc/Vol.1/index.htm).

[568] R.G. Newton (1994). S matrix as geometric phase factor, *Phys. Rev. Lett.* **72**(7), 954–956.

[569] M.V. Nezlin, E.N. Snezhkin (1993). *Rossby Vortices, Spiral Structures, Solitons*, Springer–Verlag Series in Nonlinear Dynamics.

[570] B.R. Noack, A. Banaszuk, I. Mezić (2000). Controlling vortex motion and chaotic advection, preprint.

[571] W.J. Noble (1952). A direct treatment of the Foucault pendulum, *Am. J. Phys.* **20**, 334–336.

[572] J. Norbury (1972). A steady vortex ring close to Hill's spherical vortex, *Proc. Camb. Phil. Soc.* **72**, 253.

[573] J. Norbury (1973). A family of steady rings, *J. Fluid Mech.* **57**, 417–431.

[574] E.A. Novikov (1975). Dynamics and statistics of a system of vortices, *Soviet Phys. JETP* **41**(5), 937–943.

[575] E.A. Novikov (1980). Stochastization and collapse of vortex systems, *Ann. NY Acad. Sci.* **357**, 47–54.

[576] E.A. Novikov (1983). Generalized dynamics of three–dimensional vortical singularities (vortons), *Soviet Phys. JETP* **57**, 566–569.

[577] E.A. Novikov, Yu. B. Sedov (1978). Stochastic properties of a four vortex system, *Soviet Phys. JETP* **48**, 440–444.

[578] E.A. Novikov, Yu. B. Sedov (1979a). Vortex collapse, *Soviet Phys. JETP* **50**(2), 297–301.

[579] E.A. Novikov, Yu. B. Sedov (1979b). Stochastization of vortices, *JETP Lett.* **29**, 677–679.

[580] G.C. Oates (1983). Vector Analysis, in *Handbook of Applied Mathematics*, C. Pearson, ed., 2nd ed., Van Nostrand Reinhold Co.

[581] H. Okamoto, Y. Kimura (1989). Chaotic advection by a point vortex in a semidisk, *Topological Fluid Mechanics*, Proc. IUTAM Symp., ed. H.K. Moffatt, Tsinobar, 105–113, Cambridge Press, Cambridge.

[582] W.M. Oliva (1991). On the chaotic behavior and non–integrability of the four vortices problem, Ann. Inst. H. Poincaré, **55**(21), 707–718.

[583] M. Oliver, S. Shkoller (1999). The vortex blob method as a second-grade non–Newtonian fluid, Oct. E–print, http:// xyz.lanl.gov/abs/math.AP/9910088/

[584] P.J. Olver (1982). A nonlinear Hamiltonian structure for the Euler equations, *J. Math. Anal. App.* **89**, 233.

[585] P.J. Olver (1993). *Applications of Lie Groups to Differential Equations*, 2nd edition, Graduate Texts in Mathematics, Springer–Verlag, New York.

[586] K.A. O Neil (1987). Stationary configurations of point vortices, *Trans. AMS* **302**(2), 383–425.

[587] T.M. O Neil (1999). Trapped plasmas with a single sign of charge, *Phys. Today* 24–30.

[588] L. Onsager (1949). Statistical hydrodynamics, *Il Nuovo Cimento* **9**(6), supp. no. 2, 279–287.

[589] J. Oprea (1995). Geometry and the Foucault pendulum, *Amer. Math. Month.* **102**(6), 515–522.

[590] S. Orszag (1970). Analytical theories of turbulence, *J. Fluid Mech.* **41**, 363–386.

[591] J.M. Ottino (1989). *The Kinematics of Mixing: Stretching, Chaos, and Transport*, Cambridge Texts in Applied Math., Cambridge Press, Cambridge.

[592] J.M. Ottino (1990). Mixing, chaotic advection, and turbulence, *Ann. Rev. Fluid Mech.* **22**, 207–253.

[593] E.A. Overman, N.J. Zabusky (1982). Evolution and merger of isolated vortex structures, *Phys. Fluids* **25**(8).

[594] J.I. Palmore (1982). Relative equilibria of vortices in two dimensions, *Proc. Nat l. Acad. Sci. USA* **79**, 716–718.

[595] S. Pancharatnam (1956). *Proc. Ind. Acad. Sci.* **A 44**, 247.

[596] R.A. Pasmanter (1994). On long lived vortices in 2D viscous flows, most probable states of inviscid 2D flows and a soliton equation, *Phys. Fluids* **6**, 1236–1241.

[597] V.E. Paul (1934). Bewegung eines wirbels in geradlinig begrenzten gebieten, *Z. Angew Math. Mech.* **14**, 105–116.

398   References

[598] J. Pedlosky (1985). The instability of continuous heton clouds, *J. Atmos. Sci.* **42**, 1477–1486.

[599] J. Pedlosky (1987). *Geophysical Fluid Dynamics* , 2nd Ed., Springer–Verlag, New York.

[600] G. Pedrizzetti, J.B. Kadtke, H. Luithardt (1993). Chaotic trapping phenomena in extended systems, *Phys. Rev. E* **48**(5), 3299–3308.

[601] S. Pekarsky, J.E. Marsden (1998). Point vortices on a sphere: stability of relative equilibria, *J. Math. Phys.* **39**, 5894–5907.

[602] R. Pelz (1997). Locally self–similar, finite–time collapse in a high–symmetry vortex filament model, *Phys. Rev. E* **55**(2), 1617–1626.

[603] A. Péntek, T. Tél, Z. Toroczkai (1995a). Chaotic advection in the velocity field of leapfrogging vortex pairs, *J. Phys. A: Math. Gen.* **28**, 2191–2216.

[604] A. Péntek, T. Tél, Z. Toroczkai (1995b). Fractal tracer patterns in open hydrodynamical flows: the case of leapfrogging vortex pairs, *Fractals* **3**(1), 33–53.

[605] A. Péntek, T. Tél, Z. Toroczkai (1996). Transient chaotic mixing in open hydrodynamical flows, *Int. J. Bif. and Chaos* **6**(12B), 2619–2625.

[606] I. Percival, D. Richards (1982). *Introduction to Dynamics*, Cambridge Press, Cambridge.

[607] L. Perko (1996). *Differential Equations and Dynamical Systems*, **TAM 7**, 2nd ed., Springer–Verlag, New York.

[608] A.E. Perry, M.S. Chong (1986). A series expansion study of the Navier–Stokes equations with applications to three–dimensional separation patterns, *J. Fluid Mech.* **173**, 207–223.

[609] A.E. Perry, B.D. Farlie (1974). Critical points in flow patterns, *Adv. Geophysics* **B 18**, 299–315.

[610] K. Petersen (1982). *Ergodic Theory*, Cambridge University Press, Cambridge.

[611] R.T. Pierrehumbert (1980). A family of steady translating vortex pairs with distributed vorticity, *J. Fluid Mech.* **99**, 129–144.

[612] H. Poincaré (1890). Sur la probléme des trois corps et les équations de la dynamique , *Acta Math.* **13**, 1–271.

[613] H. Poincaré (1892). *Les Méthods Nouvelles de la Méchanique Céleste*, Gauthiers–Villars, Paris.

[614] H. Poincaré (1893). *Théorie des Tourbillons*, George Carré, Editeur.

[615] Y. B. Pointin, T.S. Lundgren (1976). Statistical mechanics of two-dimensional vortices in a bounded container, *Phys. Fluids* **19**(10), 1459–1470.

[616] H. Pollard (1962). *Mathematical Introduction to Celestial Mechanics*, Prentice–Hall, New York.

[617] L.M. Polvani, D.G. Dritschel (1993). Wave and vortex dynamics on the surface of a sphere, *J. Fluid Mech.* **255**, 35–64.

[618] L.M. Polvani, J. Wisdom (1990a). On chaotic flow around the Kida vortex, in *Topological Fluid Mechanics*, H.K. Moffatt, A. Tsinobar, eds., 34–44, Cambridge Press, Cambridge.

[619] L.M. Polvani, J. Wisdom (1990b). Chaotic Lagrangian trajectories around an elliptical vortex patch embedded in a constant and uniform background shear flow, *Phys. Fluids* **A2**(2), 123–126.

[620] G. Ponce (1985). Remarks on a paper by J.T. Beale, T. Kato, and A. Majda, *Comm. Math. Phys.* **98**, 349–353.

[621] J. Pöschel (1982). Integrability of Hamiltonian systems on Cantor sets, *Comm. Pure App. Math.* **35**, 653–695.

[622] C. Pozrikidis (1986). The nonlinear instability of Hill's spherical vortex, *J. Fluid Mech.* **168**, 337–367.

[623] C. Pozrikidis (1997). *Introduction to Theoretical and Computational Fluid Dynamics*, Oxford U. Press, Oxford.

[624] E.G. Puckett (1993). *Vortex Methods: An Introduction and Survey of Several Research Topics*, Incompressible Computational Fluid Dynamics: Trends and Advances, Cambridge Press, Cambridge.

[625] A. Pumir, E. Siggia (1992a). Finite time singularities in the axisymmetric three dimensional Euler equations, *Phys. Rev. Lett.* **68**, 1511.

[626] A. Pumir, E. Siggia (1992b). Development of singular solutions to the axisymmetric Euler equations, *Phys. Fluids* **A4**, 1472.

[627] T. Ratiu (1980). Involution theorems, in *Geometric Methods in Mathematical Physics*, G. Kaiser, J. Marsden, eds., Springer Lecture Notes **775**, 219–257.

[628] R.L. Ricca (1991). Rediscovery of DaRios equations, *Nature* **352**, 561–562.

[629] R.L. Ricca (1992). Physical interpretation of certain invariants for vortex filament motion under LIA, *Phys. Fluids* **A4**, 938–944.

[630] R.L. Ricca (1994). The effect of torsion on the motion of a helical vortex filament, *J. Fluid Mech.* **273**, 241–259.

[631] R.L. Ricca (1993, 1995a). Torus knots and polynomial invariants for a class of soliton equations, *Chaos* **3**(1), 83–91, Erratum: *Chaos* **5**(1) 346.

[632] R.L. Ricca (1995b). Geometric and topological aspects of vortex filament dynamics under LIA, in *Small Scale Structues in Three-Dimensional Hydro and Magnetohydro-dynamic Turbulence*, M. Meneguzzi ed., 99–107, Lecture Notes in Physics **462**, Springer–Verlag, New York.

[633] R.L. Ricca (1996). The contributions of DaRios and Levi–Civita to asymptotic potential theory and vortex filament dynamics, *Fluid Dyn. Res.* **18**, 245–268.

[634] R.L. Ricca, M.A. Berger (1996). Topological ideas and fluid mechanics, *Phys. Today* **49**(12), 28–34.

[635] R.L. Ricca, D.C. Samuels, C.F. Barenghi (1999). Evolution of vortex knots, *J. Fluid Mech.* **391**, 29–44.

[636] S. Richardson (1973). On the no–slip boundary condition, *J. Fluid Mech.* **59**, 707–719.

[637] G.F. Roach (1982). *Green's Functions*, 2nd edition, Cambridge Press, Cambridge.

[638] J.M. Robbins (1994). The geometric phase for chaotic unitary families, *J. Phys. A: Math. Gen.* **27**(4), 1179–1189.

[639] J.M. Robbins, M.V. Berry (1992). The geometric phase for chaotic systems, *Proc. Roy. Soc. Lond.* **A 436**, 631–661.

[640] J.M. Robbins, M.V. Berry (1994). A geometric phase for $m = 0$ spins, *J. Phys. A: Math. Gen.* **27**(12), L435–L438.

[641] R. Robert (1991). A maximum entropy principle for two-dimensional perfect fluid dynamics, *J. Stat. Phys.* **65**(3–4), 531–553.

[642] R. Robert, C. Rosier (1997). The modeling of small scales in two–dimensional turbulent flows: a statistical mechanics approach, *J. Stat. Phys.* **86**(3–4), 481–515.

[643] R. Robert, J. Sommeria (1991). Statistical equilibrium states for two dimensional flows, *J. Fluid Mech.* **229**, 291–310.

[644] R. Robert, J. Sommeria (1992). Relaxation towards a statistical equilibrium state in two-dimensional perfect fluid dynamics, *Phys. Rev. Lett.* **69**, 2776–2779.

[645] P. Roberts (1972). Hamiltonian theory for weakly interacting vortices, *Mathematika* **19**, 169–179.

[646] B. Robertson (1979). Application of maximum entropy to non–equilibrium statistical mechanics, in R.D. Levine, M. Tribus, eds., *The Maximum Entropy Formalism*, MIT Press, Cambridge.

[647] C. Robinson (1988). Horseshoes for autonomous Hamiltonian systems using the Melnikov integral, *Erg. Theory Dyn. Sys.* **8**, 395–409.

[648] V. Rom-Kedar, A. Leonard, S. Wiggins (1990). An analytical study of transport, mixing and chaos in an unsteady vortical flow, *J. Fluid Mech.* **214**, 347–394.

[649] L. Rosenhead (1930). The spread of vorticity in the wake behind a cylinder, *Proc. R. Soc. Lond.* **A 127**, 590–612.

[650] N. Rott (1958). On the viscous core of a line vortex, *ZAMP* **9**, 543–553.

[651] N. Rott (1989). Three vortex motion with zero total circulation, *J. App. Math. Phys. (ZAMP)* **40**, 473–494.

[652] N. Rott (1990). Constrained three and four vortex problems, *Phys. Fluids* **A2**(8), 1477–1480.

[653] N. Rott (1994). Four vortices on doubly periodic paths, *Phys. Fluids* **6**(2), 760–764.

[654] E.J. Routh (1881). Some applications of conjugate functions, *Proc. Lond. Math. Soc.* **12**, 73–89.

[655] D. Ruelle (1969). *Statistical Mechanics: Rigorous Results*, W.A. Benjamin, New York.

[656] S.M. Rytov (1989). Dokl. Akad. Nauk. USSR, 18, 263, 1938; reprinted in B. Markovski, V.I. Vinitsky, *Topological Phases in Quantum Theory*, World Scientific, Singapore.

[657] E.B. Saff, A.B.J. Kuijlaars (1997). Distributing many points on a sphere, *Math. Intelligencer* **19**(1), 5–11.

402   References

[658] P.G. Saffman (1967). The self propulsion of a deformable body in a perfect fluid, *J. Fluid Mech.* **28**, 385–389.

[659] P.G. Saffman (1970). The velocity of viscous vortex rings, *Stud. App. Maths.* **49**, 371–380.

[660] P.G. Saffman (1979). The approach of a vortex pair to a plane surface in inviscid fluid, *J. Fluid Mech.* **92**, 497–503.

[661] P.G. Saffman (1992). *Vortex Dynamics*, Cambridge University Press, Cambridge.

[662] P.G. Saffman, G. Baker (1979). Vortex interactions, *Ann. Rev. Fluid Mech.* **11**, 95–122.

[663] P.G. Saffman, D.I. Meiron (1986). Difficulties with three-dimensional weak solutions for inviscid incompressible flow, *Phys. Fluids* **29**, 2373–2375.

[664] P.G. Saffman, J.S. Sheffield (1977). Flow over a wing with an attached free vortex, *Stud. in App. Math.* **57**, 107–117.

[665] G. Salmon (1854). *A Treatise on Conic Sections*, 6th ed. Chelsea, New York.

[666] R. Salmon (1988). Hamiltonian Fluid Mechanics, *Ann. Rev. Fluid Mech.* **20**, 225–256.

[667] J.A. Sanders (1982). Melnikov's method and averaging, *Celestial Mech.* **28**, 171–181.

[668] J.A. Sanders, F. Verhulst (1985). *Averaging methods in nonlinear dynamical systems*, **AMS 59**, Springer–Verlag, New York.

[669] S. Schochet (1995). The weak vorticity formulation of the 2D Euler equations and concentration–cancellation, *Comm. Part. Diff. Eqn.* **20**(5–6), 1077–1104.

[670] S. Schochet (1996). The point-vortex method for periodic weak solutions of the 2D Euler equations, *Comm. Pure and App. Math.* **49**(9), 911–965.

[671] P. Schwarz (1994). *Topology for Physicists*, Springer–Verlag, New York.

[672] D. Serre (1984). Invariants et dégénérescence symplectique de l'équation d'Euler des fluides parfaits incompressibles, *C.R. Acad. Sci. Paris* **298**, 349–352.

[673] J. Serrin (1959). Mathematical principles of classical fluid dynamics, in *Handbuch der Physik 8*, 125–263, Springer–Verlag, New York.

[674] J.A. Sethian (1991). A Brief Overview of Vortex Methods, in *Vortex Methods and Vortex Motion*, eds. K.E. Gustafson, J.A. Sethian, Society for Industrial and Applied Math., Philadelphia.

[675] C.E. Seyler Jr. (1976). Thermodynamics of two-dimensional plasmas or discrete line vortex fluids, *Phys. Fluids* **19**, 1336–1341.

[676] A. Shapere, F. Wilczek (1987). Self-propulsion at low Reynolds number, *Phys. Rev. Lett.* **58**, 2051–2054.

[677] A. Shapere, F. Wilczek (1989a). Geometry of self-propulsion at low Reynolds number, *J. Fluid Mech.* **198**, 557–585.

[678] A. Shapere, F. Wilczek, eds. (1989b). *Geometric Phases in Physics*, World Scientific, Singapore.

[679] K. Shariff, A. Leonard (1992). Vortex rings, *Ann. Rev. Fluid Mech.* **24**, 235–279.

[680] B.N. Shashikanth (1997). Vortex dynamics and the geometric phase, Ph.D. thesis, Department of Aerospace Engineering, University of Southern California.

[681] B.N. Shashikanth, J.E. Marsden, J. Burdick, S. Kelly (2000). The Hamiltonian structure of a 2–D rigid cylinder interacting dynamically with N point vortices, CDS preprint, Caltech.

[682] B.N. Shashikanth, P.K. Newton (1998). Vortex motion and the geometric phase. Part I. Basic configurations and asymptotics, *J. Nonlinear Sci.* **8**, 183–214.

[683] B.N. Shashikanth, P.K. Newton (1999). Vortex motion and the geometric phase. Part II. Slowly varying spiral structures, *J. Nonlinear Sci.* **9**(2), 233–254.

[684] B.N. Shashikanth, P.K. Newton (2000). Geometric phases for corotating elliptical vortex patches, *J. Math. Phys.* **41**(12), 8148–8162.

[685] J.S. Sheffield (1977). Trajectories of an ideal vortex pair near an orifice, *Phys. Fluids*, **20**, 543–545.

[686] T.G. Shepherd (1987). Non–ergodicity of inviscid two–dimensional flow on a beta–plane and on the surface of a rotating sphere, *J. Fluid Mech.* **184**, 289–302.

[687] T.G. Shepherd (1990). Symmetries, conservation laws, and Hamiltonian structure in geophysical fluid dynamics, *Adv. in Geophys.* **32**, 287–339.

[688] S. Shkoller (1999). The geometry and analysis of non–Newtonian fluids and vortex methods, UC Davis preprint.

[689] S. Shkoller (2000). On averaged incompressible Lagrangian hydrodynamics, preprint. E–print, http://xyz.lanl.gov/abs/math.AP/9908109/

[690] C.L. Siegel, J. Moser (1971). *Lectures in Celestial Mechanics*, Springer–Verlag, New York.

[691] E. Siggia (1981). Numerical study of small scale intermittency in three–dimensional turbulence, *J. Fluid Mech.* **107**, 375–406.

[692] E. Siggia (1985). Collapse and amplification of a vortex filament, *Phys. Fluids* **28**, 794–805.

[693] B. Simon (1983). Holonomy, the quantum adiabatic theorem, and Berry's phase, *Phys. Rev. Lett.* **51**(24), 2167–2170.

[694] K.R. Singh (1954). Path of a vortex round the rectangular bend of a channel with uniform flow, *Z. Angew. Math. Mech.* **34**, 432–435.

[695] C. Sire, P.H. Chavanis (2000). Numerical renormalization group of vortex aggregation in 2D decaying turbulence: The role of three–body interactions, *Phys. Rev. E*, **61**(6), 6644–6653.

[696] S. Smale (1970). Topology and Mechanics, *Inv. Math.* **10**, 305–331, **11**, 45–64.

[697] R.A. Smith (1991). Maximization of vortex entropy as an organizing principle in intermitent decaying two-dimensional turbulence, *Phys. Rev. A* **43**, 1126.

[698] A. Sommerfeld (1964). *Mechanics of Deformable Bodies* (Chapt. IV), Academic Press, New York.

[699] A. Sommerfeld (1964). *Thermodynamics and Statistical Mechanics*, Academic Press, New York.

[700] I. Stakgold (1979). *Green s Functions and Boundary Value Problems*, Wiley, New York.

[701] H.J. Stewart (1943). Periodic properties of the semi–permanent atmospheric pressure systems, *Quart. App. Math.* **1**(3), 262–267.

[702] M.A. Stremler, H. Aref (1999). Motion of three point vortices in a periodic parallelogram, *J. Fluid Mech.* **392**, 101–128.

[703] J.T. Stuart (1967). On finite amplitude oscillations in laminar mixing layers, *J. Fluid Mech.* **29**, 417–440.

[704] Y.K. Suh (1993). Periodic motion of a point vortex in a corner subject to a potential flow, *J. Phys. Soc. Jap.* **62**(10), 3441–3445.

[705] C. Sulem, P. Sulem (1983). The well–posedness of two–dimensional ideal flow, *Journal de Mécanique Théorique et Appliquée, Special Issue*, 217–242.

[706] C. Sulem, P.L. Sulem (1999). *The Nonlinear Schrödinger Equation: Self Focusing and Wave Collapse*, **AMS 139**, Springer–Verlag, New York.

[707] J.L. Synge (1949). On the motion of three vortices, *Can. J. Math.* **1**, 257–270.

[708] J.L. Synge, C.C. Lin (1943). On a statistical model of isotropic turbulence, *Trans. R. Soc. Can.* **37**, 45–63.

[709] V. Szebehely (1967). *Theory of Orbits – The Restricted Problem of Three Bodies*, Academic Press, New York.

[710] M. Tabor (1989). *Chaos and Integrability in Nonlinear Dynamics: An Introduction*, John Wiley and Sons, New York.

[711] R. Takaki, M. Utsumi (1992). Middle–scale structure of point vortex clouds, *Forma* **7**, 107–120.

[712] C.M. Tang, S. Orszag (1978). Two-dimensional turbulence on the surface of a sphere, *J. Fluid Mech.* **87**, 305–319.

[713] J. Tavantzis, L. Ting (1988). The dynamics of three vortices revisited, *Phys. Fluids* **31** (6), 1392–1409.

[714] J.B. Taylor (1972). Negative temperatures in two-dimensional vortex motion, *Phys. Lett.* **40A**, 1–2.

[715] R. Temam (1975). On the Euler equation of incompressible perfect fluids, *J. Funct. Anal.* **20**, 32–43.

[716] R. Temam (1976). *Local Existence of C^∞ Solutions of the Euler Equations of Incompressible Perfect Fluids*, Lecture Notes in Math., **565**, Springer–Verlag, New York.

[717] A. Timofeev (1978). On the constancy of the adiabatic invariant when the nature of the motion changes, *Sov. Phys. JETP* **48**, 656–659.

[718] L. Ting, R. Klein (1991). *Viscous Vortical Flows*, Springer Lecture Notes in Physics, Springer–Verlag, New York.

[719] L. Ting, C. Tung (1965). Motion and decay of a vortex in a nonuniform stream, *Phys. Fluids* **8**, 1309–1351.

[720] F. Tisserand (1889). *Traité de Mécanique Céleste*, Gauthiers–Villars, Paris.

[721] V.K. Tkachenko (1966). Stability of vortex lattices, *Soviet Phys. JETP* **23**, 1049–1056.

[722] D.V. Treshchev (1991). The mechanism of destruction of resonant tori of Hamiltonian systems, *Math. USSR Sb.* **68**, 181.

[723] C. Truesdell (1954). *The Kinematics of Vorticity*, Indiana U. Press, Bloomington.

[724] C.Y. Tsai, S.E. Widnall (1976). The stability of short waves on a straight vortex filament in a weak externally imposed strain field, *J. Fluid Mech.* **73**, 721–733.

[725] B. Turkington (1983a, b). On steady vortex flows in two–dimensions, *Comm. Part. Diff. Eqns.* **8**(9) 999–1071.

[726] B. Turkington (1987). On the evolution of a concentrated vortex in an ideal fluid, *Arch. Rat. Mech. Anal.* **97**, 57–87.

[727] B. Turkington (1998). Statistical equilibrium measures and coherent states in two dimensional turbulence, *Comm. Pure App. Math.* to appear.

[728] B. Turkington, N. Whittaker (1996). Statistical equilibrium computations of coherent structures in turbulent shear layers, *SIAM J. Sci. Comp.* **17**, 1414–1433.

[729] J.H.G.M.van Geffen, V.V. Meleshko, G.J.F. van Heijst (1996). Motion of a two–dimensional monopolar vortex in a bounded rectangular domain, *Phys. Fluids* **8**(9), 2393–2399.

[730] J.A. Viecelli (1990). Dynamics of 2D turbulence, *Phys. Fluids* **A2**(11), 2036–2045.

[731] J.A. Viecelli (1993). Statistical mechanics and correlation properties of a rotating two–dimensional flow of like signed vortices, *Phys. Fluids* **A** **5**(10), 2484–2501.

[732] H. Villat (1930). *Lecons sur la Théorie des Tourbillons*, Gauthier–Villars, Paris.

[733] V.V. Vladimirskii (1941). *Dokl. Akad. Nauk. USSR* **21**, 222, 1941; reprinted in *Topological Phases in Quantum Theory*, B. Markovski, V.I. Vinitsky, eds., World Scientific, Singapore.

[734] V.M. Volosov (1963). Averaging in systems of ordinary differential equations, *Russ. Math. Surveys* **17**, 1–126.

[735] Y.W. Wan (1986). The stability of rotating vortex patches, *Comm. Math. Phys.* **107**, 1–20.

[736] Y.H. Wan, M. Pulvirenti (1985). Nonlinear stability of circular vortex patches, *Comm. Math. Phys.* **99**, 435–450.

[737] S.E. Warschawski (1945). On Theodorsen's method of conformal mapping of nearly circular regions, *Quart. App. Math.* **3**, 12–28.

[738] W. Wasow (1973). Adiabatic invariance of a simple oscillator, *SIAM J. Math. Anal.* **4**, 78–88.

[739] C.E. Wayne (1990). Periodic and quasi–periodic solutions of nonlinear wave equations via KAM theory, *Comm. Math. Phys.* **127**, 479–528.

[740] C.E. Wayne (1996). An introduction to KAM theory , in *Dynamical Systems and Probabilistic Methods in PDE*, Lectures in Applied Math, **31**, P. Deift, C.D. Levermore, C.E. Wayne, eds., American Mathematical Society, Providence.

[741] A. Weinstein (1983). The local structure of Poisson manifolds, *J. Diff. Geom.* **18**, 523–537.

[742] J.B. Weiss, J.C. McWilliams (1991). Nonergodicity of point vortices, *Phys. Fluids* **A** **3**(5), 835–844.

[743] J.B. Weiss, J.C. McWilliams (1993). Temporal scaling behavior of decaying two–dimensional turbulence, *Phys. Fluids* **A** **5**, 608.

[744] J.B. Weiss, A. Provenzale, J.C. McWilliams (1998). Lagrangian dynamics in high–dimensional point vortex systems, *Phys. Fluids* **10**, 1929.

[745] A.L. Whipple, V. Szebehely (1984). The restricted problem of $n + \nu$ bodies, *Cel. Mech.* **32**(2), 137–144.

[746] E.T. Whittaker (1904). *A Treatise on the Analytical Dynamics of Particles and Rigid Bodies*, Cambridge Press, Cambridge.

[747] S.E. Widnall (1972). The stability of a helical vortex filament, *J. Fluid Mech.* **54**, 641–663.

[748] S.E. Widnall (1975). The structure and dynamics of vortex filaments, *Ann. Rev. Fluid Mech.* **7**, 141–165.

[749] S.E. Widnall, J.P. Sullivan (1973). On the stability of vortex rings, *Proc. R. Soc. Lond.* **A 332**, 335–353.

[750] S. Wiggins (1988). *Global Bifurcations and Chaos: Analytical Methods*, **AMS 73**, Springer–Verlag, New York.

[751] S. Wiggins (1990). *Introduction to Applied Nonlinear Dynamical Systems and Chaos*, **TAM 2**, Springer–Verlag, New York.

[752] S. Wiggins (1992). *Chaotic Transport in Dynamical Systems*, IAM 2, Springer–Verlag, New York.

[753] S. Wiggins (1993). *Global Dynamics, Phase Space Transport, Orbits Homoclinic to Resonances, and Applications*, Fields Inst. Monographs, American Mathematical Society, Providence.

[754] F. Wilczek, A. Shapere (1989). Efficiencies of self–propulsion at low Reynolds number, *J. Fluid Mech.* **198**, 587–599.

[755] C.D. Winant, F.K. Browand (1974). Vortex pairing: the mechanism of turbulent mixing layer growth at moderate Reynolds number, *J. Fluid Mech.* **63**, 237–255.

[756] G. Winckelmans, A. Leonard (1988). Weak solutions of the three–dimensional vorticity equation with vortex singularities, *Phys. Fluids* **31**, 1838–1839.

[757] G. Winckelmans, A. Leonard (1993). Contributions to vortex particle methods for the computation of three-dimensional incompressible unsteady flow, *J. Comp. Phys.* **109**, 247–273.

[758] A. Wintner (1941). *Analytical Foundations of Celestial Mechanics*, Princeton Press, Princeton.

[759] W. Wolibner (1933). Un theorém sur l'existence du mouvement plan d'un fluide parfait, homogéne, incompressible, pendent un temps, infinitent long, *Math. Z.* **37**, 698–726.

[760] H. Yamada, T. Matsui (1978). Preliminary study of mutual slip–through of a pair of vortices, *Phys. Fluids* **21**, 292–294.

[761] S. Yoden, M. Yamada (1993). A numerical experiment on two–dimensional decaying turbulence on a rotating sphere, *J. Atmos. Sci.* **50**, 631–643.

[762] W.R. Young (1985). Some interactions between small numbers of baroclinic geostrophic vortices, *Geo. Astro. Fluid Dyn.* **33**, 35–61.

[763] V.I. Yudovich (1963). Non-stationary flow of an ideal incompressible liquid, *Zh. Vych. Mat.* **3**, 1032–1066.

[764] N.J. Zabusky, M.H. Hughes, K.V. Roberts (1979). Contour dynamics for the Euler equations in two-dimensions, *J. Comp. Phys.* **30**, 96–106.

[765] V.E. Zakharov, A.B. Shabat (1972). Exact theory of two-dimensional self–focusing and one–dimensional self-modulation of waves in nonlinear media, *Sov. Phys. JETP* **34**(1), 62.

[766] L. Zannetti, P. Franzese (1993). Advection by a point vortex in closed domains, *Eur. J. Mech. B/Fluids* **12**(1), 43–67.

[767] L. Zannetti, P. Franzese (1994). The non-integrability of the restricted problem of two vortices in closed domains, *Physica D* **76**, 99–109.

[768] G. Zaslavsky (1999). Chaotic dynamics and the origin of statisitical laws, *Physics Today*, 39–45.

[769] I. Zawadzki, H. Aref (1991). Mixing during vortex ring collision, *Phys. Fluids* **3**(5), part 2, 1405–1410.

[770] H. Zhou (1997). On the motion of slender vortex filaments, *Phys. Fluids* **9**(4), 970–981.

[771] S.L. Ziglin (1980). Non-integrability of a problem on the motion of four point vortices, *Sov. Math. Dokl.* **21**(1), 296–299.

[772] S.L. Ziglin (1982). Addendum to K.M. Khanin, Quasi-periodic motions of vortex systems, *Physica D* **4**, 261– 269.

[773] S.L. Ziglin (1987). Concerning the paper Horseshoes in Perturbations of Hamiltonian Systems with Two Degrees of Freedom by P.J. Holmes and J.E. Marsden, *Comm. Math. Phys.* **108**, 527–528.

[774] J.W. Zwanziger, M. Koenig, A. Pines (1990). Berry's phase, *Ann. Rev. Phys. Chem.* **41**, 601.

Index

Applied Mathematical Sciences

Applied Mathematical Sciences

Applied Mathematical Sciences

(continued from previous page)